T0406591

Schooling for Sustainable Development

Volume 8

Series editors
John Chi-Kin Lee, Education University of Hong Kong, Tai Po, New Territories, Hong Kong
Michael Williams, Emeritus Professor of Education, Swansea University, UK
Philip Stimpson, Formerly Associate Professor, Faculty of Education, University of Hong Kong

This book series addresses issues associated with sustainability with a strong focus on the need for educational policy and action. Current attention and initiatives assume that Education for Sustainable Development (ESD) can be introduced successfully and gradually into schools worldwide. This series explores the issues that arise from the substantial and sustainable changes to be implemented in schools and education systems.

The series aims to counter the prevailing Western character of current research and enable cross-cultural comparisons of educational policy, practice, and project development. As a whole, it provides authoritative and comprehensive global coverage, with each volume providing regional/continental coverage. The volumes present data and insights that contribute to research, policy and practice in ESD-related curriculum development, school organization and school-community partnerships. They are based on ESD-related project experiences, empirical studies that focus on ESD implementation and teachers' perceptions as well as childhood studies that examine children's geographies, cultural characteristics and behaviours.

More information about this series at http://www.springer.com/series/8635

Heila Lotz-Sisitka • Overson Shumba
Justin Lupele • Di Wilmot
Editors

Schooling for Sustainable Development in Africa

 Springer

Editors
Heila Lotz-Sisitka
Environmental Learning Research
  Centre (ELRC)
Education Department, Rhodes University
Grahamstown, South Africa

Overson Shumba
School of Mathematics and Natural
  Sciences
Copperbelt University
Kitwe, Zambia

Justin Lupele
Chief of Party: USAID, Strengthening
  Educational Performance Up (STEP-UP)
Lusaka, Zambia

Di Wilmot
Faculty of Education
Rhodes University
Grahamstown, South Africa

Schooling for Sustainable Development
ISBN 978-3-319-45987-5   ISBN 978-3-319-45989-9   (eBook)
DOI 10.1007/978-3-319-45989-9

Library of Congress Control Number: 2016957748

Printed on acid-free paper

This Springer imprint is published by Springer Nature
The registered company is Springer International Publishing AG
The registered company address is: Gewerbestrasse 11, 6330 Cham, Switzerland

Contents

Contributors

Romakala Banda is a Commonwealth scholar and a final year master of science education for sustainability student at London South Bank University. She has 4 years of experience in the WASH sector and has specialised in the integration of gender in WASH (water, sanitation and hygiene) with a focus on equity and inclusion in WASH as well as the integration of WASH in national policy and official school curriculum including WASH material development. She has wider interests in sustainable development particularly with regard to participatory approaches, advocacy for equality and climate change responses. Social research and capacity building are among her key competencies. She currently serves as technical advisor for the Reform of the Water Sector Programme with Deutsche Gesellschaft fur Internationale Zusammenarbeit – GIZ-Zambia – for both rural and urban water programmes including water resource management for climate change resilient agriculture.

Ravhee Bholah A Fulbrighter and Tertiary Education Commission scholar, Dr. Ravhee Bholah is associate professor in the Department of Science Education and coordinator of the education for sustainable development (ESD) at the Mauritius Institute of Education (MIE). He has been coordinating various ESD and climate change education (CCE)-related projects and has also been actively involved in curriculum and learning resource development at institutional and national levels. He has been a consultant for UNESCO, UNDP and Government of Japan-funded projects related to ESD, CCESD and Africa Adaptation Programme. He has published in and reviewed various scientific papers for international peer-reviewed journals and conference proceedings.

Wolfgang Brunner has studied Earth sciences, zoology, botany and geography at Swedish Universities in Stockholm, Uppsala and Umeå and has a Master of Science degree in teacher education. He has taught biology, chemistry and natural sciences within the senior level of compulsory school in Sweden. He also worked part-time for the Swedish National Boards of Education and Environment, developing methods, materials and approaches concerning environmental and sustainability education. He has published textbooks for teachers and students and worked as a lecturer and course developer on environmental issues at all levels of the Swedish education system – from pre-school to teacher education institutions at universities. Between 2010 and 2015, Wolfgang worked as programme specialist at SWEDESD (Swedish International Centre of Education for Sustainable Development, Uppsala University) with teachers and teacher educators in India and southern Africa. Through the ESSA programme (Education for Strong Sustainability and Agency), Wolfgang has developed strategies, courses and teaching materials used at 42 universities and teacher education institutions in 14 SADC countries in southern Africa. Wolfgang has also produced online courses on sustainability and CSR (corporate social responsibility) for Uppsala University.

Charles Chikunda is a senior researcher with the Association for Water and Rural Development in the Olifants River Basin in South Africa, a project that aims to reduce vulnerability to climate change through building improved transboundary water and biodiversity governance and management of the Olifants River Basin. His work focuses on stakeholder engagement, professional development and training in natural resources management and basin resilience building. Charles holds a Ph.D. in education with a focus on science education and education for sustainable development. He has over 15 years' experience in teacher education focusing on these fields. Charles has been actively involved in the Southern Africa Development Community Regional Environmental Education Programme over the years. His research interests centre around equity, social justice and sustainability as well as integrated natural resources management. He has published in and reviewed various scientific papers for international peer-reviewed journals and conference proceedings.

Raviro Chineka An early career researcher and scholar, Raviro Chineka is a lecturer in physics education in the Department of Science and Mathematics Education at the University of Zimbabwe. She is currently studying towards a Ph.D. in climate change education with the University of Technology, Sydney. Raviro's past and present research passion lies in sustainability science and in conducting change-oriented research with a focus on community innovations for transformation towards a more livable, egalitarian and just world. She researches real problems affecting real people in real-life settings. Raviro served as the focal person at the University of Zimbabwe (UZ) for the SADC-REEP EE/ESD research initiative and UNEP-MESA programmes, which sought to reorient curricula towards sustainability. She has participated in regional and international training programmes on greening the curriculum and has shared her research outputs at a number of conferences.

Carolina Dube taught school geography in Zimbabwe for many years before studying academic geography, specialising in environmental courses. At Ph.D. level, her research focused on teachers' experiences of implementing EE and ESD through the geography curriculum at secondary level in the Western Cape in South Africa. Her research revealed the contextual and structural challenges facing the implementation of EE and ESD in the school context. In 2013–2014, she was a postdoctoral fellow in the Department of Education at Rhodes University. Working with Professor Di Wilmot, they researched geography teachers' experiences of implementing the geography curriculum at secondary level in Grahamstown, in the Eastern Cape. She is currently based in Zimbabwe where she is a tutor of a number of geography courses that include environmental education at the Zimbabwe Open University. Her publications are mainly in the field of school geography and environmental education.

Mantoetse Jobo holds a master's degree in agronomy and systems from the University of Adelaide, Australia. She holds B.Sc. honours in science education from the University of Limerick in Ireland. Her interests are in education for sustainable development. She has been an active member of the Lesotho Environmental Education Support Project (LEESP), working in the monitoring and evaluation team. She participated in the writing of *Environmental Education Strategy Toward 2014: A Strategic Plan for Education for Sustainable Development in Lesotho*. She has coordinated education for sustainable development (ESD) and environmental education activities at Lesotho College of Education working in collaboration with the National University of Lesotho. Mantoetse has participated in UNESCO, UNDP and Government of Lesotho ESD and climate change activities.

Bridget Kakuwa is a Ph.D. student of communication science at the University of South Africa. She has published on communication strategies used to promote regional integration in Africa and runs a family life and parenting blog on WordPress. Her professional interests are in participatory communication; educational child development and psychology; information management; water, sanitation and hygiene education; knowledge management and organisational learning; and children broadcasting/television and film production. She has worked as knowledge management, partnerships and communication advisor for FHI 360. She previously worked as an electronic librarian for the Regional Integration Support Programme at COMESA and as school teacher. She currently works as the regional communications officer for Feed the Future Harmonized Seed Regulations Project in southern Africa.

Evaristo Kalumba is an environmental educator, practitioner, writer, course developer and a senior lecturer of biology and science teaching methods in Zambia. He wrote the *Environmental Education Module for Primary School Colleges* supported by the CDN SADC REES Project, SADC-REEP and WESSA and the *Mathematics and Science Education Module 4: Conservation of the Environment* supported by USAID. He has contributed chapters to books and articles to the *Zambia Journal of Teacher Professional Growth* (ZJTPG). He has presented papers at local, regional and international environmental education attachment programmes, conferences, WWF ZEP EE courses and workshops supported by SADC-REEP. He had been a member of ZANEEP and EEASA. He was formerly a lecturer at Mufulira College of Education and is currently working at Josmak College of Education in Zambia.

Royda Kampamba is a lecturer in the School of Mathematics and Natural Sciences at the Department of Mathematics and Science Education at the University of Copperbelt (Zambia). She has worked as a secondary school science teacher and science teacher educator in the College of Education. She is a co-establisher of the Department of Mathematics and Science Education and of Sustainability (2007, 2009). She has participated in the southern African environmental education, Vrije University in the Netherlands and UNESCO short courses, workshops, research and conferences in ESD-related teaching, learning and research. She has presented papers and published peer-reviewed articles. She is a member of the Southern African Development Community (SADC) research network based at Rhodes University, South Africa. Her research interests include science innovative teaching and learning in secondary and primary schools; chemistry/science teacher education, interdisciplinary and transdisciplinary research; and precolonial Zambian indigenous science and technology knowledge.

Heila Lotz-Sisitka holds a national research foundation chair in transformative social learning and green skills learning pathways. She previously held the Murray & Roberts chair of environmental education and sustainability for 15 years and was also director of the Centre for Postgraduate Studies at Rhodes University. She has been involved in national and international education and environmental education research and policy development since 1992. She has coordinated the master's and Ph.D. programmes in environmental education at Rhodes University and has served on a number of international scientific and international policy committees, including as international reference group member for the UN Decade on Education for Sustainable Development and scientific chair of the World Environmental Education Congress in 2007. She is editor-in-chief of the *Southern African Journal of Environmental Education* and a coeditor of the journal *Learning, Culture and Social Interaction*. She recently coedited a Routledge book with Leigh Price, entitled *Critical Realism, Environmental Learning and Social-Ecological Change*. Her research interests include learning, agency and social change in contexts of risk and vulnerability and critical research methodologies.

Justin Lupele is a distinguished Rhodes and Commonwealth scholar with a string of academic papers and chapters in refereed journals to his credit. His professional interests are in education research; curriculum development; climate change; water, sanitation and hygiene education; participatory educational materials development and institutional/organisational capacity building. He has provided leadership in education through USAID contractors (Academy for Education Development, FHI 360 and Chemonics International) where he has served as a chief of party. He previously worked as programme manager for SADC Regional Environmental Education Programme, education officer for WWF and a school teacher. He holds a Ph.D. (environmental education) from Rhodes University in South Africa. He is currently chief of party for Chemonics International under a USAID-funded project – STEP UP (Strengthening Educational Performance Up) in Zambia.

Caleb Mandikonza holds a Ph.D. in environmental education from Rhodes University, South Africa. He has a long history in environment and sustainability education in southern Africa. He was involved with capacity development of teacher educators for environmental education through the Secondary Teacher Training Environmental Education Programme (STTEEP) in Zimbabwe prior to joining the Southern African Development Community Regional Environmental Education Programme (SADC-REEP) as training coordinator. As training coordinator, Caleb supported capacity building for environment and sustainability in the SADC subregion. He has researched and wrote up the UNESCO Climate Change Education case study for South Africa. Caleb's professional and research interests lie in the notion of mediation for academic professional development in environment and sustainability education, climate change education, science education and indigenous knowledge practices in relation to science education and sustainability.

Tšepo Mokuku is a senior lecturer in science and environmental education in the Department of Science Education at the National University of Lesotho. His research and curriculum development interests include school-community interface in respect of epistemological integration, biodiversity conservation and outdoor learning. He has coordinated and participated in a number of local and international projects in environmental sustainability. These include curriculum development and training focused projects under the aegis of Development Partnership in Higher Education (DelPHE), Erasmus Mundus (STETTIN) programme and SADC-REEP. Community-based projects he has coordinated include those funded by the United Nations Development Programme (UNDP), UNDP-Global Environment Facility (UNDP-GEF) and Lesotho Highlands Development Project (LHDP). He has published a number of articles in peer-reviewed international journals.

Manoah Muchanga is a Southern Africa Science Service Centre for Climate Change and Adaptive Land-use (SASSCAL) Ph.D. scholar. His Ph.D. focuses on water and the problem of sedimentation. He is a lecturer in the Environmental Education Unit at the University of Zambia, School of Education. He is currently coordinating the development of a community of practice around learning for sustainable management of the sediment problem in small reservoir catchments in southern Zambia. His current research interests include learning around sediment management, philosophy of science, environmental education and modelling as well as water and climate change.

Cecilia Mukundu is an experienced biology/science educator with the University of Zimbabwe with a keen interest in education for sustainable development (ESD), particularly mainstreaming environmental and sustainability issues in the curriculum. She was involved in quality and relevance of education research projects with a focus on children living in contexts of risk and vulnerability including marginalised groups for the last 10 years. She has benefited from capacity development programmes through the Southern African Development Community Regional Environmental Education Programme (SADC-REEP), Swedish International Training Programme (ITP) in ESD and Mainstreaming Environment and Sustainability in Africa (MESA) Universities. She was instrumental in the formation of the Regional Centre of Expertise on ESD (Harare) and continues to coordinate activities in this network. Cecilia is currently a Ph.D. scholar with the University of Witwatersrand, South Africa, where she is researching the teaching of biodiversity conservation in biology classrooms in the Zimbabwean secondary school curriculum.

Charles Namafe is currently associate professor of environmental education at the University of Zambia, School of Education. He is also the SADC-MESA chair for teacher education to address sustainability. He holds a master's degree in geographical education (with an orientation in environmental education) from McGill University, Montreal, Canada. His doctor of philosophy obtained from the University of London, Institute of Education, was in the same field. His current research interests are in water education and conflict resolution as well as indigenous education.

Kenneth Mlungisi Ngcoza is a senior lecturer in science education and deputy head of the Department in the Faculty of Education at Rhodes University. Kenneth is a mathematics and science teacher by profession. At the Education Department, he is involved with science teacher education (both pre-and in-service) and is the coordinator of the B.Ed. (hons) science elective and M.Ed. in science education courses. He has supervised a number of honours, master's and doctoral students. He is involved in community engagement activities, and he is the Faculty of Education community engagement representative. He is the former chairperson and now treasurer of the **Southern African Association for Research in Mathematics, Science and Technology Education** (SAARMSTE), Eastern Cape Chapter, as well as a chapter representative in SAARMSTE. His research interests include professional development, science curriculum, education for sustainable development (ESD), cultural contexts and indigenous knowledge.

Rob O'Donoghue is a professor at Rhodes University and director of the Environmental Learning Research Centre. In his research on environmental education processes of learning-led change in the Fundisa for Change Teacher Education programme, he has developed models of process for contemplating concerns in the environment, for working with the sustainable development goals and for mapping pathways to future sustainability. This work has given close attention to indigenous knowledge practices to frame environmental learning in post-colonial curricula and community contexts.

Andrew Petersen is a science education specialist who currently works in the field of teacher professional development at the Schools Development Unit at the University of Cape Town. His main focus is biodiversity and climate change and sustainability education. He has been involved in various education for sustainable development (ESD) projects involving both collaborative resource development for teachers and networking locally and internationally. He was a coresearcher in an international project on ESD as part of the UN Decade of Education for Sustainable Development which culminated in his participation in a seminar presentation at the world conference on ESD held in Kanoya, Japan. He has a special interest in eLearning and is currently engaged in collaborative work to integrate this approach in teacher professional development.

Ingrid Schudel is a senior lecturer involved in teaching and research in the Environmental Learning Research Centre, Rhodes University. She has an undergraduate degree in science and a master's in education (environmental education). She is currently manager of the Fundisa for Change Teacher Education programme. Her Ph.D. research focused on environmental education in a teacher professional development programme and examined the emergence of active learning as transformative praxis in foundation phase classrooms. She aims to strengthen quality learning, through environmental education processes which equip learners with the competence and confidence to actively reimagine, try out and critique new ways of developing sustainable people-environment relationships in local and global contexts. Her research interests include the exploration of change and change processes, learning and transformative praxis and knowledge and knowledge building.

Soul Shava is currently an associate professor in environmental education in the Department of Science and Technology Education at the University of South Africa (UNISA). He holds a Ph.D. in environmental education from Rhodes University, South Africa. Soul has research interests in indigenous knowledges, particularly their representation in knowledge generation processes at the interface of modern institutions and local communities and their application in community development contexts and environmental education processes in southern Africa, indigenous epistemologies and decoloniality. He also has interests in community-based natural resources management (CBNRM) and socio-ecological resilience; climate change; green economy; intangible cultural heritage; sustainable agriculture and traditional agrobiodiversity conservation; and the use of traditional crops and indigenous food plants by local communities to achieve food security and resilience to climate change impacts.

Overson Shumba is a professor of science education at the Copperbelt University, Zambia, where he is research and innovations coordinator in the School of Mathematics and Natural Sciences. He is a former W.K. Kellogg Foundation fellow (at Iowa State University, USA) and German Academic Exchange Services fellow (at the University of Muenster, Germany). He served as a member of the UNESCO Monitoring and Evaluation Expert Group (MEEG) during the UN Decade of Education for Sustainable Development. His research interests and publications are in science teacher education, environmental science and ESD, curriculum and instruction, research methods and monitoring and evaluation.

Nthalivi Silo is a senior lecturer in environmental education in the Department of Primary Education of the University of Botswana. She holds a Ph.D. in environmental education from Rhodes University, South Africa. Her areas of interest are in curriculum issues in environmental education and development of children's civic agency and competences in environmental-related and sustainability issues. She has used her educational experience to work for social justice in formal and informal setups with minority children. She is a council member of the Southern African Association of Environmental Education of which she is the editor of the Association's bulletin and is a regular reviewer of the articles of the Association's journal. She is also a key participant in the SWEDESD/SADC-REEP teacher education programme in which she is, with other teacher educators in the region, developing an online course on Education for Strong Sustainability and Agency (ESSA) for Southern African Development Community (SADC) teacher educators and teachers.

Zintle Songqwaru is a lecturer and a Ph.D. scholar at the Education Department at Rhodes University in South Africa, with an interest in teacher professional development. Before joining Rhodes University, she was a high school science teacher, teaching natural sciences, life sciences and physical sciences from 2001 to 2012. During her years as a teacher, she was an eco-schools coordinator in her school, through which she became involved with the Rhodes University Environmental Learning Research Centre (ELRC). In 2013, she became the first coordinator of the Fundisa for Change Teacher Education programme until 2015. She currently teaches pre-service teachers in the postgraduate certificate in education at Rhodes University and in-service teachers in the Fundisa for Change Teacher Education programme.

Sirkka Tshiningayamwe is a postdoctoral fellow at Rhodes University in the Environmental Learning and Research Centre (ELRC). Her research interests are in teacher professional development, and she thus helps to coordinate Fundisa for Change Teacher Education programme, a national teacher professional development programme. This is a partnership programme that aims to enhance transformative environmental learning through teacher education in South Africa. She has been involved in teacher education resource development focusing on environmental education in Namibia and South Africa. She has peer-reviewed journal articles for the *Southern African Journal of Environmental Education*.

Shepherd Urenje is a programme specialist in education for sustainable development with the Swedish International Centre of Education for Sustainable Development (SWEDESD). His current work supports processes of transformation in education and creating opportunities for quality and relevance in education within and among countries in Scandinavia, southern Africa and south Asia. His expertise includes developing and implementing learning for change strategies that develop in learners relevant skills for a changing world. He studied developmental education at the University of London. His professional background includes teaching environmental science and development education in Zimbabwe, SADC Regional

Environmental Education Programme manager for education and training and principal examiner of environmental science in the UK. He is the current chair of partners in the UNESCO Global Action Programme Priority Action Area 3 Network, responsible for increasing the capacity of educators and trainers in mainstreaming ESD in their practice.

Di Wilmot is an experienced geography teacher educator and researcher and is currently dean of education at Rhodes University, South Africa. Her Ph.D. research focused on social sciences teacher professional development in a context of national curriculum transformation and policy overload. Her current research interests include curriculum and active learning pedagogical approaches. She coordinates and teaches initial and postgraduate geography education courses in South Africa and Namibia. As a member of the International Network of Teacher Education Institutions associated with the UNESCO Chair on reorienting teacher education to address sustainability, Di piloted and contributed to the further development of UNESCO's climate change course for secondary teachers. She has presented papers on her research at international conferences in the UK, China, Finland, the USA, Germany and Canada and has published and reviewed papers for national and international education journals.

Cryton Zazu holds a doctorate degree in environment and sustainability education from Rhodes University, South Africa. Born into a family of eight, Zazu was brought up in a typical African traditional family context, a background that influenced his research interest around the integration of indigenous knowledge systems into mainstream policy and practice as a way of improving the relevance and quality of education and sustainable development practices in post-colonial southern Africa. He is an active environment and sustainability practitioner in the SADC region and is currently based in Zimbabwe.

Part I
Orientation to Education for Sustainable Development and Schools in Africa and Education for Sustainable Development Learning Processes

Chapter 1
ESD, Learning and Quality Education in Africa: Learning Today for Tomorrow

Heila Lotz-Sisitka and Justin Lupele

As the twenty-first century continues to unfold, African societies are characterised by the continuing effects of a long history of colonial intrusion, the challenges associated with establishing new societies and governance structures following the post-1950s' independence period and a complex array of risks and uncertainties associated with the more recent spread of hyper-capitalism, globalisation and earth system degradation. African societies, like societies elsewhere, are in the process of working out what the full meaning of educational quality might be in such a world. In this chapter we suggest that in working towards a fuller understanding of the meaning of educational quality, there is need to consider insights provided by environment and sustainability education (ESE) and education for sustainable development (ESD).[1] These perspectives can potentially also frame and provide perspective on emerging research agendas for ESD in Africa, as demonstrated in the chapters of this book. We propose the need to place emphasis on ESD learning processes in

---

[1] In this chapter we acknowledge that ESD is not an uncontested discourse on the African continent and more widely in the developing world (Gonzalez-Gaudiano, 2005, 2009; Gonzalez-Gaudiano & Silva, 2015; Leff, 2009; Lotz-Sisitka, 2004). Normally we would prefer to use ESE (environment and sustainability education), but in keeping with the title of the book series, we use the notion of ESD. In southern Africa, environment and sustainability education has taken account of development challenges from the start, but we also recognise that notions of development under neo-liberalism and globalisation are not without serious problems with various performative impacts on societies in the Global South that need to be troubled. In using ESD in this paper, we include a strong focus on politics and social justice (i.e. a strong political ecological perspective).

H. Lotz-Sisitka (✉)
Environmental Learning Research Centre (ELRC), Education Department,
Rhodes University, Grahamstown, South Africa
e-mail: h.lotz-sisitka@ru.ac.za

J. Lupele
Chief of Party: USAID, Strengthening Educational Performance Up (STEP-UP),
Lusaka, Zambia
e-mail: lupelejustin@yahoo.com

© Springer International Publishing Switzerland 2017
H. Lotz-Sisitka et al. (eds.), *Schooling for Sustainable Development in Africa*,
Schooling for Sustainable Development 8, DOI 10.1007/978-3-319-45989-9_1

conceptions of educational quality and, in turn, probe how such ESD learning processes are conceptualised.

This chapter provides insights into some emerging conceptual framings of ESE and quality education, with an emphasis on ESE learning processes. The chapter draws on a recent review of literature and case examples of ESE learning processes in Africa, published in a Southern African Development Community Study undertaken for the Association for the Development of Education in Africa (Lupele & Lotz-Sisitka, 2012). *It links this literature and case study review to the recent emphasis on ESD and educational quality in the global sustainable development goals and to wider imperatives for decolonisation of education in Africa.*

1.1   The African Context

Dussel (1998) insightfully commented on key intersecting challenges facing all developing countries affected by colonial histories and modern environmental degradation. He framed the intersecting challenges for former colonised communities and nation states as follows:

1. Ecological destruction of the planet based on a view of nature as an exploitable object
2. Poverty and inequality based on ongoing exploitation and accumulation of wealth
3. Narrow rationalities epitomised by colonial and imperialist thinking

Writing from a broader global perspective, Mythen (2004, p. 1) contextualised the African situation further within the wider transformations and challenges of the twenty-first century when she suggested that increasing portions of our everyday lives are spent negotiating change, dealing with uncertainty and assessing risk and impacts of situations that are complex and that often appear to be out of control.

Thus, the contemporary context poses many challenges for societies in transformation. In most African societies, there is a need to address a range of intersecting issues such as social inequality and poverty (economy); cultural change, social justice and health risks (society); natural resource depletion, biodiversity loss and global climate change (environment); and governance, democracy and peace (politics). Besides the key objectives for societal transformation such as food and water security, peace, well-being and sustainability, there is also a need for communities in Africa to learn to deal with the 'slow emergence' of climate change impacts (IPCC, 2014).

Although these commonly noted problems affect development on the African continent, Africa has the immense asset of a young population. It is one of the most youthful continents on the planet, and as Mbembe (2001) stated a few years ago, Africa is most well-known for its lacks or what it is not. Too little is known about what Africa *is* or *can be*. To give adequate attention to the potential of Africa's youth, and to their futures, there is a need to develop knowledge, skills, values and capabilities to maximise the wealth of resources and cultures that exist on the continent and to ensure that future generations are able to live sustainable lives free of poverty and other factors that impede quality of life. To achieve this, ESD learning processes and new concepts of educational quality, as outlined below, are needed to guide education on the continent.

Framing this need from an African policy perspective, the African Union recently released the Africa Agenda 2063 (African Union [AU], 2014). This is an endogenous, shared strategic framework for inclusive growth and sustainable development aimed at guiding Africa's transformation. It represents a continuation of the pan-African drive for self-determination, freedom, progress and collective prosperity (AU). Education, access to education and access to *quality and relevant education* is a key focus of this vision, as is the commitment to people-centred, sustainable development and gender equality. The AU 2063 foregrounds people's participation in the transformation of the continent and the building of caring and inclusive societies through empowering women and providing an enabling environment for its children and young people to flourish and reach their full potential (ibid). Key to this is the promoting, strengthening and valuing of African cultural identity, values and ethics. This vision has implications for the way in which educational quality is framed and pursued and for the way in which ESD learning processes are conceptualised.

To address historical challenges related to the provisioning and quality of education, sustainable development goal (SDG) four, one of the recently proclaimed global goals, focuses on the provision of quality lifelong learning and education for all (United Nations [UN], 2015). Following the years of emphasis on access, the SDG global goal also emphasises *quality education*, stating that ESD and global citizenship education must form part of the conceptualisation of quality education (ibid.). Yet, as shown by the recently completed international educational quality research programme led by Tikly and Barrett (2011), *quality education* remains poorly defined on the African continent, and there is even less definition of the relationship that exists between ESD and educational quality. It is this question, and especially the relationship that exists between educational quality and ESD, that inspires the work of this book.

1.2 Methodological Note

In 2012 we were requested by the Southern African Development Community and the Association for the Development of Education in Africa to review ESD learning processes. This study involved a comprehensive literature review of research published in Africa on ESD and learning processes. It also involved collecting a number of case studies of ESD learning processes. To conduct the review, we reviewed 36 published studies of ESD learning processes and solicited further 12 case studies of ESD learning processes in Africa (Lupele & Lotz-Sisitka, 2012). In this work, we found that there was indeed a strong interest in ESD learning processes in Africa, and an emerging body of published research was arising that reflected on ESD learning processes. To interpret these ESD learning processes in relation to the central question of the relationship between ESD and educational quality and relevance, we:

- Developed an argument for bringing ESD learning processes into focus in debates on educational quality and relevance
- Critically reviewed and developed a reframed notion of educational quality and relevance which provides for perspectives on ESD learning processes and their potential contribution to educational quality
- Developed a set of propositions from the cases reviewed which provides a framework for further research on ESD and educational quality and relevance (Lupele & Lotz-Sisitka, 2012)

We share a summary of this work in this chapter. Since we undertook this conceptual framing research in 2012, a number of additional studies into ESD and educational quality and relevance have emerged in Africa, many of which are reported on in this book. This chapter therefore does not provide an 'update' on the case-based research undertaken since our earlier conceptual framing work, but rather cross references to the more recent work as reported on across the chapters of this book. In many ways the research that has emerged since 2012 in the field of ESD in Africa demonstrates the importance of producing conceptual frameworks for generating research into ESD learning processes in ways that can contribute critically and cumulatively to our knowledge of ESD in Africa.

1.3   ESD Learning Processes in Focus

Learning processes are at the heart of education, including ESD.[2] ESD learning processes involve engagement with matters of concern that arise at the social-ecological-economic-political interface. These matters of concern often involve engagement with risk, uncertainty and wicked or difficult-to-resolve problems. They also involve envisioning new futures and engagement in actions and practices that model and contribute to the emergence of a more sustainable, inclusive and socially just society. ESD learning should be understood as a process of discovery that allows children in schools to participate in and generate a new understanding about themselves and the world around them. As part of the midterm evaluation of the United Nations Decade of Education for Sustainable Development (2005–2014), UNESCO released an expert study on ESD learning processes (Tilbury, 2011)

---

[2] The following sections provide a summary/synthesis of the conceptual framework developed out of the Lupele and Lotz-Sisitka (2012) study on ESD learning processes, and as such they also serve to provide a framework for reading the remaining chapters of the book.

which suggested that ESD learning, in addition to the gaining of knowledge, values and theories related to sustainable development, also refers to learning:

- To ask critical questions
- To clarify one's own values
- To envision more positive and sustainable futures
- To think systemically
- To respond through applied learning
- To explore the dialectic between tradition and innovation (Tilbury, 2011, p. 104)

In order to understand ESD learning, there is a need to develop and work with theories of learning that meet the challenges of social transformation and to consider the knowledge and learning implications of the complexity and wicked nature of often intractable environmental and social challenges. ESD learning links *being* to *becoming*. The distinction between knowledge, understanding and expression of valued beings and doings deepens our grasp of the layering of learning. As children are given a chance to express their valued beings and doings, and expand their knowledge, values and action competence through ESD learning, they can engage in a wider range of debates, think more critically and participate in the world differently. This is what an ESD focus in schooling allows.

ESD learning processes and associated educational practices often emphasise new ethics, norms and understandings and therefore require engagement with new concepts, knowledge and teaching approaches and practices. Yet, how does learning and culture interact in African schools when it comes to ESD, especially if we have an understanding that ESD learning processes engage people in processes of social change? If Tilbury's (2011) point about the dialectic between tradition and innovation is to be taken seriously in ESD learning, then the pedagogies needed in schools should not only take cognisance of children's sociocultural histories and contexts but should also expand and challenge cultural practices that need to change. Such a form of education not only involves acculturation into existing sociocultural heritages and histories but *also* requires a challenging and changing of these cultures, heritages and histories in the face of new challenges outlined above. Commenting on the legacy of Vygotsky's research on learning, Holland, Skinner, Lachicotte and Cain (2003, p. 176) claimed that 'individuals have access to the cultural legacy of the collective through others. Social interaction is the context in which cultural forms come to individuals and individuals come to use cultural forms'. It is in this sense that 'learning is situated'.

1.4 ESD and Situated Learning

ESD brings the importance of *situated learning* to the fore, an approach to learning that focuses on the cultural, social and socio-material 'figured worlds' (Holland et al., 2003) in which individuals act as members of social groups and interact with material and linguistic resources that are situated in and emerge from historical and cultural contexts (Gerstenmaier & Mandl, 2001; Holland et al.). ESD learning processes are, within a situated learning framework, seen as active and constructive meaning-making processes in which participation in a system of distributed knowledge and practice emerges in ways that are also contextually located and situated in the real socio-material world, out of which learning praxis emerges. In this way, ESD learning processes can therefore also be described as 'adaptation' to constraints and affordances, but they are also potentially transformative of these situations (Chabay et al., 2011; Holland et al.; Vygotsky, 1978).

In many of Africa's schools and communities, it is necessary to conceptualise what ESD means at the interface of poverty, environmental degradation and health conditions. Findings from research undertaken into ESD at the interface of these issues suggest that such education involves a complex combination of learning processes that involve (1) proactive engagement with risk reduction and mitigation, (2) reactive engagement with the consequences and impacts of risk (e.g. the impacts of the HIV/AIDS pandemic) and (3) critical, advocacy-based education that strengthens institutional efficacy and state accountability (Lotz-Sisitka, 2012; see Fig. 1.1 below). According to authors such as Namafe (2008), education should also be based on and emerge out of existing strengths at individual, community and societal level and people's valued beings and doings.

ESD learning processes, therefore, introduce engagement with substantive learning theory in African educational contexts. They also usher in new ways of thinking about educational quality and relevance in African schools. Researchers working on these issues, as shown in the chapters of this book, are providing useful insights into the relationship between knowledge, learning and meaning making. These issues have somewhat been neglected in the drive to attain *Education for All*, which, in many cases, has tended to focus on structural aspects of education, at the expense of *process* aspects of education (United Nations Educational, Scientific and Cultural Organisation [UNESCO], 2015). The book and its chapters focus on the process aspects of education and on how ESD can shape these processes in new and exciting ways that also strengthen the quality and relevance of education.

ESD learning processes are not only concerned with issues of participation in education or with situated learning approaches. They are also interested in critical engagement with the new forms of knowledge; new values, skills and dispositions; and cognitive and social development of learners. This is one of the key roles defined for schools, i.e. the responsibility for ensuring that children acquire new knowledge. Sfard (1998) argued that education should be concerned with both acquisition and participation. There is much new knowledge that children have to learn in today's world. Climate change, biodiversity loss, environmental health, disaster risk reduction and sustainable development are only a few of the topics they encounter and need to make sense of. Such topics carry universal meaning, but they are also locally imbued with contextual meaning, as the issues they represent differ in different contexts. Teachers, too, need to acquire a deep understanding of these new concepts and contexts so that they are able mediate them effectively to children. So, while acquisition of knowledge remains an important aspect of ESD learning in schools in

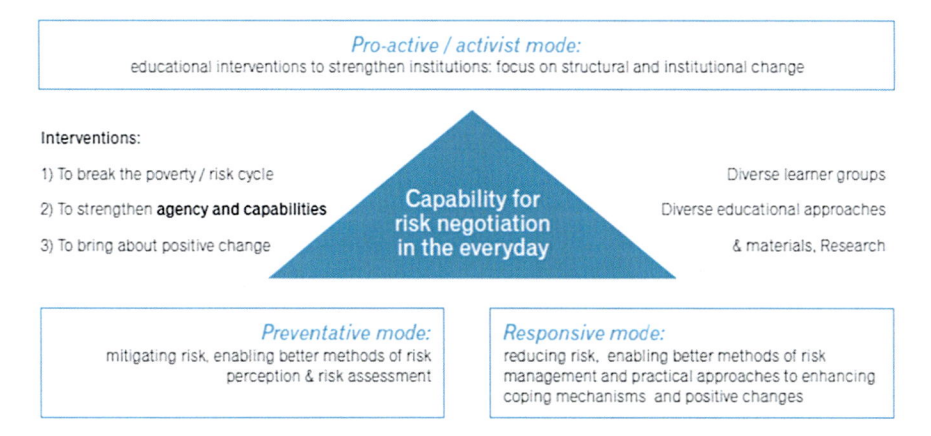

Fig. 1.1 New orientations to education that foreground capability and agency (Lotz-Sisitka, 2012)

Africa, the participation interest in ESD learning processes cannot be neglected *and needs to be foregrounded as a means to strengthen critical and reflexive engagement with concepts, knowledge and values*, as is shown in many of the chapters in the book. Chabay et al. (2011, p. 28, citing Reid & Nikel, 2007) suggested that 'people's capacity for participation in societal change processes is *learnt, constructed and dynamic* – and that this *can be enhanced* (rather than being regarded as something that is, for example, fixed, largely inherited, or stable)' (our emphasis). Holland et al. (2003, p. 177) reminded us here of the importance of dialogue in the learning process noting that Bakhtin emphasised the 'always continuing activity of producing meaning in dialogue', which takes place in relation to and in the context of everyday activity and practice.

Chabay et al. (2011, p. 27) suggested too that current ESD learning process research approaches tend to 'strongly underestimate the importance of local and indigenous knowledge, as they mostly focus on its contents, but not on its importance for value systems, local theories-in-use, and hence its role in the learning process'. Similar findings are expressed in the work of indigenous knowledge researchers interested in ESD learning processes (Asafo-Adjei, 2004; Mokuku & Mokuku, 2004; Shava, 2005), an interest that is carried forward in a number of different chapters in the book (e.g. see Chaps. 5, 6 and 15). Shava (2010) explained that

> a critical aspect in the development of indigenous knowledge is the resilience of indigenous knowledge systems as evidenced by the continued sustenance of traditional medicinal practice and traditional cultural practices even in urban settings […] This shows that indigenous knowledge is not entirely lost […] however, these practices are often not represented in formal educational settings, and if so, it is normally the researchers' anthropological eye that dominates the representations. (p. 40)

O'Donoghue, Lotz-Sisitka, Asafo-Adjei, Kota and Hanisi (2007) argued that mobilising indigenous and local knowledge in ESD learning processes is an act of *enabling epistemological access* to abstract forms of knowledge that tend to circulate in schools and universities. These authors draw their claims back to an understanding of mediation in education. Shava (2010, p. 40) went on to argue that 'we need to constantly evaluate what knowledge is represented, and most importantly, how it is represented and applied'.

Zaalouk (2004) writing from North Africa explained such forms of ESD learning more directly as a 'pedagogy of empowerment', while writers from southern Africa (Ketlhoilwe, 2008; Namafe, 2008; Shumba, Kasembe, Mukundu, & Muzenda, 2008) have noted that ESD learning processes are about not only pedagogies of empowerment but also about *epistemological changes* in the way in which knowledge is viewed and worked within educational settings (where local and indigenous forms of knowledge are also 'counted' as being valid in the educational context and where they are 'brought into' processes of enabling epistemological access to new/more abstract forms of knowledge that circulate in formal educational settings). A number of the chapters in the book address these aspects of ESD learning in African schools.

1.5 ESD and Quality Education

Quality education has its roots deep in the history of the United Nations and international declarations. The right to education is mentioned in the Universal Declaration of Human Rights and the Convention on the Rights of the Child. The importance of education is also mentioned in the World Declaration on Education for All, the Dakar Framework for Action and the Millennium Development Goals, and most recently the sustainable development goals as mentioned above. Education is crucial for the well-being of individuals, nations and the world.

In 1990, the World Declaration on Education for All noted that the generally poor quality of education needed to be improved and recommended that education be made both universally available and more relevant. The Declaration also identified quality as a prerequisite for achieving the fundamental goal of equity. While the notion of quality was not fully developed, it was recognised that expanding access alone would be insufficient for education to contribute fully to the development of the individual and society. Emphasis was accordingly placed on assuring an increase in children's cognitive development by improving the quality of their education.

A decade later, the Dakar Framework for Action (World Education Forum, 2000) declared that access to quality education was the right of every child. It affirmed that quality was 'at the heart of education' – a fundamental determinant of enrolment, retention and achievement. Its expanded definition of quality set out the desirable characteristics of learners (healthy, motivated, students), processes (competent teachers using active pedagogies), content (relevant curricula) and systems (good governance and equitable resource allocation). Although this established agenda for achieving good educational quality, it did not ascribe any relative weighting to the various dimensions identified nor did it provide demonstrations on *how* this was to be achieved. Similarly, the UN DESD Implementation Plan has failed to provide in-depth guidance on *how* quality is to be achieved.

Since the first Education for All Conference at Jomtien (UNESCO, 1990), African governments, through the support of international donors, have invested billions of dollars in programmes and reforms designed to improve access to quality education. It is generally accepted that quality education is an effective means to fight poverty, build democracies and foster peaceful societies. Quality education empowers individuals, gives them voice, unlocks their potential, opens pathways to self-actualisation and broadens perspectives to open minds to a pluralist world. There is no one definition, list of criteria, definitive curriculum or list of topics for a quality education. Quality education is a dynamic concept that changes and evolves with time and changes in the social, economic and environmental contexts of place. It is often locally relevant and culturally appropriate and thus takes many forms around the world. There is no consensus amongst the different national stakeholders on what quality education is and what is prioritised to achieve it.

In many sub-Saharan African countries, conceptualisation of what quality of education entails is informed by the international development community. The discourse on quality education has mainly focused on the improvements in quality and retention, educational efficiency and ability of learners to benefit from their educational opportunity, high drop-out and repetition rates and internal efficiency of an education system. The main focus of ESD is on how education can contribute to a sustainable development and a sustainable future. Without close attention to the acquisition of literacy and numeracy of sufficient quality, societies are unlikely to achieve sustainability in a modernising and globalising world. ESD provides a real-life context for the learning of literacy and numeracy skills. With its focus on culture, ESD also emphasises education-community links, use of local and indigenous knowledge in learning and connections with the local cultures and languages in the learning process as is discussed above and demonstrated by the different chapters in this book.

Quality education is understood differently in different countries in sub-Saharan Africa. This chapter draws on three conceptions of quality education (Barrett, Chawla-Duggan, Lowe, Nikel, & Ukpo, 2006; Tikly, Barrett, Nikel, & Lowe, 2010; Tikly & Barrett, 2011; Lotz-Sisitka, 2008, 2012):

1. *An efficiency/mastery discourse of educational quality.* This discourse seeks out mastery, efficiency and learner achievement and performance against set standards and expectations as measures of quality.
2. *An inclusivity/participatory discourse of educational quality.* This discourse seeks inclusion in the education system as its measure of quality – for example, if girls are included in school system, the quality of the system is seen to be higher or better; or if learners' views are included in pedagogical processes through learner-centred approaches, then the quality of the education is seen to be higher or better.

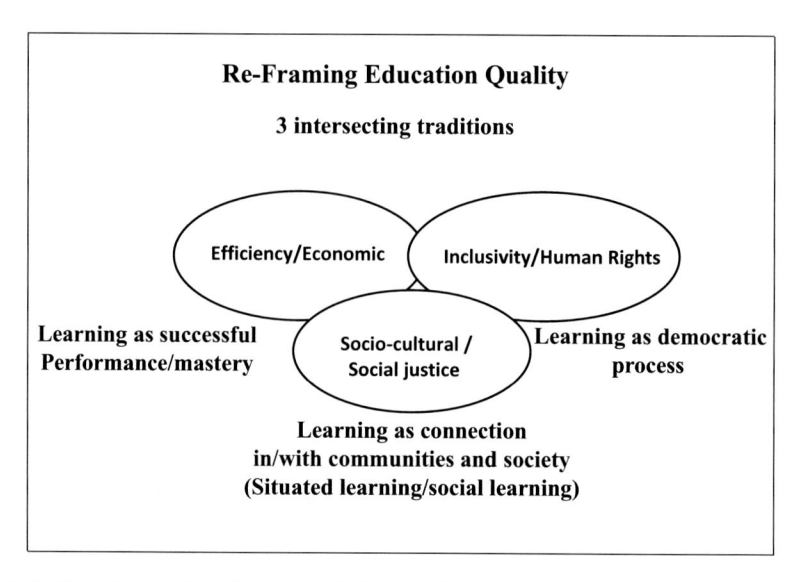

Fig. 1.2 Three intersecting discourses of educational quality which reflect both the acquisition, participation and relevance interests of ESD learning processes (Lotz-Sisitka, 2008; adapted from Barrett et al., 2006)

3. *A concept of quality that favours or emphasises the sociocultural* or the processes of meaning making that occur at the interface of existing experience and context and more abstract forms of representation most used in the symbolic practices in schools. To put it more simply, it is concerned with the meaning that occurs at the interface between context and concept. In the quality education literature, this is being expanded to include the notions of well-being and capabilities (Tikly et al., 2010; Tikly & Barrett, 2011), which also foregrounds a *social justice* discourse on educational quality. This concept also draws on sociocultural traditions of learning theory. In our work in the southern African region, we have come to refer to this using the concept of 'learning as connection' (Lotz-Sisitka, 2008, 2012; see Chap. 14, Shumba & Kampamba).

Lotz-Sisitka (2008) represented these the three conceptions for *Reframing Education Quality*, as shown in Fig. 1.2.

To further develop research trajectories and applied interpretations of this framework for ESD learning research, a review of ESD research in southern Africa (Lupele & Lotz-Sisitka, 2012) suggested that it was possible to identify the three key propositions that relate to ESD learning processes and educational quality. These are outlined below.

1.5.1 ESD Enables Learning as Connection

Learning as a connection expresses the relationship between meaning making, context and concept. For example, if a child is inducted into a complex scientific concept as fermentation through reference to his or her cultural experience of local beverage making, the scientific concept is easier to learn and grasp, particularly in contexts where learners are receiving instructions in second or third language (O'Donoghue et al., 2007). Most children in Africa are taught in western languages (as part of the colonial inheritance), which are in many cases difficult to learn. This requires that teachers make an extra effort to ensure that learners understand the vocabulary and concepts used in classrooms. Culturally located concepts and experiences can assist with enabling learners' ability to come to know more complex concepts presented in abstract languages.

One example of the notion of learning as a connection can be found in most African traditional education. Before missionary and colonial education, for the vast majority of African societies, education was an integral part of everyday life-long learning (Irwin, 1993; Mwanakatwe, 1968; World Wide Fund for Nature Zambia Education Project (WWF ZEP), 1999). Traditional education involved a detailed understanding of the local biological resources and sociocultural contexts of relational engagement and helped to develop knowledge and skills that enabled people to adapt to and manipulate their land, flora and fauna. By the time a child became a teenager, he/she had been exposed to an educational process facilitating the acquisition of knowledge and skills for survival and adulthood (Lupele, 2002).

Wals, van der Hoeven and Blanken (2009) argued that ecosystems are based on networks, mutual dependency, flexibility, resilience and sustainability. They observed that in order to cope with issues of a risk society (after Beck, 1992), we need learning processes that lead to a more reflexive society in which creativity, flexibility and diversity are released and used as tools to deal with the challenges of 'wicked' problems. A reflexive society requires reflexive citizens who critically review and alter everyday systems that we live by and that we often take for granted (Wals et al., 2009). This requires learners to constantly evaluate what they are learning in relation to the real world and to situations that they have experience of. This is *not* a conservative form of contextualism, but rather a sophisticated dialectical relationship involving everyday knowledge and school knowledge or formal learning and experiential learning. Relevant to this relational, dialectical thinking about learning in ESD, Capra (citied in Wals, 2007) argued that in trying to create a more sustainable world, we need to have a better look at how ecosystems work and become competent system thinkers. He described systems thinking as 'seeing connections and interrelationships, fine tuning functions and roles, utilising diversity, and creating synergies' (cited in Wals, p. 37).

In formal educational settings, learning as a connection has curriculum implications – how can the curriculum be developed and/or practised so that it allows learners to 'make connections'? Most African countries have started to work with curriculum reform that allows for localisation beyond setting test scores or examinations, as is seen by some of the chapters in this book (see all chapters in Parts II and III). Localised curriculum attempts to bring local knowledge into schools so that it can be recognised and so that it can be given respect. However, in some African countries, localised curriculum means superficial integration of local content into the existing curriculum. As shown in this book, curriculum localisation can mean much more than that if implemented via ESD learning processes.

Learning as connection also has teaching implications. It raises questions about the ways in which teachers can make connections and how learners make connections between school knowledge and everyday/indigenous knowledge. In localisation of curriculum, caution must be taken to avoid a conservative overemphasis on local context, which may lead to graduates failing to engage with the challenges of global market or the nature of global issues and the structural features of globalisation. Localisation of the curriculum requires critical engagement with local issues as these relate to wider systems of emergence and divergence. Sinyama (2012), in his study on curriculum localisation in Zambian primary schools, found that one of the challenges of implementing curriculum localisation approaches using ESD learning processes was teachers' abilities to formulate critical questions. This leads us to the second key proposition that relates to ESD learning processes and educational quality.

1.5.2 ESD Involves Critical Thinking, Action Competence, Agency and Developing Capabilities

The notions of critical thinking, capabilities (making choices about what is valuable to be and do), action competence (abilities to act) and agency (evidence of actions) are intertwined with power relations and with the struggles of how to bring out what is within us. They are also linked to language and how we express ourselves, what is said and how it is said. They are visible in actions and in what is done by whom and why.

Capabilities are those things that people value being and doing (Sen, 2001). For example, people might value being educated; or people might value being healthy; or people might value aspects of their communal or cultural lives. Few education systems in Africa allow children and parents the time and opportunity to fully conceptualise and express what they value being and doing and to work through what this means for education. Instead, it is simply assumed that what education offers is what is to be valued in society. Walker (2005), drawing on the work of Sen, defined the capability approach as an approach that is based on the notion of life and living as a combination of various valued doings and beings, with quality of life to be assessed in terms of our capability to achieve valuable functionings. ESD learning processes engage learners in questions of valuing and can therefore help to develop capabilities and valuable functionings. Walker argued that education in itself is a basic capability that affects the development and expansion of other capabilities. For example, education can expand a child's ability to swim or the ability to add up numbers correctly. Acquiring basic numeracy skills makes it possible for the child to develop more complex mathematical abilities, which in turn allows the child to participate in a wider variety of social practices. However, children can also deliberate what they value being and doing, and ESD learning processes allow children to engage in discussions about normative issues in society, for example, why it is important to recycle waste or why resources ought to be shared equitably amongst people and so forth. This helps learners to develop their abilities to choose what constitutes *valued beings and doings* at a broader level.

Action competence is defined as the ability to act (Jensen & Schnack, 1997). It develops 'learners' abilities to act at the personal and at the societal level' (1997, p. 6). Jensen and Schnack further argued that education is not about simple behaviour modification without understanding, but about creating a democratic process of participation in which students decide for themselves what action they will take. An action competence approach points to democratic, participatory and action-orientated teaching and learning that can help students develop their ability, motivation and desire to play an active role in finding democratic solutions to problems and issues connected to sustainability development (Finn & Karsten, 2010, p. 5).

Human agency, according to Walker (2005), is having the capacity to make informed and reflexive choices. She explained that educational action research is potentially a powerful tool in the capability approach because it requires us to evaluate how we are *doing well-being* and *doing fairness* in education (2005, p. 109). This requires critical thinking skills.

A review of ESD learning process literature and associated cases (Lupele & Lotz-Sisitka, 2012) showed strong support for ESD practices taking place in different contexts in sub-Saharan Africa that draw on the notions inherent to the capability approach, action competence and agency. These include:

- Undertaking local environmental action projects
- Developing and implementing school environmental policies and management plans through consultative processes and communities of practice
- Involving learners in action competence processes to strengthen their participation in decision-making
- Researching (including action research projects)
- Greening schools and communities (including permaculture projects)
- Celebrating commemorative days (e.g. World Wetlands Day, World Environment Day, HIV/AIDS Day, Youth Day, Human Rights Day)
- Supporting school clubs and competitions

A study involving 14 southern African countries showed that there is wide support amongst ESD practitioners in southern Africa for the use of participatory, active and learner-centred approaches to teaching and learning (Lotz-Sisitka, Gumede, Olvitt, & Pesanayi, 2006). The rationale behind the use of participatory approaches in ESD initiatives is that they contribute to capacity building and ownership of ESD initiatives amongst stakeholders and can lead to local action taking. They also encourage the use of methods which ensure effective implementation of ESD programmes and objectives and contribute to changes in educational practice, such as group work, research projects, experiential learning, presentations or theatre, and they help to situate learning. However, these approaches to ESD learning processes in African school settings are also hampered by shortages of suitable teaching and learning materials, low levels of literacy amongst participants, difficulties in accessing relevant information and different interpretations of *participatory* and *active* learning amongst participants (Lotz-Sisitka et al., 2006). This is not only the case in southern Africa, as Ndaruga (2003), writing from East Africa, reported that active learning pedagogies are critical for learner participation in ESD for improved wetland management. Similarly, Babikwa (2003) wrote from Uganda that deficient interpretation of educational concepts hampers the manner in which active critical approaches to learning can be developed and implemented in farmers' learning of sustainability practices.

Other examples of a commitment to active, participatory approaches to learning emerge from work undertaken in a Schools and Sustainability Professional Development Programme in South Africa that uses an active learning framework developed by O'Donoghue (2001) as a guide for supporting active learning and action competence development (Lotz-Sisitka & Schudel, 2007; Schudel, 2014); see Fig. 1.3). This active learning framework proposes the following important ESD learning processes: mobilising learners' prior knowledge and experience; identifying possible focus, risks or concerns that need to be investigated; seeking out new information on the issues or risks; undertaking enquiries; taking action; and reporting on findings.

Using this pedagogical framework, many teachers have been supported to implement ESD learning processes that are action centred and that help to develop learners' agency and their critical thinking skills. Research undertaken by Schudel (2014) showed how the use of this framework can lead to transformative praxis in environmental education. In working with this active learning framework in analysing empirical cases of active learning, Lotz-Sisitka and Schudel (2007) reported on various areas where ESD learning processes can be enhanced and deepened, all of which have implications for teacher education. They stated that

> reflections on the active learning lessons indicate that in each case, the focus ... was on gaining an understanding of the issue / risks, with less focus on deeper, more critical *analysis* of the issues. Causes and consequences were not probed, nor were relationships between a healthy environment and social justice issues explicitly probed There was generally a heavy reliance on learners getting information from community members, with little checking on the accuracy or validity of the information. Engaging in issues of local concern and pedagogical changes (i.e. use of different kinds of activities) are perhaps the most striking features of the lesson sequences. Available information used by the teachers was general in nature, and not necessarily always appropriate to the context. (Lotz-Sisitka & Schudel, 2007, p. 253)

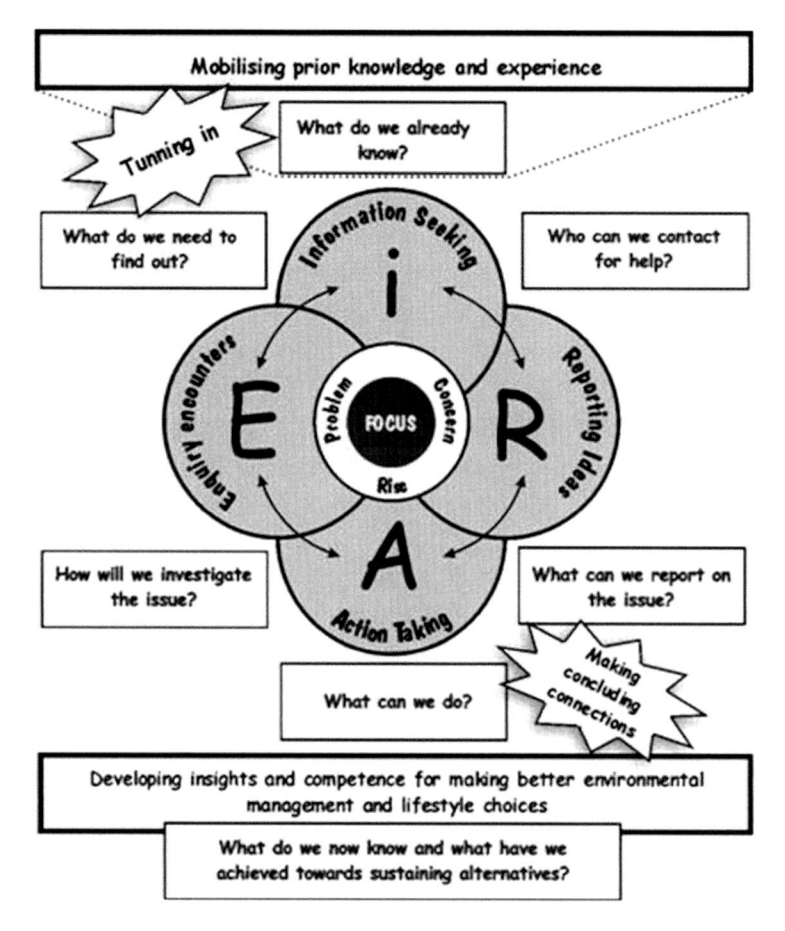

Fig. 1.3 Active learning framework (O'Donoghue, 2001)

This brings us to our third proposition on ESD learning processes and educational quality.

1.5.3 ESD Learning Processes Help to Make Education Relevant

In discussing education relevancy, Hawes and Stephen (1990) mentioned that education must be rooted in a society and a culture that learners can comprehend. An alien education is both unproductive and psychologically disturbing, often leading to a dangerous form of half learning. Sadly, many children in schools in Africa today are exposed to this dangerous form of half learning. Children can answer questions on the content, yet do not fully understand what they are being asked or why they are answering, because it has little connection with their daily lives and experiences. Along similar lines Shava argued that

> formal education usually suffers the setback that it is out of context with the learners 'lived environment'. This set-up creates two separate worldviews for the learner: the school world and the world in which they live. Indigenous knowledge should be integrated into mainstream education to enrich the learning environment and to put learning processes into context with the learners' living environment. (2005, p. 80)

To address these issues, various research projects have been undertaken into the context of schooling in Africa, with all researchers showing an interest in the relationship that exists between ESD learning processes and relevance, which forms part of a wider interest in educational quality. In their research on the relevance and quality of the VaRemba initiation school curriculum and its impact on formal schooling in a rural district in Zimbabwe, Chikunda and Shoko (2009) concluded that education would be considered to be relevant and of good quality if it develops learners' cognitive skills as well as their values and attitudes in a way that contributes to a wider reorientation of society towards equity and sustainability.

Hogan (2008) consulted with parents, learners, villagers and teachers to identify ways of contextualising and localising a module of the formal curriculum in Tanzania. Her research found that allowing local issues, such as the control of forest logging or the marketing of mushrooms, into formal education curriculum provided openings for students to 'get the 'insight and knowledge' and 'social skills' needed for their engagement in 'concrete action' with their communities for the environment' (Hogan, p. 55). Hogan's study provides useful findings on what happens when a curriculum is localised or contextualised. The issue of curriculum contextualisation and relevance is also demonstrated through an example of learners' experiences of foreign concepts in school, when ESD learning processes could so easily make these same concepts visible and accessible to learners. Lotz-Sisitka (2009) documented a case in which a child and his entire class failed to understand the scientific concepts of mutualism and commensalism, when asked to do so without substantive mediation. Simple forms of scaffolding demonstrated that the concepts in fact were obvious in the local fields where children played, and could be observed through watching birds benefit from the movement of cow's feet and lichen growing on trees.

Farrington (2008) also wrote about the benefits and relevance of place-based education in ESD, noting that in Africa little has been done to include youth in decision-making at the local, regional and global level. Firstly, Farrington identified that youth have multiple and multilayered identifications in place, which are influenced by their mode of transport (i.e. walking in this instance), their environmental concerns (linked to the well-being of the inhabitants of their communities in this instance), their responsibilities, their peer interactions and desires for solitude or company; and their cultural experiences of place. Secondly, Farrington identified that youth have changing (sometimes paradoxical, but not irreconcilable) attachments to place influenced by the hybrid intersection of global, urban and capital aspirations and fashions, attachments to local social and physical environments and growing freedoms of choice. Farrington reflected that 'the ease with which the young adults were able to maintain multiple identifications within global cultures (which stress change, opportunity and flexibility) and local cultures (which offer security and stability) appeared relatively effortless' (2008, p. 200).

These studies, and their emphasis on relevance, allow for engagement with hybrid influences of the global and the local. Farrington argued that learners' identifications with contexts and places are *socially situated* and include influences of global media and ideologies, but are simultaneously 'grounded in a general sense by their social and embodied interactions within their communities, families and peer groups' (2008, p. 183). As such, she argued that youth in South Africa are 'not the passive victims of the structural forces of globalisation, but are actively engaged in the world and with the circumstances and conditions that surround them' (Farrington, 2008, p. 183).

There is an increased understanding that a relevant quality education system should contribute to socio-economic development and wider development goals, such as equality, controlling the HIV/AIDS pandemic, social justice and democracy of any country. Relevance is critical for deliberating the applicability and value of knowledge, the applicability for the subject (himself/herself) and the applicability to society (see Schudel, Chap. 3). Engagement with knowledge questions in education is not simply about learning externally produced powerful knowledge, but more about how knowledge becomes powerful in the minds of learners as they engage with and act on critical concerns in their societies (Zipin, Fataar, & Brennan, 2015). However, relevance also goes beyond applicability and involves engaging learners in hybrid processes, such as dialogue and navigating between the local and the global, and in deliberating what solutions are needed in the future.

1.5.4 Conclusion: Towards a Programme of Action and Research

The SADC Regional Environmental Education Programme (SADC REEP, 2002, p. 21) noted that environmental issues are complex and context specific and arise in a range of different contexts. According to this principle, for any education to be relevant, it must aim to respond to contextual needs of a given society. Learning theory in the twenty-first century is also increasingly acknowledging the importance of taking cognisance of prior knowledge and the disposition, cultural context and needs of the learner, as well as the learners' identity formation and the relationship between knowledge, learning and societal change (Chabay et al., 2011; Holland et al., 2003). There are interesting questions that can be analysed further and researched in relation to ESD learning processes and the enhancing of quality and relevance in education. For example, are there contextual limits to applicability? What scope of contextuality is involved in establishing relevance? How do ESD issues, risks and contexts provide new frameworks for considering how relevance can be established in education? And what role does knowledge and pedagogy play in establishing relevance?

This chapter sought to synthesise some of what is known about ESD learning processes in sub-Saharan Africa as interpreted through a contextually situated literature and case analysis review (Lupele & Lotz-Sisitka, 2012). Our analysis provides a framework for further research on ESD learning processes and quality and relevance of education in schools in Africa. It also provides some key features or propositions that can provide a way forward for further research and praxis in the area of ESD learning processes and can help to develop theory and practice for a more fully understanding of ESD learning processes and their contributions to educational quality in schools in Africa.

The chapter has shown that ESD learning processes introduce a number of important propositions into educational thinking and particularly educational thinking about quality, namely, that:

- *ESD learning processes change the focus on learning to 'learning as connection' and a foregrounding of meaning making in the learning process as basis for efficacy and inclusivity.* This requires a wider engagement with different ways of knowing, or inter-epistemological dialogues, and changes in pedagogy.
- *ESD learning processes can enhance learners' capabilities (abilities to express and make choices about valued beings and doings), critical thinking and action competence (abilities to act) and agency (evidence of action).* This requires different forms of pedagogy, but it is notable that there is a high level of interest in these forms of pedagogy and these outcomes of education amongst ESD practitioners on the continent. This is because these approaches all foster change or change oriented learning. Pedagogical models exist to support such approaches to learning, and these have been tried out and tested in a number of settings.
- *ESD learning processes can enhance relevance of education.* This can take place, for example, through various strategies such as including ESD topics in localised curricula and through place-based educational activities or use of local audits. Relevance of education is not simply local, but involves learners in hybrid relationships between the local and the global and in identity formations and negotiating of directions and purposes for engaging in educational activities. ESD learning processes support such initiatives.

Finally, we put forward an argument that it is possible to improve educational quality in schools in Africa through ESD learning processes. ESD learning processes emphasise a broader notion of educational quality than that which has traditionally shaped educational quality thinking and practice in Africa. They bring *learning as connection* or sociocultural/social-ecological and social justice perspectives of quality into focus, to complement and extend notions of quality as efficiency and quality as inclusion. Together, a new framing of educational quality that takes account of local epistemology, cultures and meaningful practices, which help learners to make sense of abstract concepts and knowledge, is possible. Transformations are taking place in African education systems that take account of this notion of quality, although it has not been explicitly stated as such. It seems to be more driven by a bottom-up interest in relevance and the meaning of education in Africa, as can be seen in the different chapters in this book.

The propositions put forward in this book constitute a possible way forward to guide further research and practice, shaping the contribution and role of ESD learning processes in schools in Africa. We suggest that the perspectives raised in this chaptercan expand the debate on education quality perspectives and frameworks guiding schooling in Africa. This debate should be opened up to more key stakeholders including researchers, teachers, learners and communities. This will help in exposing any hidden ideologies that underpin current educational quality thinking in schools in Africa. It will also help in the design of education quality frameworks that address the specific needs of the subregion and that give adequate attention to the futures that children in African schools will have to inhabit, and how children are being prepared for these futures, shining a light on what learners are learning today, for tomorrow.

Acknowledgement This paper draws on and summarises key aspects of a more extensive report on ESD learning processes in Africa produced for the Southern African Development Community Regional Environmental Education Programme (Lupele & Lotz-Sisitka, 2012).

References

African Union (AU). (2014). *Africa agenda 2063. The Africa we want.* Retrieved from http://agenda2063.au.int/en/sites/default/files/agenda2063_popular_version_05092014_EN.pdf.

Asafo-Adjei, R. T. (2004). *From imifino to umfuno: A case study foregrounding indigenous agricultural knowledge in school-based curriculum development.* Unpublished master's thesis. Grahamstown, South Africa: Rhodes University.

Babikwa, D. (2003). *Environmental action to community action: methodology and approaches in community-based environmental education programmes in Uganda.* Unpublished PhD thesis, Grahamstown, South Africa: Rhodes University.

Barrett, A., Chawla-Duggan, R., Lowe, J., Nikel, J., & Ukpo, E. (2006). *The concept of quality in education: A review of 'international' literature on the concept of quality in education.* EdQuality Working Paper no.2. Retrieved July 1, 2010 from http://edqual.org/publications.

Beck, U. (1992). *Risk society. Towards a new modernity.* New Delhi, India: Sage.

Chabay, I., Collins, K., Gutscher, H., Pfeiffer, E., Schmidt, F., Schreurs, M., Sibenhuner, B., & van Eijndhoven, J. (2011). *Knowledge, learning and societal change: Finding paths to a sustainable future.* Draft Science Plan for a cross-cutting core project of the International Human Dimensions Programme on Global Environmental Change (IHDP).

Chikunda, C., & Shoko, P. (2009). Exploring the relevance and quality of the VaRemba initiation school curriculum and its impact on formal schooling in a rural district in Zimbabwe. *Southern African Journal of Environmental Education, 26*, 193–209.

Dussel, E. (1998). Beyond Eurocentrism: The world-system and the limits of modernity. In F. Jameson & M. Myoshi (Eds.), *The cultures of globalization* (pp. 3–31). Durham, NC: Duke University Press.

Farrington, K. (2008). Exploring place in a changing social context. In D. Wiley (Ed.), *Toxic belonging* (pp. 179–205). Cambridge, UK: Cambridge Scholars Press.

Finn, M., & Karsten, S. (2010). The action competence approach and the 'new' discourses of education for sustainable development, competence and quality criteria. *Environmental Education Research, 16*(1), 59–74.

Gerstenmaier, J., & Mandl, H. (2001). *Methodologie und Empirie zum Situierten Leren, research report no.137*. Munchen, Germany: Ludwig-Maximilians Universitat.

Gonzalez-Gaudiano, E. (2005). Education for sustainable development. *Policy Futures in Education, 3*(3), 243–250.

Gonzalez-Gaudiano, E., & Silva, E. (2015). Education: A road to nowhere or a path to a more sustainable future? A southern perspective. In D. Selby & F. Kagawa (Eds.), *Sustainability frontiers: Critical and transformative voices from the borderlands of sustainability education* (pp. 43–58). Opladen, Germany: Barbara Budrich Publishers.

Hawes, H., & Stephens, D. (1990). Questions of quality. In M. J. Kelly (Ed.), *The origins and development of education in Zambia* (pp. 146–149). Lusaka, Zambia: Image Publishers.

Hogan, R. (2008). Contextualising formal education for improved relevance: A case of the Rufiji wetlands, Tanzania. *Southern African Journal of Environmental Education, 25*, 44–58.

Holland, D., Skinner, D., Lachiacotte, W., & Cain, C. (2003). *Identity and agency in cultural worlds*. Cambridge, UK: Harvard University Press.

Intergovernmental Panel on Climate Change (IPCC). (2014). *Climate change 2014: Impacts, adaptation, vulnerability*. Cambridge, UK: Cambridge University Press.

Irwin, P. R. (1993). *Environmental education in Botswana with particular reference to pre-service primary teacher education*. Unpublished PhD thesis. Pretoria, South Africa: University of South Africa.

Jensen, B. B., & Schnack, K. (1997). The action competence approach in environmental education. *Environmental Education Research, 3*(2), 163–178.

Ketlhoilwe, M. J. (2008). *Supporting environmental education and education for sustainable development in higher education institutions in Southern Africa*. Howick, South Africa: SADC Regional Environmental Education Programme, Share-Net.

Leff, E. (2009). *Latin American environmental thought*. Paper presented at the 6th Latin American congress of environmental education. San Clemente de Tuyu, Argentina, September 19, 2009.

Lotz-Sisitka, H. (2004). *Positioning southern African environmental education in a changing context*. Howick, South Africa: SADC REEP/Share-Net.

Lotz-Sisitka, H. (2008). Environmental education and educational quality and relevance: Opening the debate. *Southern African Journal of Environmental Education, 25*, 5–12.

Lotz-Sisitka, H. (2009). Epistemological access as an open question in education. *Journal of Education, 46*, 57–79.

Lotz-Sisitka, H., Gumede, M., Olvitt, L., & Pesanayi, T. (2006). *ESD practice in Southern Africa: Supporting participation in the UN decade of education for sustainable development*. Howick, South Africa: SADC REEP.

Lotz-Sisitka, H., & Schudel, I. (2007). Exploring the practical adequacy of the normative framework guiding South Africa's national curriculum statement. *Environmental Education Research, 13*(2), 245–263.

Lotz-Sisitka, H. B., & Southern African Development Community Regional Environmental Education Programme (SADC REEP). (2012). *Learning together for a sustainable future: 15 years of Swedish-SADC co-operation*. Howick, South Africa: SADC REEP.

Lupele, J. K. (2002). *Action research case studies of participatory materials development in two community contexts in Zambia*. Unpublished master's thesis. Grahamstown, South Africa: Rhodes University.

Lupele, J. K., & Lotz-Sisitka, H. (2012). *Learning today for tomorrow. Sustainable development learning processes in sub-Saharan Africa*. Howick, South Africa: SADC Regional Enviornmental Education Programme.

Mbembe, A. (2001). *On the postcolony: Studies on the history of society and culture*. Los Angeles: University of California.

Mokuku, T., & Mokuku, C. (2004). The role of indigenous knowledge in biodiversity conservation in Lesotho Highlands: Exploring indigenous epistemology. *Southern African Journal of Environmental Education, 21*, 37–48.

Mwanakatwe, J. M. (1968). *The growth of education in Zambia since independence.* Lusaka, Zambia: Oxford University Press.

Mythen, G. (2004). *Ulrich Beck: A critical introduction to risk society.* London: Pluto Press.

Namafe, C. (2008). What selected basic schools in western Zambia are best at in environmental and sustainable education? *Southern African Journal of Environmental Education, 25*, 59–80.

Ndaruga, A. M. (2003). *An exploration of teacher perceptions and actions to conserve wetlands in Kenya.* Unpublished PhD thesis. Grahamstown, South Africa: Rhodes University.

O'Donoghue, R. (2001). *Environment and active learning in OBE.* Howick, South Africa: Share-Net.

O'Donoghue, R., Lotz-Sisitka, H., Asafo-Adjei, R., Kota, L., & Hanisi, N. (2007). Exploring learning interactions arising in school-community contexts of socio-ecological risk. In A. Wals (Ed.), *Social learning towards a sustainable future* (pp. 435–448). Wageningen, The Netherlands: Wageningen University Press.

SADC Regional Environmental Education Programme (SADC REEP). (2002). *SADC regional environmental education programme: Programme document 2002.* Howick, South Africa: SADC REEP/Share-Net.

Schudel, I. (2014). Exploring a knowledge-focused trajectory for researching environmental learning in the South African curriculum. *Southern African Journal of Environmental Education, 30*, 96–117.

Sen, A. (2001). *Development as freedom.* New York: Oxford University Press.

Sfard, A. (1998). On two metaphors for learning and the dangers of choosing just one. *Educational Researcher, 27*(2), 4–13.

Shava, S. (2005). Research on indigenous knowledge and its application: A case of wild food plants of Zimbabwe. *Southern African Journal of Environmental Education, 22*, 73–81.

Shava, S. (2010). *Indigenous knowledges: A genealogy of representations and applications in developing contexts of environmental education and development in southern Africa.* Unpublished PhD thesis. Grahamstown, South Africa: Rhodes University.

Shumba, O., Kasembe, R., Mukundu, C., & Muzenda, C. (2008). Environmental sustainability and quality education: Perspectives from a community living in a context of poverty. *Southern African Journal of Environmental Education, 25*, 81–97.

Sinyama. I. (2012). *Enabling social learning as a response to environmental issues through teaching of localised curriculum in Zambian schools.* Unpublished master's thesis. Grahamstown, South Afric: Rhodes University.

Tikly, L., & Barrett, A. M. (2011). Social justice, capabilities and the quality of education in low income countries. *International Journal of Educational Development, 31*(1), 3–14.

Tikly, L., Barrett, A.M., Nikel, J., & Lowe, J. (2010). *Understanding quality.* EdQual research programme consortium on implementing education quality in low income countries. EdQual Working paper no. 18b. University of Bristol and University of Bath. Retrieved October 2010 from www.edqual.com.

Tilbury, D. (2011). *Education for sustainable development: An expert review of processes and learning.* Paris: UNESCO.

United Nations (UN). (2015). Sustainable development goals. 17 goals to transform our world. Retrieved from http://www.un.org/sustainabledevelopment/sustainable-development-goals/

UNESCO. (1990). World declaration on education for all. Paris: UNESCO.

United Nations Educational, Scientific and Cultural Organisation (UNESCO). (2015). *Rethinking education. Towards a global common good?* Paris: United Nations Educational, Scientific and Cultural Organisation.

Vygotsky, L. (1978). *Mind in society.* London: Harvard University Press.

Walker, M. (2005). Amartya Sen's capability approach and education. *Educational Action Research, 13*, 103–110.

Wals, A. (2007). Learning in a changing world and changing in a learning world: Reflexively fumbling towards sustainability. *Southern Africa Journal of Environmental Education, 24*, 35–45.

Wals, A. E. J., van der Hoeven, N., & Blanken, H. (2009). *The acoustics of social learning: Designing learning processes that contribute to a more sustainable world.* Wageningen, The Netherlands: Wageningen Academic Publishers.

World Education Forum. (2000). *Dakar framework for action.* Dakar, West Africa: Senegal. 26–28 April.

World Wide Fund for Nature Zambia Education Project (WWF ZEP). (1999). *Environmental education manual for teacher educators.* Lusaka, Zambia: WWF ZEP.

Zaalouk, M. (2004). *The pedagogy of empowerment: Community schools as a social movement in Egypt.* Cairo, Egypt: The American University in Cairo Press.

Zipin, L., Fataar, A., & Brennan, M. (2015). Can social realism do social justice? Debating the warrants for curriculum knowledge selection. *Education as Change, 19*(2), 9–36.

Chapter 2
Situated Learning in Relation to Human Conduct and Social-Ecological Change

Rob O'Donoghue

This chapter traces how education has developed to provide orientation in a modern world that is characterised by emerging risk. It examines how ESD initially developed as a modernist process to enable social reorientation and has been centred on problem-solving engagement in relation to issues and risk. The intractable complexity of most social-ecological problems has meant that change-orientated and transformative imaginaries arising in learning are not easily realised in tangible change to resolve the problems at hand. The chapter thus poses the question, "Is ESD as situated learning with transgressive social-ecological reorientation possible?" To address this question, the study reviews ESD as a reflexive social process in modernity and tracks some of the expansive trajectories in the developing field over the last 10 years of the UN Decade of Education for Sustainable Development in Southern Africa.

To situate and contemplate learning interactions, the issues of waste, climate change and biodiversity loss are briefly examined as learning-led processes of situated social change. In each case the focus is on the interplay of context, life experience and new environmental knowledge to probe how ethics-led knowledge work is foundational in reflexive learning processes that might enable change. The three cases are then contemplated in a teacher professional development and school curriculum context as learning processes where knowledge acquisition with participation and reflexive learning are contemplated in and as open-ended learning progressions. The study concludes that an overemphasis on problems and problem-solving in ESD pedagogy needs reframing in relation to matters of concern and purposeful human conduct, so that situated learning progressions might more directly enable learning-led change that is satisfying and contributes to the common good.

R. O'Donoghue (✉)
Environmental Learning Research Centre, Department of Education, Rhodes University, Grahamstown, South Africa
e-mail: r.odonoghue@ru.ac.za

© Springer International Publishing Switzerland 2017
H. Lotz-Sisitka et al. (eds.), *Schooling for Sustainable Development in Africa*, Schooling for Sustainable Development 8, DOI 10.1007/978-3-319-45989-9_2

2.1 ESD and Modernity

Modern education is characterised by an increasing diversity of learning and change in imperatives that now include Education for Sustainable Development (ESD). The concept of ESD emerged as a proposition that, if successfully implemented, would address pressing issues and risks to enable humanity to mediate social life towards more sustainable ways of knowing and doing things together in a finite world. ESD is thus commonly centred on resolving social-ecological problems and sustainability in the modern world. To this end, ESD is usually directed at creating the awareness that will enable learning and change that resolves problems and their associated risks (UNESCO, 2014).

Popkewitz (2008) noted how modern education practices have developed with an instrumental character and have emerged as 'salvation narratives' that are characterised by 'a double gesture of hope and fear'. ESD thus commonly highlights the fearful prospect of social-ecological, economic and political risk as threatening problems of future sustainability, to be solved with education. As an education process for enabling society to solve problems, education practices have been relatively blind to how modernity is producing successively more wicked and global problems, an insight that led Ulrich Beck to conclude that modernity has developed as a 'risk society'. Beck (2009) noted how problems have been compounded by emerging solutions that have often, in turn, produced more and more complex and intractable problems. He noted how this complex has shaped a mechanism of 'reflexive modernisation' that is emerging as a process of reorientation in our modern world of 'wicked'[1] problems.

2.2 Waste as a Wicked Problem for Problem-Solving Education Practices

The issue of waste commonly includes a call for education practices to create awareness of the need to recycle paper, plastic and glass and so on, so that the problem can be resolved. Legislation and education have thus been initiated so that recycling is becoming 'the right thing to do' in the modern world. Waste education in ESD practices seldom probes how recycling is both energy intensive and polluting or that, with the exception of the recycling of organic matter, waste recycling is an unsustainable feature of modern-day technologies and cultural practices that needs to change. By treating waste as a problem to be solved with recycling, some ESD practices have not addressed the need for us to come up with new ways of doing

---

[1] Beck did not use the concept of wicked problems but this was developed in the sciences to note how a change to seemingly resolve a problem produces effects that change the game and shape other problems in a complex system.

things that transgress the modernist legislative norm by contributing to change that might avoid perpetuating the wasteful and polluting recycling of modernity.

Read in this way, it is apparent that early ESD in relation to the waste problem developed as a modernist education process of mediated reorientation that engaged the problem within a unitary modern order of recycling – even though this does not meet modernising ideals of being a resolution of the problem that is coherent, transparent and applicable to all. Bauman (1993) noted how contradictions such as this have bedeviled modernism and have shaped modernity as a culture that is in need of continuous reorientation as latent contradictions lead to ambivalence in the ordering and regulating of social life. Modern education practices (ESD) that promote recycling as the solution to a modernity-induced waste problem are engaging learners in a process of social change but are not producing tenable solutions with the necessary transgressive reorientation.

Where the intractable problems of waste and recycling have been addressed as a wicked problem in some ESD initiatives we have been involved in, they have primarily shaped creative social imaginaries that are *suggestive of* transgressive change but are difficult to translate into reorientated practices. This is common in activities that encourage creative reimagining and often includes deliberative ethics in relation to intractable problems, but with little successful translation into different ways of doing things that foster future sustainability. The learning is, however, engaging and useful in as much as reflexive critical deliberation opens the way to contemplating reimagined future possibilities that are outside the risk-producing normative practices of the day.

2.3   Case 1: Engaging with Waste on an ESD Short Course

In a 'changing practices' short course, we explored basket shopping and composting as practices to reduce recycling and waste towards a zero-waste ideal as a transgressive solution to modern waste production and management. This is done through a three-session learning process involving:

1. *Deliberation* on the concepts and issues, where people share and clarify their matters of concern
2. *Demonstration* of composting practices, with a creative reimagining of basket shopping to reduce waste bin contents
3. *Intervention* project reporting as learner-led shopping innovation and organic recycling towards a sustainability ideal of zero waste

Throughout the situated learning cycle of deliberation, demonstration and innovation, participants engaged in matters of concern and developed intervention experiments towards change projects that were sustained, narrated and shared as transgressive learning and innovation towards a sustainable future.

Given the brief overview of this first case of course-based co-engaged social innovation that can be read against the emerging history of our present ESD practices (above), this chapter attempts to probe how reflexive education processes might be initiated to enable *transgressive social-ecological reorientation* that might play out in more learner-led change. The foundations for this *oeuvre* in ESD developed through the recent review of the concept of ESD by UNESCO.

2.4 An Expanding Perspective on ESD in the UN Decade of ESD

In a recent clarification of the concept of ESD at the Aichi-Nagoya Global Conference (2014) as part of a review of the UN Decade of Education for Sustainable Development (UN-DESD), we noted how ESD had expanded since its formal inception a decade earlier. Four intermeshed dimensions were identified in the modern concept of ESD as processes of co-engaged learning for social change (see Fig. 2.1):

- Situated knowledge and systems thinking (*knowledge*)
- An ethics-led process in emergent context (*ethics*)
- A valuing and purposeful process of learning with and from others (*valued purpose*)
- Developing agency and skills in stewardship practices that bring about change (*practices*)

These reflect a change in the concept of ESD from its inception as a structural functionalist process to 'create awareness' and to 'get the message across' so as to change behaviour, into a more co-engaged and participatory process of learning to change (refer to Fig. 2.1). What was not sufficiently explored in the review was the character of ESD as a response to unsustainability and risk where education processes tend to develop as learning towards resolving intermeshed global problems like biodiversity loss, poverty, climate change, social injustice and global warming. A subtle shift from *problems* to a more explicit focus on *human conduct and situated matters of concern* emerged in a position paper developed to contemplate education practices into the Global Action Programme (GAP) in our southern African context of work (see O'Donoghue, 2014).

2.5 A Critical Review Towards ESD as a Process of Transgressive Reorientation

In the UNESCO DESD concept review process (O'Donoghue, 2014), emerging insights and evidence of ESD practice in its current form suggest that ESD is not fully realising its intent. This chapter is thus an attempt to probe ESD with a view to

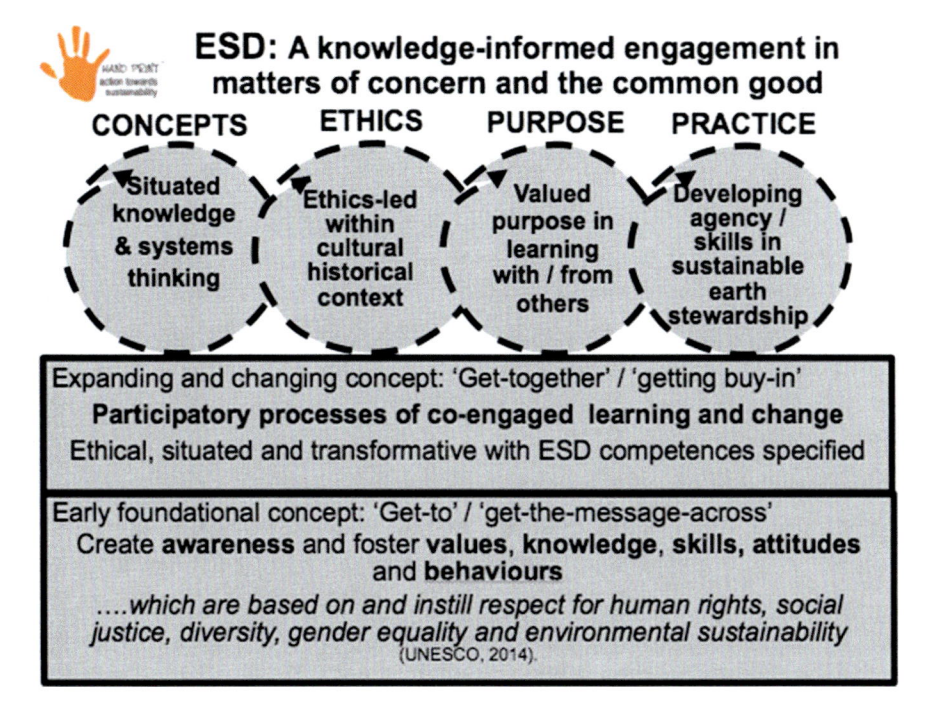

Fig. 2.1 Foundational and expansive trajectories in the concept of ESD (Adapted from O'Donoghue, 2014)

reframing the game within education practices that are more likely to produce reflexive learning and enable a transgressive reorientation towards a sustainable future.

ESD with transgressive reorientation is not an easy matter. Knowledge-led, problem-solving engagement (*structural functionalist*) has been limited because much of the knowledge engaging the complexity of wicked problems tends to circulate in contested deliberation that is, at times, all but impossible to resolve and take up into change practices. Participatory (*constructivist*) alternatives that focus on individual and collaborative change are a conventional wisdom that is playing out as social imaginaries of other possibilities that are not easily translated into change. *Competence* approaches that specify the attributes of people transitioning to sustainability have added a layer of direction to participatory processes but can be elusive and intangible in application. Similarly, a situated, *sociocultural* framing of learning can simply circulate conventional wisdom in participatory processes where prevailing dispositions are seldom overturned.

All these trends that are apparent in the twists and turns of ESD over the last 10 years have their strengths and enthusiastic promoters but the jury is out on which is likely to be best, particularly where little tangible change is evident in reviews of the UN-DESD. All that can be said at this juncture is that there is a positive enthusiasm for the concept of ESD and it has generated some momentum into the Global Action Programme and the forthcoming focus on global citizenship education.

The value of a critical review of prevailing ESD conventions is that it is pointing to many areas that need attention and it is surfacing central concerns that we are not yet adequately theorising and mediating in co-engaged learning processes. The central concern surfacing here is an inadequate grasp of how ESD might engage and enable transgressive learning practice towards future sustainability.

2.6 Sociocultural Reorientation with New Environmental Knowledge for Future Sustainability

The review has allowed us to begin to understand education responses to risk as initiatives to provide better social orientation – a long established function of education processes in the modern age. Noting this, we were able to deduce that modernity is a sustained cultural trajectory that is misaligned with the life support systems and processes of natural habitats and the global planetary systems as a whole. We are thus facing a double process of enabling better orientation in/of modern societies and reorientating global cultural practices to align better with how the intermeshed sociocultural orders and biophysical systems are functioning to sustain social-ecological life on earth.

This insight enabled us to conclude, in line with Beck's reflexive modernisation, that we are concerned with ESD as expanding processes of *sociocultural reorientation* that will have to be *transgressive* to take us beyond the modern conventional wisdom of the present day and the emerging polycrisis (Brito & Stafford Smith, 2012) that scientists have been mapping out and monitoring on a global scale. Here it is notable how many of the earlier problems like erosion, habitat degradation, waste and pollution became reinscribed as drivers of more and more intertwined issues that now extend to biodiversity loss, global warming and climate change.

As scientific research has detected and begun to monitor wider and deepening risk like climate and biodiversity, new environmental knowledge and systems thinking have been produced to represent what is going wrong and to point to some of what needs to be done to resolve emerging problems of future sustainability. The scientific literacy pathway of problem detection and monitoring has produced new environmental knowledge of the more and more intractable problems that are being taken up into education within new co-engaged pedagogy and using participatory approaches to learning and change that developed within the emerging concept of ESD during the past UN-DESD (as outlined above).

Differing knowledge streams converge in schools (see Fig. 2.2) as learners work with the funds of knowledge they bring into and encounter in the classroom. Here conventional wisdom is engaged alongside the disciplinary knowledge in the curriculum subjects. Environment and sustainability concerns are also in general circulation on the internet and in the media but most students in African contexts have limited access to these narratives. These intermeshed streams of knowledge often present the environmental problem as the focus of ESD, without adequate translation into matters of concern in a local context.

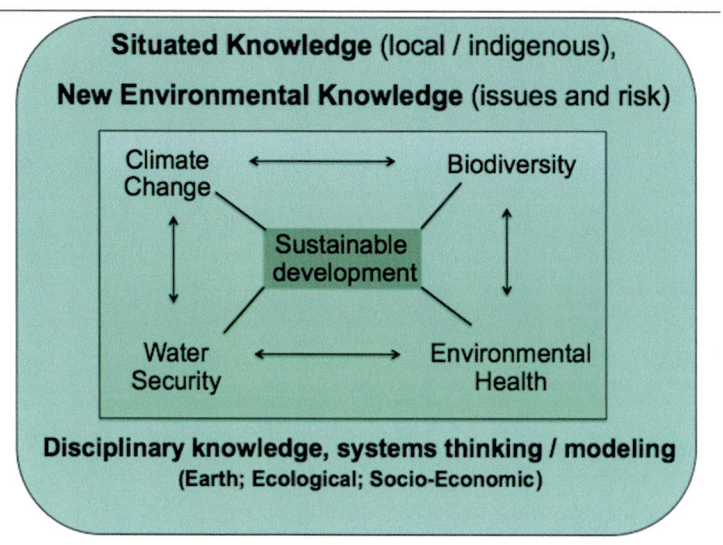

Fig. 2.2 Problems and knowledge streams in a learning context

The next case of global warming and climate change in southern Africa reflects how knowledge of a global problem was translated into matters of concern in a regional context and then related to patterns of practice and possibilities of change.

2.7   Case 2: Engaging a Global Problem as a Matter of Human Conduct and Changing Practice

Scientific knowledge and the monitoring of earth systems tell us that the inter-meshed problem of global warming and climate change is a matter of increasing CO_2 in the atmosphere that is retaining heat and powering global weather systems. Simply put and closer to home in the Eastern Cape in South Africa, these processes are:

Intensifying the southern forcing of the southern African high pressure system during the winter months to produce

… hotter and drier winters that can block the advance of some of the moist winter cold fronts from the Southern Ocean system

… and influence a later start to the summer rains in future seasons, with

… increasing seasonal variability in the weather, and

… more extreme events, with

greater frequency.

Translating this new environmental knowledge of the global system into matters of concern and human conduct, it is possible to track how a dry winter and a late start to the season have been episodic events in the Eastern Cape for hundreds of years but at a lower frequency than is now anticipated with global warming. Also in simplified overview, we now know that seasonal variability is primarily driven by how hot the subcontinent gets in the winter and how strong the southern ocean system is driven by the El Niño–La Niña phenomenon (ENSO).[2] The variation in the Southern Ocean system is primarily driven by the windblown movement of the surface waters of the southern Pacific Ocean and the episodic upwelling of cold deep waters that are drawn up or overrun and warmed in a cyclical process. This influences the moisture in the atmosphere and the drive of the ENSO phenomenon that reaches across into southern Africa.

When this knowledge is related to historical patterns of practice in the Eastern Cape, it can be noted that, primarily in dry years, the amaXhosa people migrated to the Zuurveld for around 4 months of the year when the grasses there were highly nutritious. In dry years this would allow the rolling grasslands of the Amatole to recover from a winter drought by responding to the rains. It is also notable that the amaXhosa 'broke the sod'[3] to capture water in the soil so that their croplands were ready for the summer rains, especially when these were late and the cattle would not have the strength to plough because of the reduced food in pastures after a winter drought. Today some commercial farmers still migrate their herds, but by truck, to other parts of the country during a dry winter season. Into modern times, most people also grow less food and many have forgotten how things were done in the past.

Addressing the problem of climate change through considering what change practices would enable people living in the Eastern Cape to adapt to and mitigate climate variability and extreme events enables learners to contemplate human conduct both in relation to responses to the conditions in context and in relation to the drivers of global change like fossil fuel use that increases atmospheric carbon. This experience has led us to approach learning as change practices towards a responsive *reorientation* of modern social practices in the face of global warming and climate change. Much of the learning has also been centred on planning for extreme events and the introduction of water conservation farming practices that are now being developed in a Water Research Commission project, Amanzi for Food (www.amanziforfood.co.za). Social imaginaries towards future sustainability have also included speculation on how living in a warming world will need to change with a reduction in the use of fossil fuels and the advent of a 'green economy'.

---

[2] The El Niño Southern Oscillation (ENSO).

[3] The Xhosa surface-dug and ploughed their fields, a process of *gelesha*, to enable water infiltration in anticipation of summer rains and the prospect that these could be late some seasons.

2.8 How Do We Work Less on Problems and More on Human Conduct and Situated Matters of Concern?

In problem-centred approaches, ESD is commonly approached as social-ecological and economic issues of sustainability in relation to, for example, global warming and climate change, biodiversity loss or pollution and food security. Instead of the abstracted 'surface making' of ESD as global problems in need of problem-solving responses, we are finding it educationally more useful to translate the big problems into 'matters of concern' related to current patterns of human conduct and looking to a sustainable future. The focus on situated *matters of concern* and current *human conduct* is not a trivial issue as it:

1. Locates the global issues in a life world and lived context for learners and their matters of concern
2. Decentres the fearful and intractable problem to a matter of reorientating human conduct and exploring change practices towards future sustainability

This enables a pedagogy of translation that decentres a focus on the big and scary problem so that participants engage with matters of concern that relate to the practices in their context. In this process, participants are able to develop an explanatory critique of human conduct that begins to enable them to identify absences[4] that are producing the risk. Contemplating changing practice then allows them to apply their creativity to doing things differently so that what is not there, and thus producing an undesirable condition, might be absented through some sort of creative intervention. In this way ESD is reframed as a situated and co-engaged process of identifying matters of concern that relate to the risk of global climate change. Learning in this way can enable participants to surface and to contemplate things that need to be changed. Here changing practices can simply involve identifying things that are not being done and could be done to mitigate or to 'absent' a problem in some way.

2.9 Case 3: Restorative Biodiversity and the Recovery of Lost Heritage Knowledge

In the ELRC Handprints for Change initiative with the Centre for Environmental Education, India, we began to explore practice-centred learning in ESD. An interesting project undertaken in the Eastern Cape involved knowledge related to the planting of trees in the semi-arid Eastern Cape with its seasonal variations that are influenced by oscillations in the Pacific and into the Southern Ocean (as outlined

---

[4] Roy Bhaskar (1993) explored the notion of the 'dialectic pulse of freedom' and 'absenting the absences' producing risk in his critical realism *oeuvre* in contemporary philosophy. These enable a real grasp of matters of concern and the reframing of conduct so that the undesirable conditions are not produced.

above). When the people of Grahamstown began to be concerned with the replanting of indigenous trees to maintain the diversity of wild plants and birds in the city, common knowledge was that their efforts were seldom rewarded as many indigenous plants would die after a few years, normally in the first extended period of hot dry weather after a dry winter. A scientist soon resolved the problem but, to our surprise, so had some rural herbalists who had been accustomed to transplanting saplings from the forests into their home gardens to make living hedges or to have some common medicinal plants close at hand. Their custom was to 'take the forest with the plant', in this way transferring mycorrhizae with the transplanted sapling. Put simply, mycorrhizae are a variable mix of soil flora that stimulate root growth in a mutualistic way so that both organisms benefit; the plant gets nutrients and moisture in an expanded root system and the microorganisms gain the moist nutritious conditions around the roots that they need to thrive. The trees that had been dying had been grown in heat-sterilised potting soil, a common practice in modern nurseries as the heating kills weeds and excludes some diseases that might inhibit plant growth.

Here again, the translation of the problem of biodiversity loss to a situated practice and something tangible to do at home not only related the matter to a local concern but allowed the active pursuit of an absence[5] that could be removed to make something positive possible. The Handprints for Change materials on biodiversity loss now tell this story in three parallel narratives (dying indigenous trees, indigenous knowledge practices and the modern scientific information on mycorrhizae), making these an interesting and challenging learning programme that can include some practical tree planting of indigenous trees to restore biodiversity in a town.

In this case, learning interactions developed across three funds of knowledge or literacy-enabling dialogical streams,[6] notably, everyday experience, heritage practices and disciplinary knowledge (see Fig. 2.3).

Across the cases of waste, global warming and biodiversity (above), situated matters of concern, learning engagement that draws on diverse realms of knowledge and reflexive processes of learning-led change were evident in expanding processes of ESD. These features of ESD had emerged in the Aichi-Nagoya review of the concept (above), leading us to conclude that clarifying an expanding concept of ESD might be contemplated as *knowledge and ethics-led* learning processes that are centred on achieving a changed, *valued purpose* through *changing practice*, transgressive processes involving changing literacies and changing practice. Here a reframing of ESD would have to be historically and socioculturally located (*emergent, situated and open ended*) as learning and ethics-led processes of co-engaged change initiatives (*participatory and purposeful within changing practice trajectories*).

---

[5] In this case the absence was the mycorrhizae that enable indigenous tree health and resistance to episodic drought conditions.

[6] Here I have drawn on the self-fashioning 'dialogism' after Bakhtin (Holland et al. 1998) to contemplate the situated grasp of a matter of concern as an emerging literacy steering learning-led change.

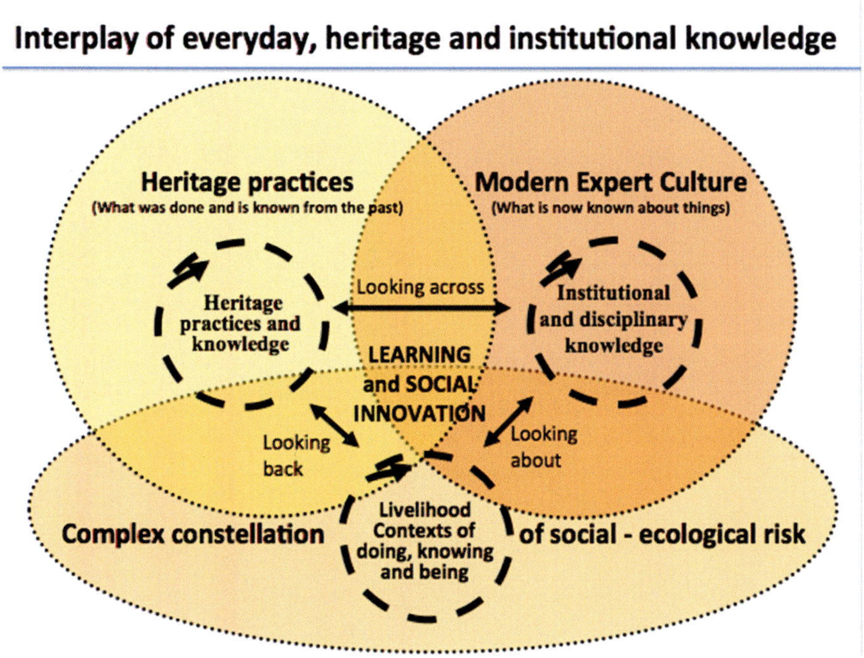

Interplay of everyday, heritage and institutional knowledge

Heritage practices
(What was done and is known from the past)

Modern Expert Culture
(What is now known about things)

Heritage practices and knowledge

Looking across

Institutional and disciplinary knowledge

LEARNING and SOCIAL INNOVATION

Looking back

Looking about

Complex constellation

Livelihood Contexts of doing, knowing and being

of social - ecological risk

Fig. 2.3 Interacting realms of knowledge

To successfully clarify, work with and integrate many of these attributes, one clearly needs a cultural-historical (CHAT) anchor that is orientated to a situated grasp of matters of concern. Here associated uncertainties and experience can surface contradictions that enable us to raise matters of concern. These reflexive processes of questioning engagement can enable learning-led change as reflexive processes within networked communities with the purpose of working towards future sustainability and the common good (ethics).

This direction-seeking overview allows one to see how ESD appears to be taking on a plural character that is more contextually situated and practice orientated in ways that draw together a variety of literacy streams.

Once again, doing something with these insights is not an easy matter. ESD has come to be positioned in critical opposition to the way that education is being enacted, particularly in relation to a knowledge-centred curriculum and the management of day-to-day schooling today. There are thus many debates around the constraints of knowledge stipulation, testing and numerous other aspects of the cultural conventions associated with modern education. Many of the deliberations have merit but what is often missed is that at the heart of most processes of social change is usually a reflexive process of transformative praxis. Recognising this in relation to both the practice of schooling and the processes of ESD within this, the approach of the Fundisa for Change (www.fundisaforchange.co.za) initiative has involved beginning to work reflexively with teachers within the prevailing culture of

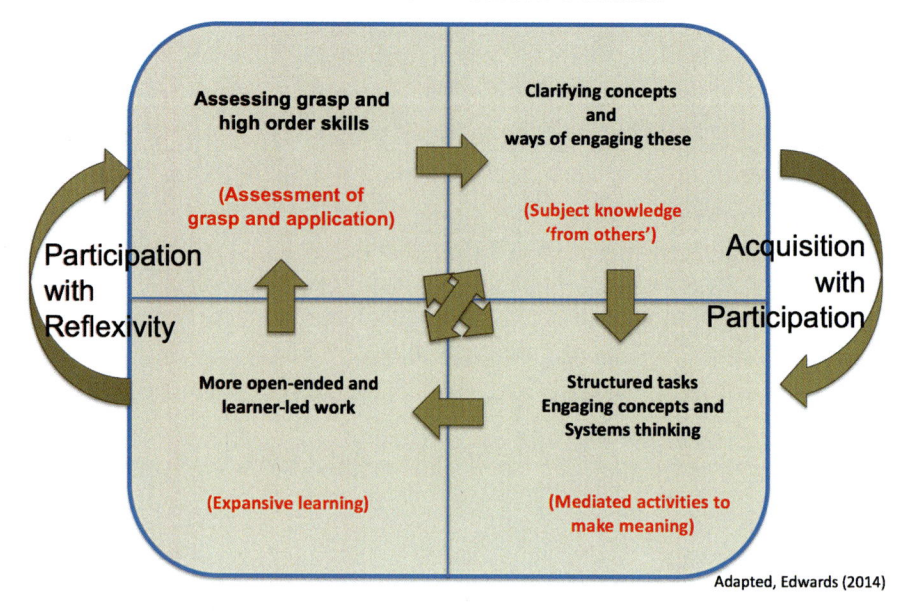

Fig. 2.4 A lesson sequence with acquisition, participation and open-ended reflexivity (Adapted from Edwards, 2014)

schooling. This required the framing of professional development activities supporting teachers to:

- Know their subject (new environmental content knowledge)
- Improve their teaching practices (methods and lesson planning sequences)
- Improve their assessment practices (marking and testing)

Engaging teacher professional development in this collaborative way also included teachers formalising their work in professional learning communities and developing ESD as expansive innovations within the knowledge-led curriculum to the South African Curriculum and Assessment Policy Statement (CAPS). The section that follows frames the emergence of ESD in a knowledge-led curriculum as a process that is centred on improving education quality and access.

2.10   Reviewing Each Case in Relation to an Open-Ended Knowledge-Led Learning Progression

Edwards (2014), working with a Vygotskian learning sequence (see Fig. 2.4), illustrated how relevant learning appears to emerge through situated teaching for concept acquisition that includes enabling learner-led participation. Here, processes of reflexive critical deliberation (higher-order skills) are made possible by the acquisition of knowledge and attendant cognitive skills developed through careful work with concepts and working with these ways to shape meaningful learning.

Vygotsky, elaborating on learning around his concept of a zone of proximal development, noted how:

> … learning awakens a variety of developmental processes that are able to operate only when the child is interacting with people in his environment and with peers. Once these processes are internalised they become part of the child's independent developmental achievement. From this point of view, learning is not development; however, **properly organised learning results in mental development and sets in motion a variety of developmental processes that would be impossible apart from learning**. (1978, p. 90, my emphasis)

Here, situated participation is a key driver of developmental learning as Bakhtin (1981) noted:

> The word in language is half someone else's. It becomes 'one's own' only when the speaker populates it with his own intentions, his own accent, when he appropriates the word, adapting it to his own semantic and expressive intention. (p. 293)

In the development of a Vygotskian-based learning sequence, Edwards (2014) emphasised how key concepts are embedded and made real in the ways they are represented in and out of a context of representation. When we teach, we not only teach the concepts alone but represent and model the situated literacies that have shaped these with a degree of reality congruence within the context of their expression. In good teaching, the words for new ideas 'come to us from others' and 'we make them our own' in dialogue (Bakhtin) with an attendant learning engagement in the company of others that sets in motion developmental processes that can, as we will endeavour to illustrate, be transgressive in relation to the adequacy of the cultural practices of the day and the risks that these are producing.

The value of the Edwards' learning sequence is as a model of process for classroom pedagogy where acquisition of new concepts as the words of others and the ideas being advanced by particular literacy trajectories can be taken up and then explored, clarified and developed in learner-led work into an ESD context of changing practice for future sustainability.

Each of the examples from African contexts briefly sketched in this chapter reflects processes of acquisition with participation. Here the call is for participation that is better situated and more learner led and exploratory. These processes can emerge together in dialogical processes at the level of individuals and groups. Here critical engagement in relation to a matter of concern (issue) and human conduct (practices) is orientated within the interacting social-ecological context of daily life.

Waste recycling has discussion and demonstration processes where acquisition and learner-led deliberative work interplay in reflexive critical review of current practices and shape transgressive innovation experiments to develop and share practices that are both realistic in context and are better orientated in and with the world.

The climate change case shifts the complexity of the global threat into a deliberative conversation in relation to the context of the Eastern Cape. Here, there is also an elaborating interplay between how the developing literacy of climate science information is able to explain climatic variation and how it is escalating with global warming and an attendant change in the seasonal cycles in southern Africa. This case is rich in intersecting literacies that resonate and build more object-congruent knowledge that might enable transgressive change that reduces risk by absenting misaligned dispositions and practices that are driving global warming.

Finally, the biodiversity example reflects a wonderfully dialectical interplay between the problem of the indigenous trees dying and the literacies of science and indigenous knowledge reflected in the uncovering of the key role of mycorrhizae in semi-arid environments.

Seen in overview, each could be framed by a creative teacher into an open-ended lesson sequence where learners are engaged in taking up the words (concepts) and ideas in/of diverse literacy threads (discourses) into deliberative meaning-making towards, where necessary, transgressive social innovation with more reality-congruent orientation in a finite social-ecological world of and at risk.

Research on transgressive processes of reorientating intervention and innovation is needed to probe how changing practices to absent risk might become characterising features of environment and sustainability education curriculum and social learning practices.

References

Bakhtin, M. (1981). The dialogic imagination. In M. Holquist (Ed.), *Four essays by M.M. Bakhtin* (C. Emerson & M. Holquist, Trans.). Austin, TX: University of Texas Press.

Bauman, Z. (1993). *Postmodern ethics*. London: Blackwell.

Beck, U. (2009). Critical theory of world risk society: A cosmopolitan vision. *Constellations, 16*(1), 3–22. Oxford: Blackwell.

Bhaskar, R. (1993). *Dialectic: The pulse of freedom*. London: Verso.

Brito, L., & Stafford Smith, M. (2012, March 26–29). *Planet under pressure*. State of the planet declaration: New knowledge towards solutions conference, London.

Edwards, A. (2014). Designing tasks which engage learners with knowledge. In I. Thompson (Ed.), *Task design, subject pedagogy and student engagement*. London: Routledge.

Holland, D., Lachicotte, W., Jr., Skinner, D., & Cain, C. (1998). *Identity and agency in cultural worlds*. London: Harvard University Press.

O'Donoghue, R. B. (2014). ThinkPiece: Re-thinking education for sustainable development as transgressive processes of educational engagement with human conduct, emerging matters of concern and the common good. *Southern African Journal of Environmental Education, 30*, 7–26.

Popkewitz, T. (2008). *Cosmopolitanism and the age of school reform: Science, education, and making society by making the child*. London/New York: Routledge.

UNESCO. (2014). *Shaping the future we want: UN decade of education for sustainable development (2005–2014)* (Final Report). Paris: UNESCO.

Vygotsky, L. (1978). *Mind in society*. London: Harvard University Press.

Chapter 3
Deliberations on a Changing Curriculum Landscape and Emergent Environmental and Sustainability Education Practices in South Africa

Ingrid Schudel

This chapter describes Environmental and Sustainability Education (ESE) in South Africa against the backdrop of a changing educational system. It discusses changing educational imperatives in post-apartheid South Africa and how these have been interpreted and applied in three curriculum revisions in South Africa since 1994: Curriculum 2005, the Revised National Curriculum Statements and the Curriculum Assessment Policy Statement. The narrative also examines a changing picture of teacher professional development in South Africa highlighting how constructivist and social realist understandings of education, development and learning have entered the discourse and influenced practice in South Africa. The chapter concludes with highlighting how a relational approach to learning, as evident in ESE practices in South Africa, can help to avoid a pendulum swing between the dichotomies of competing discourses.

South African ESE practices have evolved in a post-apartheid context of radical and regular curriculum change. The study draws on illustrative examples of research and practice from two national South African ESE programmes, namely, the National Environmental Education Programme for General Education and Training (NEEP-GET), which operated during the early post-apartheid educational changes in South African, and the Fundisa for Change teacher education programme which was started as a pilot programme in 2011. This chapter sets out to describe this evolving educational context, examining the ideological and national imperatives of the time, as well as the curriculum, pedagogical and teacher professional development trends emerging from these imperatives.

I. Schudel (✉)
Environmental Learning Research Centre, Department of Education, Rhodes University, Grahamstown, South Africa
e-mail: i.schudel@ru.ac.za

© Springer International Publishing Switzerland 2017
H. Lotz-Sisitka et al. (eds.), *Schooling for Sustainable Development in Africa*, Schooling for Sustainable Development 8, DOI 10.1007/978-3-319-45989-9_3

3.1   Educational Ideology and National Imperatives

The post-apartheid South African curriculum had strong social and economic agendas to ensure that citizens had access to economic opportunities in a prosperous new South Africa. This agenda was an important reaction against the apartheid curriculum's agenda to 'prepare different groups for dominant and subordinate positions in social, political and economic life' (Harley & Wedekind, 2004, p. 195). The response to these social and economic exclusions was taken up in the 1995 South African *White Paper on Education and Training* which called for a 'prosperous, truly united, democratic and internationally competitive country with literate, creative and critical citizens leading productive, self-fulfilled lives in a country free of violence, discrimination and prejudice' (South Africa. Department of Education [DoE], 2002a, p. 4).

One of the strategies for a more accessible and equitable system was a new National Qualifications Framework which placed all forms of learning on a single qualifications ladder to enable previously educationally and economically marginalised South Africans smoother access into the educational system, the working world and the South African economy.

In the schooling sector, educational access was also a priority, and research priorities shifted from access as indicated by enrolment statistics to access as indicated by more complex issues of attendance after enrolment (Community Agency for Social Enquiry & Joint Education Trust, 2007; Nelson Mandela Foundation, 2005; South African Human Rights Commission, 2010). School performance and education quality became another important priority in the early years of system transformation with particular concerns raised around literacy and numeracy amongst school-going learners (South Africa. Department of Basic Education, 2011a; South Africa. Department of Education, 2008). Fifteen years into democracy, South Africa is still faced with ongoing inequality in the country evident throughout social, economic and educational systems (Jahan, 2015; Makholwa, 2015; Spaull, 2013).

Also the South African education system has undergone a significant educational shift in its pedagogical thinking influenced by the above-mentioned White Paper which called for a 'shift from the traditional aims-and-objectives approach to outcomes-based education' (South Africa. Department of Education, 2002a, p. 4). Christie (1997) explained how the notion of outcomes-based education has its roots in the 'competence' trends that were being trialled in a few countries around the world at the time of transformation in South Africa. However, by the time South Africa entered the debate, the notion of competences was being heavily criticised for narrow instrumentalist applications of the idea in practice (Dwyer, 1995 in Christie, 1997). Strongly influenced by the Australian system at the time, efforts were made in South Africa to revive the ideal of competences as embracing 'the ability to apply skills to performing a task, and encompass theoretical understanding of the task, as well as the ability to transfer knowledge, skills and understanding to another context' (ibid., p. 56). The importance of this view of outcomes will become evident as the narrative of this chapter unfolds.

Closely related to competence is the concept of knowledge. From the above definition of competence, it is clear that knowledge works in tandem with competence and that knowledge of underpinning ideas as well as of different contexts is crucial to understanding why something should be done, how it should be done and how it might be done differently in different contexts. That reflects a view of knowledge as dynamic, applied and relational (Schudel, 2014) rather than static, disembodied from the knower and disembedded from the context in which it emerges.

The applied and practical nature of knowledge is highlighted in the White Paper with the argument that the integration between education and training implies 'a view of learning which rejects a rigid division between 'academic' and 'applied', 'theory' and 'practice', 'knowledge' and 'skills', and 'head' and 'hand''.

Also pertinent to the discussion on applied and practical knowledge is the argument that everyday knowledge can provide a useful context in which to apply and try out (practice) knowledge. Taylor agreed that 'the two-way relationship between everyday and school knowledge provides important pedagogical tools for inducting learners into the art of formal discourse, and for the practical application of formal knowledge to problems in the real world' (1999, p. 113). Yet at the same time, he warned of a naïve application of the emphasis on local knowledge in the hands of a teacher without strong conceptual frames, where 'school knowledge is totally submerged in an unorganized confusion of contrived realism' (Taylor, p. 121). This is an important argument warning about the dangers of extreme or naïve interpretations of educational philosophies such as constructivism.

Yet, the discussion on local and real-world knowledge takes a different turn when revolving on differences between constructivist and social realist philosophies. Since the constructivist interests of the early post-apartheid outcomes-based curriculum, South African education has more recently been influenced by the social realist imperative of 'bringing knowledge back in' (Wheelahan, 2010; Young, 2008). Tensions still remain in the country between these two perspectives and their interpretations and application. Authors such as Taylor (arguing from a social realist perspective) have been criticised for privileging of specialised disciplinary knowledge over 'everyday' knowledge (Zipin, Fataar, & Brennan, 2015) as is evident in statements from Taylor such as that 'real world examples, even where they are appropriate illustrations of formal knowledge, [can] obscure an understanding of the logical relations which constitute the ultimate goal of school knowledge' (Taylor, 1999, p. 115). While Taylor is talking specifically about schooling above, it would seem problematic to undermine the value of 'real world examples' when according, for example, to the South African National Development Plan, 'Knowledge is the systemically integrated information that allows a citizen, a worker, a manager, or a finance minister to act purposefully and intelligently in a complex and demanding world' (South Africa. National Planning Commission, 2013, p. 94).

A second problem with Taylor's argument is not acknowledging that a relational exploration between 'real world' and 'abstract knowledge' can also constitute an exercise in 'logical relations'. This appears to be typical of the dichotomous argumentation highlighted by Zipin et al. in their critique of social realism. Part of their critique revolves around social realist arguments that everyday knowledge is limited by its emphasis on 'segmentation (not integration), locality (not generality), context dependence (not context-independence), and contradiction (not coherence and commensurability) across contexts' (2015, p. 19). Instead of insisting on a choice between each of these dichotomies, environmental educators (in agreement with Zipin et al.'s argument for a dialectical approach to curriculum) embrace the possibilities for rich learning experiences enabled by exploring the relationships between each respective 'opposing' perspective. That is, encouraging 'play on difference'.

That being said, there are some useful relational perspectives on knowledge that are highlighted by social realist philosophy. For example, Schudel (2014) drew on Maton's (2014) descriptions of ontic epistemic relations (ontic relations between knowledge and its objects of study) and discursive epistemic relations (discursive relations between different knowledge) in order to develop a description of environmental knowledge in the South African curriculum. For example, this paper argues the importance of relating local to global knowledge in a way that highlights context-rich rather than context-bound knowledge, thus highlighting the relationship between local and global perspectives and context-dependent and context-independent knowledge.

In a second example of 'play on difference', O'Donoghue drew on a case of a preservice science education course to illustrate how exploring pedagogical experiences at the 'plural interface of lived world experience (for some students this included indigenous knowledge), formal institutional concepts (natural sciences) and the issues of the day (environment and sustainability concerns' can strengthen epistemological access and the relevance of learning in schools (2016, p. 168). These two particular examples are chosen in response to a contentious issue around constructivist pedagogies that guided the development of outcomes-based education in South Africa. They highlight, as does Michelson (2004, in Zipin et al., 2015), that an emphasis on locatedness does not necessarily have to compromise the rigour of knowledge.

Another important perspective on knowledge raised by those countering the practices that have emerged from the social realist emphasis on specialised knowledge is stimulated by questions around ethics and knowledge. The emphasis on specialised subject knowledge has inadvertently led to a pendulum swing in the curriculum with the prioritisation of a social-epistemological over a social-ethical basis for authorising curriculum choices (Zipin et al., 2015). From this concern I argue that ethics is in a social-ethical-epistemological relationship with knowledge in a way that enables processes of questioning the selective privileging of 'cultural capital associated with relatively powerful social-structural positions' (Zipin et al., p. 26). This call for highlighting the ethical dimensions in education is key to international aspirations for sustainable development as promoted by the United Nations Educational, Scientific and Cultural Organisation (UNESCO)'s call for 'education for the common good' underpinned by a 'humanistic approach, stressing the need for an integrated approach to education based on renewed ethical and moral foundations … a process that is inclusive and does not simply reproduce inequalities' (UNESCO, 2015, p. 17).

This humanistic ideology is evident in the South African national Fundisa for Change programme which focuses on ESE in teacher professional development. The programme states that:

> Massive over-consumption of resources and continued environmental degradation are undermining the natural systems we depend on, and impacting most severely on the poor and marginalised people in our society ... We can use our imaginations and problem-solving skills to creatively develop diverse solutions to our problems, and to change how we live for the better. (Fundisa for Change Programme, 2013, p. 3)

This next section will discuss how the persistent and changing ideologies, philosophies and imperatives described above have been taken up in an equally volatile curriculum policy environment in South Africa.

3.2 School Curriculum

A milestone for environmental educators in the transforming education and training system was a statement in the above-mentioned White Paper calling for an interdisciplinary approach to environmental learning to be included across all phases and levels of the education and training system (Lotz-Sisitka & Raven, 2001). This White Paper imperative arose from the ANC environmental desk, which actively supported an environmental focus in educational deliberations for post-apartheid curriculum transformation in the early 1990s in association with the Environmental Education Policy Initiative and its follow-up structure, the Environmental Education Curriculum Initiative (Lotz-Sisitka, 2002). The response to this in the post-1994 school curriculum – Curriculum 2005 – was to include 'environment' as one of the legislated 'phase organisers'. Phase organisers were a curriculum design tool to deal with significant crosscutting issues in society.

This cross-curricular approach was supported by the reorganisation of the South African curriculum into 'learning areas' instead of 'subjects'. A learning area was defined as 'a field of knowledge, skills and values which has unique features as well as connections with other fields of knowledge and learning areas' (South Africa. Department of Education, 2002a, p. 9). This meant a curriculum structure with more fluid boundaries and opportunities for integration across different 'fields of knowledge'. Integration became a key feature of lesson planning, and teachers were expected to integrate across learning outcomes and across learning areas. Learning outcomes were the foundation of the new outcomes-based Curriculum 2005 and referred to skills which learners were expected to master and demonstrate in order to proceed to the next curriculum level. The depth and scope of each learning outcome were detailed for each grade through a number of 'assessment standards'.

In 2000 in a ministerial committee review of Curriculum 2005, Chisholm (2000) recommended a streamlining of the design features of the curriculum. A result of the streamlining was that the revised curriculum, the 2002 Revised National Curriculum Statement (RNCS), no longer had environment as a phase organiser.

ESE work at this time was based on the environmental focus integral to each learning area. This focus was explicit in learning area statements, learning outcomes, assessment standards and core knowledge foci of each learning area (Schudel, 2014). This interest in environment as integral to curriculum was already a strength in the environmental community as South Africa underwent a third major revision of its curriculum, and the new Curriculum and Assessment Policy Statements (CAPS) reintroduced the notion of curriculum as arranged into 'subjects' rather than 'learning areas'.

CAPS was implemented in South Africa in 2012 and included an emphasis on content as an important foundation for learning. This was clearly influenced by the 2009 national curriculum review which highlighted that 'a key dimension related to the successful implementation of curriculum relates to the detail and clarity provided by policy in relation to what to teach' (Dada et al., 2009, p. 47). The highlighting of content in the latest curriculum change has been strongly supported by the social realist imperative of 'bringing knowledge back in' and the increased emphasis on specialised knowledge as discussed in the previous section (Wheelahan, 2010; Young, 2008). 'High knowledge' was already identified as an important principle in the RNCS, was an explicit and fundamental departure from Curriculum 2005 (Dada et al.) and is still an important principle of the CAPS (South Africa. Department of Basic Education, 2011b). The strategy identified for the development of such knowledge was to develop a 'stronger base from which to enable the development of a high level of skills and knowledge by all. It does so by specifying the combination of minimum knowledge and skills' in different learning areas and grades (South Africa. Department of Education, 2002a, p. 12).

One aspect of 'high knowledge' is, for example, the CAPS curriculum insistence that the call for detail and clarity in terms of what to teach does not mean an atomistic approach to content knowledge. This is illustrated in the point made in the Natural Science Intermediate Phase CAPS curriculum which states that 'The main task of teaching is to build a framework of knowledge for learners and to help them make connections between the ideas and concepts in their minds – this is different to learners just knowing a lot of facts' (South Africa. Department of Basic Education, 2011c, p. 11).

This new focus on content, the implications of 'making connections' and the meaning of knowledge became a key interest for environmental educators (Fundisa for Change Programme, 2013; Schudel, 2014). Yet this making of connections, from an environmental perspective, is more than simply making connections between ideas and concepts presented amongst stated topics in the curriculum, but includes making connections by highlighting relationships such as those between 'local contexts, while being sensitive to global imperatives' (South Africa. Department of Basic Education, 2011b, p. 4), 'the relationship between indigenous knowledge and science' (South Africa. Department of Basic Education, 2011, p. 13) and the relationship between 'topics and content [and] a particular discovery or a particular scientist' and between school science and lives outside the school (South Africa. Department of Basic Education, 2011d, p. 17). These extracts from the subject of Life Sciences illustrate that the CAPS curriculum has not lost the constructivist influences of early South African curriculum reform despite the renewed emphasis on specialised content knowledge.

Besides 'making connections', a second important perspective on curriculum content knowledge is to see it in the *critical and active* light as emphasised in the RNCS's mission to develop the 'critical and active citizen' (South Africa. Department of Education, 2002a, p. 8). The same principle of critical and active approaches has been carried through into the CAPS, which call for 'encouraging an active and critical approach to learning, rather than rote and uncritical learning of given truths' (South Africa. Department of Basic Education, 2011e, p. 3).

The coupling of the words 'critical' and 'active' is significant for environmental educators with the implication that curriculum interpretation should not simply be the internalisation of the foundations of knowledge, but one that promotes a reflexive approach emerging from a willingness and ability to 'question (and break away from) existing routines, norms, values and interests' (Wals, 2007, p. 38). Such a reflexive approach is important for ESE and the humanistic approach described above which promotes strong counter-hegemonic (going against) and expansive (going beyond) imperatives as evident in a call for a 'stand against violence, intolerance, discrimination and exclusion. Regarding education and learning, it means going beyond narrow utilitarianism and economism to integrate the multiple dimensions of human existence' (UNESCO, 2015, p. 10).

A third important perspective on curriculum is influenced by the concern raised in the previous section regarding the privileging of the social-epistemological over the social-ethical. Such concerns have already been raised by other authors even within the time of the RNCS during which Carrim and Keet (2005, p. 107) argued against what they called 'a minimum infusion option of human rights ... propelled by the contradictory "pulls" and "pushes" of human rights and democracy, and capitalist development simultaneously'. I argue that teacher professional development needs to highlight ethics in the CAPS context in a way that brings into question prevailing perspectives and thus questioning of the selective privileging of certain more powerful knowledge (Zipin et al., 2015). One of the ways of doing this is working with and emphasising the principle (persistent across all post-apartheid versions of the curriculum) highlighting the relationship between human rights, a healthy environment, social justice and inclusivity (South Africa. Department of Basic Education, 2011b; South Africa. Department of Education, 2002a). This principle has consistently provided an important anchor for an ethical foundation to ESE work across the changing curriculum landscape. It has also formed an important foundation for ensuring that the transformative ideologies of the post-apartheid curriculum agenda have been consistently part of the South African reflexive educational practices.

Another significant feature of curricula is assessment. Changes in emphasis in assessment can be illustrated through considering the evolution of the skill of 'decision making' in life orientation in the foundation phase. In Curriculum 2005, a specific learning outcome designed for the General Education and Training (GET) band was that learners should 'Practise acquired life and decision-making skills' (Ministry of Education, 1997, p. 246). This was elaborated in assessment criteria for the foundation phase regarding decision-making, problem solving and planning based on knowledge of safety and drawing on a variety of information methods (ibid.). This minimal content knowledge was elaborated slightly as differentiating between safe and unsafe situation, coping with hazards and handling conflict with friends (ibid.). Assessment in this system required a flexible, learner-centred approach that was 'holistic, tacit and inferential' (Harley & Parker, 1999, p. 184) and was in stark contrast to a pre-apartheid content-based system where knowledge was simply 'right or wrong' within assessment frameworks that were 'atomistic, explicit and measurable' (Harley & Parker, p. 184).

In the RNCS, after the emphasis on 'high knowledge' as discussed above, the content knowledge in association with each learning outcome was elaborated in much more detail. The life orientation learning outcome stated that learners should be able to 'make informed decisions regarding personal, community and environmental health' (South Africa. Department of Education, 2002b, p. 5). For foundation phase learners, the types of issues to focus on for each grade were more specific. For example, Grade 1 specified issues regarding nutrition, personal hygiene, communicable diseases, dangers en route to school and vulnerabilities to sexual abuse and also highlighted skills in relation to this knowledge – including identifying, explaining, distinguishing and recognising (South Africa. Department of Education). This elaboration demonstrates the intention of the RNCS to support high knowledge through a stronger knowledge base and associated skills.

The CAPS document now refers to 'decision-making' as part of the general introduction to the subject of life skills (which now integrates the old 'life orientation'). Decision-making is part of the knowledge focus of 'personal and social wellbeing', and learners are expected to 'make informed, morally responsible and accountable decisions about their health and the environment' (South Africa. Department of Basic Education, 2011f, p. 9). The knowledge foci regarding nutrition, diseases (including HIV/AIDS), safety, violence, abuse and environmental health are elaborated in much more detail than in the RNCS with each of these fleshed out with specific content and topics to be covered.

In the foundation phase CAPS, assessment is treated in an open-ended manner. However, in the intermediate phase, the change in assessment from the RNCS to the CAPS is much more significant with specified assessment tasks and topics for each grade and term. With the CAPS insistence that content knowledge should not be treated 'atomistically' (as discussed above), it remains to be seen whether the explicit detailing of content knowledge and tasks and tendencies towards more 'measurable' assessment will be treated in a way that returns to a pre-apartheid tendency towards atomistic assessment as described above.

In an effort to avoid a pendulum swing towards a reductionist approach to assessment, teacher researchers involved in the Fundisa for Change research programme are investigating how to work in a non-atomistic/holistic manner with assessment by focusing on learner competence as a functionally linked complex of knowledge, skills and attitudes that enable successful task performance and problem-solving (Wiek, Withycombe, & Redman, 2001) in ways that promote 'core competencies, such as critical and systemic thinking, collaborative decision-making, and taking responsibility for present and future generations' (UNESCO, 2014, p. 12).

To do this, Fundisa researchers are drawing on a set of sustainability competencies identified by Wiek et al., namely, systems thinking and modelling, strategic initiatives and change practices, normative reflexivity and re-imaging, interpersonal engagement and communication and anticipatory tracing and imaging (Wiek et al., 2001). Focusing on these competencies in assessment means focusing on learners' (environmental) content knowledge as well as more complex, dynamic, applied and relational knowledge with respect to 'active and critical thinking' described above.

These reflections on curriculum content, knowledge and assessment are important considerations for teacher professional development, as it is the development of classroom pedagogies and meaningful assessment practices that can help teachers to bring curriculum content alive and to develop learners' knowledge in the dynamic, applied and relational way proposed above. The next section explores two particularly pertinent debates regarding teacher professional development over the same time period discussed in this section, namely, the development of teacher competence and the development of pedagogical content knowledge.

3.3 Competence and Knowledge-Centred Teacher Professional Development

Earlier in this chapter, the changing discourse around competence- and outcomes-based education was explained. In the school system, although the principles and the approach of competence-based education remained, ultimately 'outcomes-based education' became the notion around which the new school curriculum was built. Yet, the notion of competence still could be seen in the field of teacher professional development which was developing parallel to the changing school curriculum. The *Norms and Standards for Educators* document, which was the guiding national policy for teacher professional development at the time, defined reflexive competence as a combination of practical (ability to 'do') and foundational (ability to understand) competence in which the learner demonstrates 'an ability to adapt to change and unforeseen circumstances and to explain the reasons behind these adaptations' (South Africa. Department of Education, 2000, p. 10). Teachers were expected to integrate reflexive, practical and foundational competencies in a demonstration of 'applied competence' (ibid.).

While the strong relationship between competence and knowledge has been highlighted in a previous section, the changes in teacher professional development in South Africa since 1995 represent changes in *emphasis* or *foregrounding* of these two concepts. Competences were not foregrounded as much in the new teacher development policy – the *Minimum Requirements for Teacher Education Qualifications* – which followed from the *Norms and Standards* document. The new teacher professional development policy recognises the value of its predecessor in highlighting integrated and applied competence. Remnants of the competency turn are seen in the emphasis on the 'importance of inter-connections between different types of knowledge and practices into the foreground, as well as the ability of teachers to draw reflexively from integrated and applied knowledge, so as to work flexibly and effectively in a variety of contexts' (South Africa. Department of Higher Education and Training, 2011, p. 8). Also a number of competences for beginner teachers are outlined in an appendix to this document highlighting subject knowledge; knowledge of curriculum; planning for teaching; knowledge of individual learners and how they learn and diversity of learning needs; communication and mediation of learning; highly developed literacy, numeracy and information technology skills; knowledge of assessment; a sense of professionalism; and reflexive capability (ibid.).

What *is* foregrounded in the most recent teacher education policy is different teacher knowledge. In this chapter, I will focus particularly on the development of pedagogical content knowledge and how it has gained traction over the time period of educational change covered in this chapter.

This means returning to early research on the implementation of the post-apartheid curriculum which revealed a persistence of transmission and rote learning teaching methods and implied authoritarian ideologies (to which teachers had been exposed as learners in schools and as student teachers during the apartheid era). These methods contradicted the democratic and rights-based ideologies presented in new curriculum policy (Chisholm, 2000). The contradiction here was created by persistent apartheid-rooted authoritarian ideologies consistent with a performance-based curriculum model of tight control as described by Taylor (1999), coexisting with democratic rights-based ideologies in post-apartheid curriculum (ibid.).

Environmental educators took on the challenge of helping to develop more innovative teaching methods in relation to the more learner-centred constructivist philosophies called for in the early curriculum changes. In South Africa, one of the significant responses to this need was the development of an 'active learning framework' which was used in the National Environmental Education Programme (NEEP). This framework proposed a number of important dimensions of planning for ESE aimed at 'developing insights and competence for making better environmental management and lifestyle choices' (O'Donoghue, 2001, p. 8).

With the implementation of phase organisers in Curriculum 2005, an effective way of planning classroom activities was through cross-curricular work (NEEP-GET, 2005), and this suited environmental educators who were then able to focus on themes such as water use, recycling and food gardening which could be addressed across a number of learning areas. This theme-based approach also persisted with the implementation of the RNCS despite the removal of phase organisers in the RNCS. However, a theme-based approach to learning was seen in some cases to be compromising 'conceptual learning and progression within subjects' (Dada et al., 2009, p. 24). Similarly, ESE research revealed that while focusing on cross-cutting themes, ESE classroom realities revealed limitations in furthering the 'aims of the learning area' and 'deepening of knowledge or process skills from specific learning areas' (Lotz-Sisitka & Raven, 2001, p. 67). This may partly be attributed to experiences where 'teachers appeared to have had little opportunity to engage with knowledge around specific issues and so attempted to engage learners in exploring these issues without having developed their own understandings and perspectives' (ibid., p. 27). This raised questions about teachers' own foundational knowledge and how they find, adapt and use materials that are relevant and meaningful to contemporary environmental issues and risks (Lotz-Sisitka, 2011; Mbanjwa, 2002; Nduna, 2003).

This tendency of teachers to depend too much on learners' prior knowledge also contributed to problems with conceptual learning and progression. This tendency can be attributed to naïve interpretations of constructivism and learner-centred education within outcomes-based education (Moll, 2002) where learner-centred education was favoured over teacher-centred education (Chisholm, 2000; Harley & Wedekind, 2004). For example, the 2009 curriculum review reported that overemphasis on group work was leaving learners to construct their own knowledge without acknowledging the need for the mediation of new knowledge by teachers (Dada et al., 2009). This finding also emerged in early ESE research which revealed that an overemphasis on group work actually distracted learners from core conceptual learning, 'consolidating knowledge and developing new understandings' (Lotz-Sisitka, 2000, p. 108). Together with another dichotomy, where outcomes were favoured over content (Chisholm; Harley & Wedekind), group work created the potential for learners to end up working in a knowledge vacuum and for what Lotz-Sisitka described as 'learner-centred emptiness' (Lotz-Sisitka, 2002, p. 114).

Another problem that can be associated with naïve interpretations of constructivism is an overemphasis of 'hands-on' learning at the expense of 'minds-on' learning. Practical investigations are key methods in ESE. Rosenberg, O'Donoghue and Olvitt (2013) drew attention to the possibility of sidelining meaning-making when noting that those involved in projects and practical actions may be so absorbed in the practical activity that they neglect to reflect on what they are learning. This echoes the concern of Vinjevold and Taylor that:

> Learning must involve cognitive and affective activity and not merely movement and speech. Thus … if not carefully structured and guided by the teacher, [educational activity] succeeds only in passing the time without engaging the cognitive faculties of participants, and thus results in little or no learning. (Vinjevold & Taylor, 1999, p. 65)

In essence the demand being made here is for strengthening the relationship between hands-on and minds-on learning as opposed to creating an impression of the primacy of hands-on learning only.

Another factor that may have contributed to learner-centred emptiness described above was teachers' own subject content knowledge. A further potential dichotomy in teacher professional development was the risk that teacher professional development might focus 'only on teaching methods, and ignore the need to provide trainee teachers with the content knowledge they also require' (Sayed, 2004, p. 257). With the danger of perpetual pendulum swinging from emphasis on teacher knowledge to emphasis on teacher's teaching methods, a useful development has been a move in teacher professional development to a focus on teachers' pedagogical content knowledge. Shulman described this knowledge as representing the 'blending of content and pedagogy into an understanding of how particular topics, problems or issues are organised, represented and adapted to the diverse interests and abilities of learners, and presented for instruction' (Shulman, 1987, p. 8). This highlighted that not only is a focus on both teacher knowledge and pedagogy important in teacher professional development, but also a focus on the *relationship* between these.

The recent South African *Minimum Requirements for Teacher Education Qualifications* emphasises pedagogical content knowledge, which includes 'knowing how to represent the concepts, methods and rules of a discipline in order to create appropriate learning opportunities for diverse learners, as well as how to evaluate their progress' as one of its important five aspects of teacher training (South Africa. Department of Higher Education and Training, 2011, p. 11).

However, since the move away from outcomes-based education, there has been little emphasis evident in policy documents on teaching methodologies and classroom pedagogy, for example, considering how the 'active and critical' approach to learning specified in the CAPS curriculum (described above) might be achieved. The CAPS specifically state that the curriculum 'does not prescribe particular instructional strategies or methodologies. Instead, educators have the freedom to expand concepts and to design and organise learning experiences according to their local circumstances, including the availability of resources' (South Africa. Department of Basic Education, 2011d, p. 10). This highlights the important role of teacher professional development in exploring and developing both teachers' pedagogical content knowledge as well as their capacity for working with a variety of teaching strategies, so as to best prepared for deepening and strengthening learners' knowledge.

In order to do this, the Fundisa for Change programme started by reviewing the CAPS curriculum and identified a 'wide range of environment and sustainability content and concepts in a wide range of subjects' (Lotz-Sisitka, 2011, p. 28) and developed a set of materials for teachers to strengthen and expand their foundational knowledge of and teaching and assessment strategies in relation to subject and phase-specific environmental topics such as climate change and sustainable development (in Geography), biodiversity (in Life Sciences), water management and security (in Natural Sciences, Social Sciences and Life Sciences) and healthy living and waste management (in Life Orientation) (Fundisa for Change Programme, 2013, p. 10).

3.4   Summary and Way Forward

This chapter has described the educational context in which South African ESE and teacher professional development initiatives have played out. The narrative has been one of 'pushes' against historical and contemporary inequalities in South African and 'pulls' towards more inclusive access to education, the economic fruits of the country and quality education for all. Other influences in the developing story have been the ongoing intellectual debates between underpinning constructivist and social realist philosophies and their interpretations and implications for curriculum design and implementation. Also in the narrative, stark dichotomies around competence, knowledge, curriculum planning and organisation, locatedness, teaching methods and assessment practices were identified as the differing philosophies gained and lost influence in the unfolding educational challenges in South Africa.

Through the illustrations of ESE research and practices during this turbulent time, the chapter has given examples of how ESE has the potential to transcend what can be unproductive dichotomies and naïve interpretations amongst educational ideals. The recommendation emerging from this chapter is to find ways to strengthen and take forward the relational dynamics evident in ESE practices discussed here. This means embracing both competence and knowledge in school and teacher professional development contexts and embracing both foundational content knowledge and more dynamic and processual views of knowledge as content knowledge is embodied through application and relational processes of engagement. It also means embracing an ethical perspective on knowledge and embracing a relational perspective on teacher content knowledge and pedagogical knowledge. Furthermore, it means embracing more rigorous but less reductionist or atomistic approaches to assessment in teacher training programmes. Finally, it means embracing the notion of learner centeredness without compromising access to the powerful, specialised subject knowledge that will give learners access to opportunities in the world and the capacity to make changes in that world.

References

Carrim, N., & Keet, A. (2005). Infusing human rights into the curriculum: The case of the South African Revised National Curriculum Statement. *Perspectives in Education, 23*(2), 99–110.

Chisholm, L. (2000). *South African curriculum for the twenty first century: Report of the review committee on Curriculum 2005: Report presented to the Minister of Education, Professor Kader Asmal*. Pretoria, South Africa: South Africa. Department of Education.

Christie, P. (1997). Global trends in local contexts: A South African perspective on competence debates. *Discourse: Studies in the Cultural Politics of Education, 18*(1), 55–69.

Community Agency for Social Enquiry, & Joint Education Trust. (2007). *Learner absenteeism in the South African schooling system*. Pretoria, South Africa: South Africa. Department of Education.

Dada, F., Dipholo, T., Hoadley, U., Khembo, E., Muller, S., & Volmink, J. (2009). *Report of the task team for the review of the implementation of the National Curriculum Statement*. Pretoria, South Africa: South Africa. Department of Basic Education.

Fundisa for Change Programme. (2013). *Fundisa for change, p. Introductory core text.* Grahamstown: Rhodes University. p. Environmental Learning Research Centre.

Harley, K., & Parker, B. (1999). Integrating differences: Implications of an outcomes-based national qualifications framework for the roles and competencies of teachers. In J. Jansen & P. Christie (Eds.), *Changing curriculum: Studies on outcomes-based education in South Africa* (pp. 181–202). Kenwyn, South Africa: Juta.

Harley, K., & Wedekind, V. (2004). Political change, curriculum change and social formation, 1990 to 2002. In L. Chisholm (Ed.), *Changing class: Education and social change in post-apartheid South Africa* (pp. 195–220). Cape Town, South Africa: HSRC Press.

Jahan, S. (Ed.). (2015). *Human development report 2015: Work for human development*. New York: United Nations Development Programme.

Lotz-Sisitka, H. (2000). The learning for sustainability project: Insights into policy implementation in South Africa. In E. Janse van Rensburg & H. Lotz-Sisitka (Eds.), *Monograph: Learning for sustainability: An environmental education professional development case study informing education policy and practice* (pp. 103–113). Johannesburg, South Africa: Learning for Sustainability Project.

Lotz-Sisitka, H. (2002). Curriculum patterning in environmental education: A review of developments in formal education in South Africa. In E. Janse van Rensburg, J. Hattingh, H. Lotz-Sisitka, & R. O'Donoghue (Eds.), *EEASA monograph, p. Environmental education, ethics and action in southern Africa* (pp. 97–120). Pretoria, South Africa: EEASA/HSRC.

Lotz-Sisitka, H. (2011). Teacher professional development with an Education for Sustainable Development focus in South Africa: Development of a network, curriculum framework and resources for teacher education. *Southern African Journal of Environmental Education, 28*, 30–71.

Lotz-Sisitka, H., & Raven, G. (2001). *Active learning in OBE: Environmental learning in South African schools*. Pretoria, South Africa: National Environmental Education Project. Department of Education.

Makholwa, A. (2015). *Is South Africa the most unequal society in the world? Interview with Haroon Bhorat*. Retrieved April 16, 2016, from http.p.//mg.co.za/article/2015-09-30-is-south-africa-the-most-unequal-society-in-the-world.

Maton, K. (2014). *Knowledge and knowers: Towards a realist sociology of education*. London: Routledge.

Mbanjwa, S. G. (2002). *The use of environmental education learning support materials in OBE: The case of the Creative Solutions to Waste project*. Unpublished Masters in Education study, Rhodes University, Grahamstown.

Ministry of Education. (1997). *Government Gazette, 6 June 1997, no 18051. Pretoria*. Pretoria, South Africa: Government Printer.

Moll, I. (2002). Clarifying constructivism in a context of curriculum change. *Journal of Education, 27*, 1–28.

Nduna, N. R. (2003). *The use of environmental learning support materials to mediate learning in outcomes-based education: A case study in an Eastern Cape school*. Unpublished Masters in Education study, Rhodes University, Grahamstown.

NEEP-GET. (2005). *Stories of change 1, p. Action research case studies*. Howick, New Zealand: NEEP-GET/ShareNet.

Nelson Mandela Foundation. (2005). *Emerging voices: A report on education in South African rural communities*. Cape Town, South Africa: HSRC Press.

O'Donoghue, R. (2001). *Environment and active learning in OBE: NEEP guidelines for facilitating and assessing active learning in OBE*. Howick, New Zealand: ShareNet.

O'Donoghue, R. (2016). Working with critical realist perspective and tools at the interface of indigenous and scientific knowledge in a science curriculum setting. In H. Lotz-Sisitka & L. Price (Eds.), *Critical realism, environmental learning and social-ecological change* (pp. 159–177). London: Routledge.

Rosenberg, E., O'Donoghue, R., & Olvitt, L. L. (2013). *Methods and processes to support change-oriented learning* (2nd ed.). Grahamstown, South Africa: Rhodes University: Fundisa for Change.

Sayed, Y. (2004). The case of teacher education in post-apartheid South Africa: Politics and priorities. In L. Chisholm (Ed.), *Changing class, p. Education and social change in post-apartheid South Africa* (1st ed., pp. 246–265). Cape Town, South Africa: HSRC Press.

Schudel, I. (2014). Exploring a knowledge-focused trajectory for researching environmental learning in the South African curriculum. *Southern African Journal of Environmental Education, 30*, 96–117.

Shulman, L. S. (1987). Knowledge and teaching: Foundations of the new reform. *Harvard Educational Review, 57*(1), 1–21.

South Africa. Department of Basic Education. (2011a). *Report on the Annual National Assessments of 2011*. Pretoria, South Africa: Department of Basic Education.

South Africa. Department of Basic Education. (2011b). *Curriculum assessment policy statements*. Pretoria, South Africa: Department of Basic Education.

South Africa. Department of Basic Education. (2011c). *Curriculum assessment policy statement: Intermediate phase: Natural sciences and technology: Grades 4–6*. Pretoria, South Africa: Department of Basic Education.

South Africa. Department of Basic Education. (2011d). *Curriculum assessment policy statement: Grades 10–12: Life sciences*. Pretoria, South Africa: Department of Basic Education.

South Africa. Department of Basic Education. (2011e). *Curriculum assessment policy statement: Foundation phase: Mathematics: Grades R-3*. Pretoria, South Africa: Department of Basic Education.

South Africa. Department of Basic Education. (2011f). *Curriculum assessment policy statement: Foundation phase: Life skills: Grades R-3*. Pretoria, South Africa: South Africa. Department of Basic Education.

South Africa. Department of Education. (2000). *Norms and standards for educators, p. Government Notice 82 of 2000*. Pretoria, South Africa: Government Printers.

South Africa. Department of Education. (2002a). *Revised national curriculum statement Grades R-9 (schools) policy: Overview*. Pretoria, South Africa: Department of Education.

South Africa. Department of Education. (2002b). *Revised national curriculum statement Grades R-9 (schools) policy: Life Orientation*. Pretoria, South Africa: Department of Education.

South Africa. Department of Education. (2008). *Foundations for learning campaign: Government notice 306 of 2008*. Pretoria, South Africa: Government Printers.

South Africa. Department of Higher Education and Training. (2011). The minimum requirements for teacher education qualifications. *Government Gazette, 583*, 3–59.

South Africa. National Planning Commission. (2013). *National development plan 2030: Our future, make it work*. Pretoria, South Africa: South Africa. Office of the Presidency.

South African Human Rights Commission. (2010). *7th report on economic and social rights: Millennium development goals and the progressive realisation of economic and social rights in South Africa*. Johannesburg, South Africa: South African Human Rights Commission.

Spaull, N. (2013). *South Africa's education crisis: The quality of education in South Africa 1994–2011*. Johannesburg, South Africa: Centre for Development and Enterprise.

Taylor, N. (1999). Curriculum 2005: Finding a balance between school and everyday knowledges. In N. Taylor & P. Vinjevold (Eds.), *Getting learning right: Report of the President's education initiative research project* (pp. 105–130). Braamfontein, South Africa: Joint Education Trust.

UNESCO. (2014). *Roadmap for implementing the global action programme on Education for Sustainable Development*. Paris: UNESCO.

UNESCO. (2015). *Rethinking education, p. Towards a global common good?* Paris: United Nations Educational, Scientific and Cultural Organisation.

Vinjevold, P., & Taylor, N. (1999). Research methods. In N. Taylor & P. Vinjevold (Eds.), *Getting learning right, p. Report of the President's Education Initiative Research Project* (pp. 65–104). Braamfontein, South Africa: Joint Education Trust.

Wals, A. (2007). Learning in a changing world and changing in a learning world, p. Reflexively fumbling towards sustainability. *Southern African Journal of Environmental Education, 24*, 35–45.

Wheelahan, L. (2010). *Why knowledge matters in curriculum: A social realist argument.* Oxon, UK: Routledge.

Wiek, A., Withycombe, L., & Redman, C. L. (2001). Key competencies in sustainability: A reference framework for academic program development. *Sustainability Science, 6,* 213–218.

Young, M. (2008). *Bringing knowledge back in: From social constructivism to social realism in the sociology of education.* London: Routledge.

Zipin, L., Fataar, A., & Brennan, M. (2015). Can social realism do social justice? Debating the warrants for curriculum knowledge selection. *Education as Change, 19*(2), 9–36.

Part II
Curriculum Innovations: Teaching, Learning, and Assessment

Chapter 4
The Relation of Mainstreamed Environmental Education to the Modern Schooling System in Zambia

Charles Namafe and Manoah Muchanga

This chapter provides insights into issues related to mainstreaming environmental education (EE) between and across institutions. It is noted that EE would greatly be assisted in its effort to produce impact in society if its practitioners had access to principles of how best to mainstream the field across institutions. Unfortunately, no such principles that fit every situation exist at present, thereby perpetuating challenges of mainstreaming EE across institutions. The chapter challenges readers to open the 'black box' of mainstreaming and to share issues and experiences. International readers and those struggling with mainstreaming EE in higher education institutions will find this chapter illuminating.

The importance of environmental education (EE) no longer escapes anyone – according to Ziaka (2000) writing about the features and educational implications of environmental education in France around the year 2000. Ever since the early 1970s, and especially since 1975, with the launching by UNESCO of the International Environmental Education Programme (IEEP) at the Belgrade Conference, given the conceptual thinking on this type of education as well as the learning from numerous pilot projects, of multiple proposals of activities and teaching aids of all kinds, one might think everything concerned with EE was heading for success! Ziaka mentioned, however, that in most countries including Zambia, real integration of EE into the educational systems is still lacking.

In a rather forthright title – *The Failure of Environmental Education* – Saylan and Blumstein (2011, p. 1) asserted:

> Environmental education has failed to bring about the changes in attitude and behaviour necessary to stave off the detrimental effects of climate change, biodiversity loss and environmental degradation that our planet is experiencing at an alarmingly accelerating rate.

C. Namafe • M. Muchanga (✉)
School of Education, Environmental Education Unit, University of Zambia,
P.O. Box 32379, Lusaka, Zambia
e-mail: manoahmuchangaa@yahoo.co.uk

© Springer International Publishing Switzerland 2017   57
H. Lotz-Sisitka et al. (eds.), *Schooling for Sustainable Development in Africa*,
Schooling for Sustainable Development 8, DOI 10.1007/978-3-319-45989-9_4

A situation such as this – about EE not successfully penetrating educational systems of many countries and bringing about changed behaviours – is a stark reality that requires critical analysis and strategising. It denotes the existence and relationship of two entities, that is, a mainstreamed EE scenario on one hand and a non-mainstreamed EE scenario on the other. Such a condition provides a contextual background to this chapter. In this regard, the main argument of this chapter is that, during its development, EE has to a large extent failed to permeate the surrounding mosaic of the schooling system partly because EE has preferred to concentrate on the comfort zones of various educational institutions (refer, for instance, to Chapter 3 of UNESCO's (2005) *Guidelines and Recommendations for Reorienting Teacher Education to Address Sustainability* that describes 'initiatives taken by members of the international network' referring to sustainability). To appreciate this point, let us first define what we mean by the term 'mainstreamed EE'.

4.1 Mainstreamed Environmental Education

Environmental education may be mainstreamed at various scales and levels as diagrammatically shown below. The University of Zambia will serve as an example.

In Fig. 4.1, the process of mainstreaming EE in any institution starts with an individual person, i.e. one termed a change agent for our purpose. Such a person could be a pupil, school teacher, college lecturer, university lecturer, civil servant, nurse or a medical doctor. It is vital to verify the extent (scale) of mainstreaming of EE even within a mental scheme of an individual, that is, is it full-scale or partial mainstreaming? For instance, a person may conceptually mainstream EE in his/her natural mental scheme only and neglect to mainstream it fully into the economic, social and political scheme of that same individual. Such a person, for instance, in only focusing on the natural dimension may love wildlife and, hence, maintain a game ranch. But since the other dimension of his or her mental schema is insufficiently mainstreamed, he may not participate politically in the affairs of environmental conservation, or he may socially be so self-centred and selfish that he does not care about the welfare of his neighbouring communities. In short, full-scale mainstreaming of EE within individuals should cover all major dimensions of a person's mental schema, that is, the natural, economic as well as the social and political.

In the case of the University of Zambia, EE has to a large extent been mainstreamed at levels from 1 to 5, as shown in Fig. 4.1, especially within the context of the School of Education. But the scale of such mainstreaming would require systematic verification for each stage from 1 to 5. As shown in Fig. 4.1, the letter 'B' can be any institution such as a school, a college, a university, a hospital, a non-governmental organisation or an industry. Such institutions would have layered units or departments within which EE requires to be mainstreamed. It may be that EE is mainstreamed successfully within one section or department of an institution

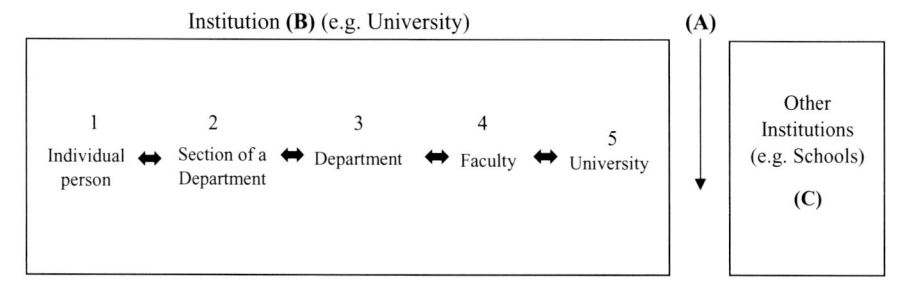

Fig. 4.1 Mainstreaming environmental education at different institutional levels

and not in other departments. Alternatively, it may be that EE is mainstreamed well in one type of institutional activity (e.g. teaching) and not in the other (e.g. research). In fact, even within a single institution, there will be a need for individual change agents at each level (e.g. department or section) in order to ensure successful EE mainstreaming.

In other cases, only individual persons represented by the number 1 in the diagram may have mainstreamed EE within their mental scheme, and not in other levels of the institution represented by numbers 2–5 in the diagram. Such individuals would go about attending conferences, processing emails or publishing on matters to do with EE while leaving levels 2–5 of an institution untouched and, hence, 'unmainstreamed'. Certificates of recognition and other forms of awards in EE may be awarded to such individuals as well. They may also be actively conducting research in EE, including disseminating their research findings in different channels. Figure 4.1 further shows that the direction of mainstreaming EE in B can be from an individual change agent (1) to the whole institution (5), or vice versa.

The above account of mainstreaming EE also applies to the other levels of an institution where, for instance, a faculty (i.e. number 4) may actively mainstream EE by way of research, publications, conference participation or even winning awards – but other levels of the institution may remain unaffected and, hence, unmainstreamed (i.e. numbers 1, 2, 3 and 5). In this regard, there is still room to interrogate the levels of mainstreaming even at B. Apart from 1, further insights on the process of mainstreaming at 2, 3, 4 and 5 need to be opened up and issues shared at these levels within C. Recommended literature on such issues can be found in the *Southern Africa Journal of Environmental Education*, volume 26 (paper by Namafe, 2009) and volume 29 (paper by Chileshe & Namafe, 2013 and paper by Shumba & Kampamba, 2013). Namafe (2005) in the *Monograph of Cases of Course Development in Environmental and Sustainability Education in southern Africa* provided rich discussion on some of the likely hurdles, opportunities and challenges of EE programme developers trying to make a positive contribution to their academic institutions in developing countries. This paper epitomises the struggles of mainstreaming EE at 'B' as presented in this chapter.

What next regarding a situation where an institution, such as the University of Zambia in this case, has largely mainstreamed all its levels from 1 to 5? The answer to this question is examined next.

4.2 Cross-Border Mainstreaming of EE

The process of winning over institution 'C' by an individual or institution from 'B' is hereby called 'crossing over'. The University of Zambia has successfully devised and mainstreamed an EE curriculum within its comfort zone represented by numbers 1–5 in Fig. 4.1. The next logical step in its mainstreaming agenda is to cross borders in order to win other institutions (i.e. represented as C) to the cause of EE. The intangible space or interface between the University of Zambia (UNZA) and other institutions is real and is represented as A in Fig. 4.1. How best can EE in institution B behave, what can it do or how best can it constitute itself in order to successfully win C to itself by crossing A? No quick, easy answers or principles exist in the literature to clarify the process. This is where the crux of the problem resides, explaining why EE is failing to attain real integration into the educational systems of many countries. Change agents wishing to cross over from B to mainstream EE into C will discover that zone A is an intersection that is a rather messy, often quite lively place full of surprises. Expeditions from B to C have been likened in some ways to travelling into unknown countries (Gibson, 1985).

In comparison to B, institutions in C exist in a world dominated by politics, pettiness and real problems. Forays into C with its various languages, cultures and values will bring rich rewards to EE practitioners in B. Without knowing these differences, Gibson (1985) warned of the danger to those moving from one institution that has mainstreamed EE to another that has not yet mainstreamed EE. Also, experience in one institution can have a detrimental effect on another institution. For instance, EE practitioners with government experience moving from institution C to B could become an undue influence in universities and threaten those values which make the university distinctive from government. Similarly, academic EE change agents moving from B may be harmful to those values which make civil service, school teaching or political appointment distinctive. In short, crossing over for purposes of mainstreaming EE between B and C is a valuable experience for EE practitioners, regardless of the direction of movement. A complete transfer of methods in either direction would be a disaster. Practitioners of EE must rethink the need and value of making such forays.

But zone A is also formed of poor relations between B and C. Attempts by change agents moving from a university setting in B, wanting to work closely with schools in C, seem to run headlong into a number of painful impediments identified and described by Barth (1990) as follows:

Scorned lovers – a university cannot begin a new EE activity with schools as if the slate is clean of old activities. School and university people make new conversations while harbouring 'antibodies' that each has built for protection against the other.

Who takes the initiative – for the academic, taking the initiative carries the risk of being prescriptive, which often causes school people to bristle defensively as many detest being responsive to university initiatives.

Muted voices – a third roadblock is to be found in the muted voices of school teachers and administrators who may feel dominated by university voices. Yet such school teachers and administrators often have valuable insider perspectives of events happening within their institutions.

Crossing boundaries – a fourth roadblock to rich exchange between the worlds of school and university is that neither adequately rewards those crossing the border between them for purposes of mainstreaming EE.

Theory versus *practice* – the locus of theory and practice is another obstacle to close engagement between universities and schools. The common saying is that "theory resides in universities and practice resides in schools". This conception is, however, simplistic and detrimental to relations between school and university.

Environmental education needs theories, principles or tested approaches that clarify the process of colonising new institutions for purposes of mainstreaming EE from its stronghold (B) to a new territory (C). At present, there appears to be too much 'in-breeding' among EE practitioners themselves, who continuously preach to themselves and to the already converted through various forms such as conferences, publications or workshops.

Such in-breeding produces different implications, one of which is a situation where potential beneficiaries of the EE message, such as school children or local communities experiencing various environmental ills such as the impact of climate change, are unaffected by such messages. The next sections of this chapter elaborate on this point by using a case study in Zambia with a climate change message.

4.3 Outlook of Climate Change Education in Zambia

Addressing climate change is now one of the major issues on the international political agenda, and it must be one of the issues that requires emphasis in environmental education (United Nations Educational Scientific and Cultural Organization (UNESCO), 2007). In Zambia, and particularly in the school system, the subject of climate change is relatively new, and consequently, learners and the general community have limited understanding of the phenomenon; such learners and even people may be aware of the localised effects of climate change but do not associate these with climate change (Muchanga, 2013). Similarly, the International Union for Conservation of Nature (IUCN) (2007) noted that some communities in Zambia are aware of climatic change but there is little understanding as to why it has occurred.

At a national level and particularly in the school system in Zambia, there is inadequate information, knowledge and skills to prepare teachers and learners to face up to the challenge of climate change. The Zambian government has opted for climate change awareness and literacy creation using different approaches such as the print and electronic media, drama and community fora. In order to go beyond awareness creation towards meaningful behavioural and social change for climate change preparedness, the actors in B, for example, at the University of Zambia initiated climate change education (under the broad EE programme)-related activities such as research at postgraduate level training, conferences and workshops (e.g. events organised with SARUA and Climate Change Education for Sustainable Development (CCESD) funded by UNESCO). Further, the institution introduced a course in Climate Change Education (CCE), among others, into the EE curriculum in anticipation of some trickledown effect into C through A (Muchanga & Nakazwe, 2015).

The envisaged key outcomes in C include updated school curricula with climate change content, updated textbooks and other learning materials with climate change content. Other expected outcomes include updating of existing courses or designing new ones with an emphasis on climate change adaptation and mitigation. This is expected to lead to the development of technical capacity in various climate change fields, engaging local administration and community leaders as well as disseminating climate change education literature in local languages for the benefit of marginalised populations and the general public who might be found in C. Education and particularly EE is a right of each individual and a means for enhancing the well-being and quality of life in the changing environment (Ministry of Education [MoE], 1996).

However, this right is not being enjoyed by the schooling system in C because of diverse obstacles at A (e.g. the Curriculum Development Centre and the MoE) maintaining the untenable argument that the curriculum is too full to include CCE conceived through the EE curriculum in B, which also faces its own internal hurdles. According to UNESCO (2000), in the *Dakar Framework for Action on Education for All* (*EFA*), the right to education imposes an obligation upon states to ensure that all citizens have opportunities to meet their basic learning needs. In contemporary society, it can be argued that the Zambian government, and particularly the Ministry of Education (MoE), is under an obligation to infuse climate change education in the school curriculum at all levels in order to meet the learning needs of learners in the context of climate change. In fact, the National Climate Change Response Strategy (NCCRS) of 2010 formed at A acknowledges CCE as one of the strategies to address diverse environmental ills that affect diverse groupings in C (Ministry of Tourism, Environment and Natural Resources (MTENR), 2010). However, this remains a mere pipedream unless a synergised approach among actors in B and C through A is adopted.

Much as we appreciate that many activities are being implemented on climate change, they do not address the needs of the Zambian schooling system. For example, with the support from the United Nations Development Programme (UNDP), the Ministry of Finance has established some climate change and mitigation projects and even frameworks, but these do not provide the required systemic and systematic content that would bring about behavioural and social change especially within school system. In fact, there are many other entities in Zambia that are involved in climate change adaptation activities whose pedagogical foci are not necessarily to influence positively behavioural and social change in the school system. The lack of a learning approach in these activities is actually what renders them inadequate to address the needs of the schooling system in Zambia (MTENR, 2010).

This lack of appropriate approaches is not only a challenge among nonacademic entities; it exists within key academic agents (especially at 2, 3, 4 and 5 in B) that are attempting to mainstream EE/ESD issues, such as climate change. For example, in their study at the Copperbelt University, Shumba and Kampamba (2013) demonstrated that although EE/ESD issues are salient in education policies in Zambia, the surveyed high school subject and the university teaching methods syllabi do not carry ESD content and change-oriented teaching and learning approaches that would foster effective mainstreaming. They are unlikely to model integration of ESD issues such as those stated in policies, for example, human rights, values education, entrepreneurship and climate change (Shumba & Kampamba, 2013). Overall, the results of policy and curriculum analysis and of the student surveys undertaken by Shumba and Kampamba point to a discrepancy between policies, school science syllabi and the university teaching methods course, which is an indicator of disconnections between B, A and C. Policies such as the National Policy on Education – *Educating our Future*, National Policy on Environment (Ministry of Education [MoE], 2013) and the latest Education Curriculum Framework (MTENR, 2007) designed at A stipulate mainstreaming of EE/ESD issues, but the syllabi do not systematically point out the content and the teaching and learning methods for integrating ESD (Shumba & Kampamba, 2013).

A paradigm shift might occur by adopting the metaphor of 'learning as connection' (Lotz-Sisitka, 2010), which implies learning that has meaning in people's lives or a notion that connects science to social and humanistic issues such as those advocated in EE/ESD discourse. Thus far, one entity in B and University of Zambia in particular is using students' educational practicum as one of the channels through which C could be mainstreamed with EE/ESD issues such as climate change. This involves a mandatory sending of all EE students for 'attachment' to various institutions in C for 3–4 months to develop a practical feel of EE/ESD and thereby mainstreaming-related issues such as climate change. Turbulent though this has been, it has proven to be one way of mainstreaming EE issues into various institutions, especially noneducational ones. However, much needs to be done to penetrate the school system due to inadequate commitment and lack of political will at A and partly due to intra-institutional inconsistencies at levels 2, 3, 4 and 5 under B.

Generally, the existence of climate change and its impacts is widely accepted in Zambia as demonstrated in various policy documents, reports and frameworks such as the National Adaptation Plan of Action, National Policy on Environment, National Climate Change Response Strategy and others. The government through these documents acknowledges that climate change has the potential to impact negatively on almost all sectors of the economy in Zambia. Using climatic data from 1960 to 2003, it has been concluded that the mean annual temperature in

Zambia has increased by 1.3 °C since 1960, at an average rate of 0.29 °C per decade (MTENR, 2010). Zambia experienced severe droughts in different agro-ecological regions during the 1990s (1991/1992, 1994/1995, 1997/1998/1999) that were partly linked to the El Nino Southern Oscillation (ENSO) and partly due to the changing climate. Such events caused national disasters, especially in 1992/1993, when Zambia was compelled to seek maize food aid from Norway due to crop failure (ibid.).

Using a survey design and hermeneutic paradigm, Muchanga (2011) noted that climate change has radiated diverse effects (water scarcity, crop failure, social disintegration) in rural areas of Luangwa District especially among women and school-going girls. This led to a strong recommendation that CCE should be responsive, topically diverse and flexible enough to address the emergent effects of climate change. This recommendation also influenced the structure of an existing course in CCE in the EE curriculum of the University of Zambia. The unfortunate situation, however, is that the knowledge which an EE teacher is obtaining from this course has not yet reached the envisaged recipients in C because of ideological contradictions at A.

As mentioned earlier, sometimes actors in C and A may argue that climate change is already mainstreamed in subjects such as geography, but the conclusion of Muchanga and Nakazwe (2015), in their critical review of the extent to which CCE is incorporated within the Zambian school curriculum from early childhood education to tertiary level, showed that what is in the curriculum is basically climate science where learners learn about weather and climate, and not CCE. Some of the key observations are that the pre-primary curriculum has little or no learning goals on climate change education. Most of the issues integrated are more to do with the general environment and not CCE. Nevertheless, the pedagogy seems favourable for the inclusion of CCE. The primary school curriculum covers a number of environment and sustainability themes under which climate change issues are currently integrated, though in a fragmented nature. At secondary school level, climate change issues are mainly integrated into geography as the main carrier subject, although not every learner chooses to study it. While other environmental issues are generally well represented, climate change themes are presented in a very simplistic way with no clear focus on the local context. Many CCE issues that can bring about community adaptive capacity and resilience are missing. The main reason for this is the same as what is already presented above, that is, CCE is still 'stuck' within the comfort zone B with actors at A being arguably the main source of EE fixation within its comfort zone (Muchanga and Nakazwe, 2015).

4.4 Conclusion

This chapter has argued that the emerging twenty-first century schooling system is experiencing various sustainability challenges largely as a result of EE practitioners focussing on 'in-breeding' messages among their already converted members to the neglect of addressing the intangible zone formed by 'A' in Fig. 4.1. As long as that interface is untouched by clear strategies, tactics and approaches, the ethos of sustainability as espoused in fields like ESD or EE will remain unreceived by people and, particularly, the schooling system it is supposed to benefit. In particular, the message of climate change will forever be trapped in universities, research centres and other high-level institutions given that its main carrier (EE) is also fixed. The University of Zambia is currently working on its next steps of implementing education for sustainable development (Namafe, forthcoming).

References

Barth, R. S. (1990). *Improving schools from within – teachers, parents and principals can make a difference.* San Francisco: Jossey-Bass.

Chileshe, B., & Namafe, C. (2013). Contextualising the curriculum through local floodplain artefacts at Lealui Basic School of Western Zambia. *Southern African Journal of Environmental Education, 29*, 167–179.

Gibson, E. M. (1985). Crossing over: Geographers in University and Government 1965–1985. *The Operational Geographer, 7*, 22–24.

International Union for Conservation of Nature-IUCN (World Heritage Conservation Union). (2007). *Zambia 'hard hit' by climate change.* Retrieved 21 May 2007, from http://www.inthe-news.co.uk/news/science/zambia-hard-hit-by-climate- change$1087447.htm

Lotz-Sisitka, H. (2010). *Conceptions of quality and 'learning as connection': Teaching for relevance.* Paper presented at the UNESCO EFA High Level Meeting, Amman.

Ministry of Education (MoE). (1996). *Educating our future: National policy on education.* Lusaka, Zambia: Zambia Educational Publishing House.

Ministry of Education (MoE). (2013). *The Zambia education curriculum framework.* Lusaka, Zambia: MoE.

Ministry of Tourism, Environment and Natural Resources (MTENR). (2007). *National policy on environment.* Lusaka, Zambia: MTENR.

Ministry of Tourism, Environment and Natural Resources (MTENR). (2010). *Climate change response strategy.* Lusaka, Zambia: MTENR.

Muchanga, M. (2011). *Perceptions of climate change and learning among selected areas in Zambia's Lusaka Province.* Lusaka, Zambia: University of Zambia.

Muchanga, M. (2013). Learning for climate change adaptation among selected communities of Lusaka Province in Zambia. *Southern African Journal of Environmental Education, 29*, 94–114.

Muchanga, M., & Nakazwe, M. K. (2015). Climate change education in the school curricula of Zambia. In G. Nhamo & S. Shava (Eds.), *Climate change education in the SADC school curricula of Zambia.* Pretoria, South Africa: Africa Institute of South Africa.

Namafe, C. M. (2005). Postgraduate course development process in geography and environmental education at the University of Zambia. In J. Lupele (Ed.), *Cases of course development in environmental and sustainability education in Southern Africa.* Howick, New Zealand: SADC-REEP.

Namafe, C. M. (2009). Sigtuna Think Piece 3: The wider context of climate change discourse. *Southern African Journal of Environmental Education, 26*, 38–48.

Namafe, C. M. (forthcoming). Next steps at the University of Zambia in implementing education for sustainable development. *Applied Environmental Education and Communication.*

Saylan, C., & Blumstein, D. T. (2011). *The failure of environmental education (and how we can fix it).* Berkeley, CA: University of California Press.

Shumba, O., & Kampamba, R. (2013). Mainstreaming ESD into science teacher education courses: A case for ESD pedagogical content knowledge and learning as connection. *Southern African Journal of Environmental Education, 29*, 151–166.

United Nations Educational Scientific and Cultural Organization (UNESCO). (2000). *Dakar framework for action*. Paris: Place de Fontenoy.

United Nations Educational Scientific Cultural Organization (UNESCO). (2005). *Guidelines and recommendations for reorienting teacher education to address sustainability* (Technical Paper no 2). Paris: Place de Fontenoy.

United Nations Educational Scientific and Cultural Organization (UNESCO). (2007). *Ahmadabad Conference Recommendations on Environmental Education*. Conference Proceedings. India.

Ziaka, Y. (Ed.). (2000). *Environmental education for 21st century: Issues and perspectives for a permanent forum*. Ermoupolis, Greece: Polis International Network in Environmental Education.

Chapter 5
The Culture Hut Concept as Curriculum Innovation: Engaging the Dialectic Nature of Heritage in Zimbabwean Schools to Support ESD Learning

Cryton Zazu

Influenced by Zimbabwe's cultural policy of 2004, the culture hut concept entails the establishment of a culture hut or village within schools as a curriculum innovation aimed at promoting the teaching and learning of culture within the country's formal education system. The culture hut is representative of what can be viewed as a 'mini cultural museum' in which cultural artefacts and objects are displayed. All schools in Zimbabwe are, therefore, as a matter of policy expected to have a culture hut; hence the culture hut concept itself as a curriculum innovation has, since 2004, became popular in the country. Despite its popularity, little research has been done to evaluate how this curriculum innovation is adding value to Zimbabwe's education system. It is against this background that this chapter interrogates the culture hut concept in Zimbabwe, pointing out some of the shortfalls in its current application and highlighting how, if carefully constituted, such curriculum innovation can foster the type of learning envisaged within the Education for Sustainable Development (ESD) framework. This chapter, as much as it provides a critique of how the culture hut concept is being used in Zimbabwean schools, also points to what could be done to reconstitute and use this curriculum innovation in ways that support more ESD learning.

This chapter is mainly motivated by my desire to share with fellow Education for Sustainable Development (ESD) practitioners some of the insights that emerged from my doctoral research on heritage and social learning (Zazu, 2013). One of the issues that I was pursuing in my studies was how to make contemporary heritage education practices in post-colonial southern Africa allow for learning that goes beyond intergenerational transmission of culture or what Malegapuru (1999) and later Cheru (2002) termed 'African renaissance' and incorporate an environment and sustainability dimension. In order to do this, I drew upon a conceptual framework

C. Zazu (✉)
Rhodes University, Environmental Learning Research Centre (ELRC),
Grahamstown 6139, South Africa
e-mail: zazucryton@gmail.com

© Springer International Publishing Switzerland 2017   67
H. Lotz-Sisitka et al. (eds.), *Schooling for Sustainable Development in Africa*,
Schooling for Sustainable Development 8, DOI 10.1007/978-3-319-45989-9_5

of heritage as inclusive of both natural and cultural elements that are dialectically interconnected (Harrison, 2013; Zazu, 2013; Lowenthal, 2005). I used this dialectic nature of heritage to point out how contemporary curriculum innovations such as the 'culture hut concept' in Zimbabwe can, if its current use is reorganised, help to reposition heritage education practices within an Education for Sustainable Development (ESD) framework and allow for a kind of learning that transcends the discourses of culture and identity.

My argument for harnessing the dialectic nature of heritage within the learning processes taking place around the culture hut concept is that it will enable learners to understand and appreciate the interconnection between their cultures and what the natural environment can provide (Daniel, Muhar, Aznar, Boyd, & Tam, 2012; Hughes, 2009; Lowenthal, 2005). Such learning is aligned with the imperatives of ESD and, as such, promotes identity, diversity and social cohesion that come with it. Furthermore, such learning processes have the potential to develop a sense of environmental responsibility and stewardship in the learner.

In this chapter I first historicise how the culture hut concept came into the Zimbabwe's formal education system followed by a brief description of what the concept entails. Drawing on field observations and interviews conducted with selected teachers from two primary schools in Zimbabwe, I describe how learning around the culture hut concept is currently constituted. Drawing on the literature, I also critically discuss what I consider were the shortfalls in how educators were organising the learning around the culture hut concept in the two schools, also pointing out what could be done to make such curriculum innovation support more ESD learning.

5.1 How Did the Culture Hut Concept Come into Zimbabwean Schools?

The 'culture hut concept' is an outcome of Zimbabwe's cultural policy of 2004 and is heavily influenced by the discourses of culture and identity, particularly the need to promote the inclusion of local cultures within the formal education system (Nyoni & Nyoni, 2010). In the country's cultural policy of 2004, it is stated:

> Zimbabwe has a rich cultural heritage built over a long period of time. Our value systems and beliefs give us an identity, as a people. The defeat of indigenous people by settler colonialists in the first Chimurenga witnessed some erosion of our traditions, values and religion. … The need to promote and preserve our cultural heritage has become more important in the face of the above factors. (Government of Zimbabwe [GoZ], 2004, p. iv)

One of the strategies for promoting the uptake of local cultures in formal education, as advocated in the cultural policy, was a proposal to establish culture huts (villages) in schools (J. Chataika, personal communication, July 17, 2011). In an interview, the teacher in charge of the culture hut project at Primary School A in Kwekwe (quoted in Zazu, 2013, p. 152) concurred with this observation:

… this cultural village was established according to a policy of government to educate the young children in urban areas about cultural life in the rural areas and villages. Here children learn about culture and social studies at the village and this helps them to understand who they are.

The rationale behind the introduction of the culture hut concept in Zimbabwe's formal education system was therefore to promote an intergenerational engagement with and transfer of culture to the learners. The idea of education as *acculturation* (Obanya, 2005) and the view of curriculum as a "selection from culture" as postulated by McKernan (2008, p. 7) seemed to have influenced and shaped this curriculum innovation. It is probably against this historical background that the kind of learning processes (with its perceived shortfalls) currently taking place around the culture hut concept can be appreciated.

5.2   What Does the Culture Hut Concept Entail?

The culture hut concept basically entails the establishment of a culture hut or village within the school premises, representative of a 'mini cultural museum' in which cultural artefacts are archived for learning purposes (Nyoni & Nyoni, 2010; Zazu, 2013). Cultural artefacts, ranging from clay pots (*hari*), yokes (*majoki*) and hunting gear such as bow and arrows (*uta nemuseve*), spears (*pfumo*) and knobkerries (*nduku*) are usually displayed in the culture hut (see Figs. 5.1 and 5.2).

Fig. 5.1   The culture huts (village) at Primary School A

Fig. 5.2 Some of the artefacts archived in the huts for learning purposes at Primary School A

By 2005 the 'culture hut concept' had become widespread, with almost all nine provinces in Zimbabwe participating (Nyoni & Nyoni, 2010). The department of National Museums and Monuments of Zimbabwe (NMMZ) was tasked to provide technical support to teachers and assist schools to get the relevant cultural artefacts for use in their culture villages (G. Chauke, personal communication, September 6, 2010).

While the 'culture hut concept' sounds a noble curriculum innovation, the extent to which it adds value to the kind of learning taking place in Zimbabwean schools is not yet fully established (Zazu, 2013). Implications of the culture hut concept for helping learners to learn about heritage as both natural and cultural, allowing them to recognise and value the dialectic interaction between their culture and the natural environment is therefore central to the discussions in this chapter.

5.3   Contemporary Use of the Culture Hut Concept in Zimbabwean Schools

Here I present an overview of how the culture hut concept is being used to promote cultural heritage education in two primary schools in Zimbabwe. To provide this overview, I drew on insights and experiences gained through both field observation and the semi-structured interviews that I conducted with teachers from the two schools.

I worked with two schools, referred to as Primary School A and B, located in the Midlands province of the country. These schools pioneered the establishment of culture huts or villages in the province.

Primary School A falls within the Kwekwe urban district and has an average enrolment of around 900 learners. The school belongs to the local Kwekwe municipality and as such all its educational activities, and the learning support materials needed are wholly financed through revenue generated from fees paid by learners. The central government only pays the salaries of the teachers.

According to Primary School A's deputy head, quoted in Zazu (2013, p.186):

> Our ministry *inoita* (does) encompass culture, and it is because most of our school kids grows up in urban areas and have very little knowledge of their culture. The ministry then said it is important to construct in the school a village that is just like the same village in the rural area where learners can learn about their culture. This is a ministry of education initiative.

The cultural village at Primary School A was established in 2010 with support from the school development authority. The village is, according to the teacher in charge, used to support learning around the culture of the Zimbabwean people. The use of the culture hut is often restricted to the teachers of social studies and Shona subjects. Other teachers seemed not to be interested in the culture village. Probed to describe the type of learning taking place at the village, the teacher was quoted in Zazu (2013, p.152):

> Here at the culture village learners learn about the type and names of huts and other infrastructures that make up a typical rural home. E.g. here we have a hut called *tsaka* (for cooking), *dura* (for storage of grains like maize and millet) and another hut in which all cultural rituals [e.g. *kudzora vafi* (bringing the dead back into the home); *kuroodza mwana* (marriage processes)] and where other family ceremonies are conducted. If you look outside there you see another structure called *dara* (for drying dishes) and you also see this one here where goats (livestock) sleep. So learners here learn about these different huts and what they represent in their culture. Learners need to know what each hut is used for, and even in the exams this is what is asked.

Interesting to note in the case of this school was the fact that the culture hut concept is actually linked to the official school curriculum and that social studies (a subject within which culture falls) and Shona teachers use the cultural village within the formal school timetable. Also worth noting is that the learning taking place around the culture village seemed to be limited to or narrowly centred on culture and identity.

During the demonstration lessons that I observed at Primary School A, it also became clear that teachers mainly focus on the naming and identification of symbolic uses of the huts and cultural artefacts that are on display. The cultural artefacts are usually labelled in the local language (see Fig. 5.2) and learners are expected to master and remember the names for the school examination. Common tools learned about at the culture village at Primary School A include clay pots and plates such as *chipfuko* (for keeping traditional drink called *mahewu*), *hadyana* (for cooking relish), *mbiya* (a plate for relish), *shinu* (for keeping cooking oils), *gate* (bigger pot for keeping water or beer), *hodzeko* (for storing milk) and *tsambakodzi* or *tsaiya* (for cooking *sadza*). Also included are tools such as *pfumo* (spear), *uta nemuseve* (bow and arrow), *demo* (axe) and *mutsi neduri* (pestle and mortar). Other tools and artefacts are made from plant materials and examples are *dende* (for storing water but which can also be used to make a musical instrument called *mbira*). As already alluded to, learning around these artefacts and tools is however more about naming and knowing what they were used for in the past.

At Primary School B, a similar culture village has also been established due to a directive from the provincial education director – the ministry of education has a policy that requires schools to incorporate culture in their lessons.

From the observations and interviews conducted at this school, it emerged that learning taking place around the culture village focuses mainly, as for Primary School A, on learning about the names of the different traditional huts, the artefacts kept in them and the cultural practices within the village. The teacher in charge of the cultural village also explained how the learning processes around the village are organised, pointing out that in the most it is the Shona and social studies teachers who find the village useful for their teaching and learning. Asked to elaborate on the cultural practices that learners learn about at the village, the teacher was quoted in Zazu (2013, p.154):

> Here we teach children about traditional practices such as *kudzura* (milling), *kuvesa moto* (starting a fire), *kudzira pasi nendove* (using cow dung to make a floor) and *kupeswa* (winnowing). We do this to make sure that children don't forget where they come from and who they are.

The teacher further explained that learning at the culture village also included building studies in which learners learn about the materials used to construct the different huts and infrastructures that make up the village.

It was interesting to note that learning at the cultural village at Primary School B is also somehow integrated into the school's main timetable, and teachers are encouraged to visit and use the village to teach different topics pertaining to culture and social studies. I participated in one of the lessons in which the teacher took the learners through a description of the village and the naming of the tools and arte-facts that are displayed in the huts. Again, the learning processes seemed to be restricted to making learners aware of what makes up a typical rural village and the cultural practices associated with Zimbabwean indignity. Even though building studies was mentioned as one of the subject areas covered at the cultural village, there was little discussion or learning around the different materials (an aspect that could have brought in the natural dimension of heritage into the learning space) that were used to construct the village.

In both Primary School A and B, learning taking place around the culture village is therefore arguably limited to the cultural dimension of heritage, thus limiting the potential that heritage has in supporting ESD learning. The following section explores this further.

5.4   The Culture Hut as a Potential Space for ESD Learning

This section presents a critical discussion of how contemporary learning around the culture hut concept, as described above, can be more oriented to ESD. To situate this discussion, I first revisit what is entailed in ESD learning. I also draw on heritage and curriculum literature to strengthen some of the claims that I make about how the culture hut concept could be used more meaningfully to support ESD learning.

5.4.1 Defining ESD and Heritage

ESD learning entails learning that fosters creative and critical thinking in learners. It also entails what Lotz-Sisitka (2010) has termed 'learning as connection', within which learners' agency to respond to real-life challenges is enhanced. ESD learning is learning that is situated within the learner's sociocultural context, but with a view to ensure sustainable futures (Tilbury, 2011). Important to note here is that learning, besides addressing political, economic and sociocultural issues, must also help learners to become environmentally literate and responsible (United Nations Educational, Scientific and Cultural Organization [UNESCO], 2010). It is in this sense that the dialectic nature of heritage which, if harnessed within the culture hut concept as a curriculum innovation, could provide valuable spaces for ESD learning.

Heritage, as defined by a number of scholars, is both natural and cultural. Lowenthal (2005, p. 81), for example, argued that the word 'heritage' denotes both 'nature' and 'culture'. Accordingly, we often talk of natural heritage, as denoting natural places such as the lands and seas we inhabit and exploit and the soils, plants and animals that form the world's ecosystems (UNESCO, 2002), while on the other hand, cultural heritage is representative of a people's ways of life (Dumbrell, 2012; Munjeri, 2004; Zazu, 2011). In emphasising the interconnectedness and dialectic nature of heritage, Lowenthal (2005, p. 85) further argued that:

> Increasingly the heritages of culture and nature came to be viewed as interconnected, and indeed indivisible. If they are twins, they are Siamese twins, separated only at high risk of demise of both.

Put simply, this implies that there is dialectic relation between people's cultures and what their natural environment can provide. Figure 5.3 illustrates this relational interaction between natural and cultural heritage.

Heritage is therefore symbolic of human-nature interaction (Munjeri, 2004). The dialectic nature of heritage has already been widely recognised (UNESCO, 1972), but its implications for shaping the kind of learning that takes place within curriculum initiatives such as the culture hut concept need further exploration (Zazu, 2013).

5.4.2 My Discomfort and What I Consider Missed ESD Opportunities

Working with the ESD narrative and the concept of heritage as inclusive of both natural and cultural relational dynamics, I began to feel uncomfortable about the way learning around the culture village, as observed at Primary Schools A and B, was organised and constituted. I also started to see how, if carefully reconstituted, the same learning processes can be reoriented to allow for learning experiences that incorporate both sociocultural and environmental sustainability perspectives.

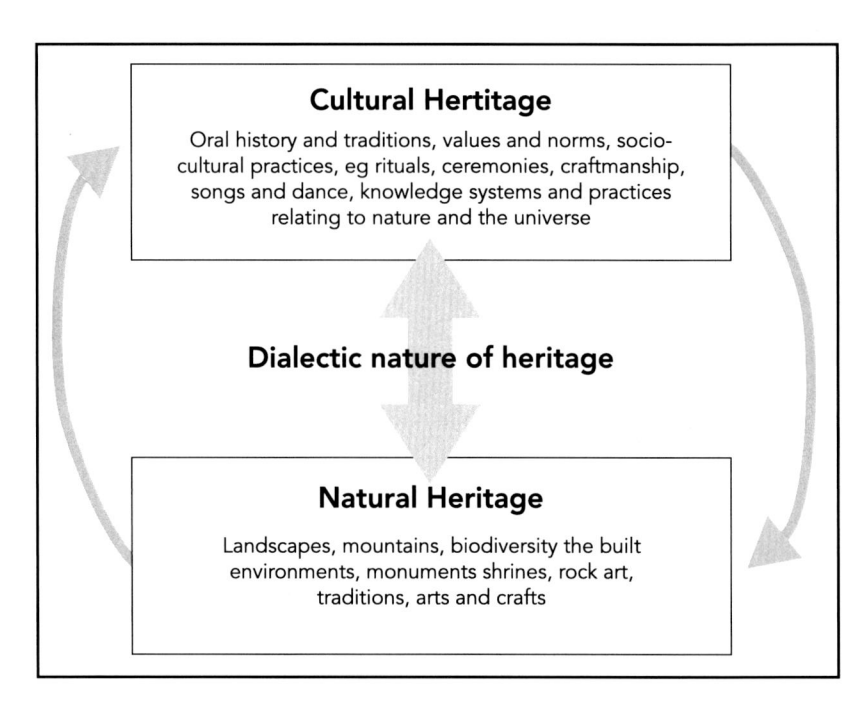

Cultural Hertitage

Oral history and traditions, values and norms, socio-cultural practices, eg rituals, ceremonies, craftmanship, songs and dance, knowledge systems and practices relating to nature and the universe

Dialectic nature of heritage

Natural Heritage

Landscapes, mountains, biodiversity the built environments, monuments shrines, rock art, traditions, arts and crafts

Fig. 5.3 Dialectic interaction between heritages of culture and nature (Zazu, 2013)

At both Primary Schools A and B, where the culture hut concept was being piloted, what emerged as a shortfall is the limiting of learning to the cultural aspects of heritage, and the lack of a realisation of the potential that the culture hut, as a learning space, has to broaden learning to include the natural aspects of heritage. A good example is how, with creative lesson planning, a teacher can provide space for learners to engage critically with questions around the type of natural resources (heritage) that are used to sustain the cultural practices in question, when learners at Primary School A engage in learning about the different types of traditional pots or pottery as a cultural practice defining a particular tribe or ethnicity (Elias and Benjamin cited in Smith & Riley, 2009). Instead of limiting learning to memorisation of names of cultural objects, the culture hut could be a space for learners to ask and engage with critical questions about how their cultural practices relate to and are shaped by their natural environment and what this means in the context of the worsening local and global socioecological crisis. Such a framing of learning is much more empowering, allowing learners to clarify their cultural values and norms, to think systematically (and see their culture as dialectically related to what their natural environment can provide) and to contribute to a sustainable future (Tilbury, 2011). It will also, as Lotz-Sisitka (2010) argued, allow for learning within local contexts but with a focus on the complex interplay amongst social and environmental sustainability risks, issues and crises.

A possible explanation for the teachers' failure to broaden contemporary learning processes within the culture village is that their understanding of heritage may be limited. Often they limit heritage to culture and forget that, as Perez, Lopez and Ferres-Listan (2010, p. 132) pointed out "the term heritage itself does not distinguish between cultural and natural manifestations".

Because of the way learning around the culture hut concept is organised and structured, learners continue to learn about culture as if it is not connected to nature. Such learning experiences may not be enough to allow the learner to make the relevant connections between meaning making, context and concept (Lotz-Sisitka, 2010; Lupele & Lotz-Sisitka, 2012) or think both critically and systematically as advocated for by ESD (Tilbury, 2011).

In emphasising the importance of recognising the dialectic nature of heritage, Hughes (2009) alerted us to the risks of treating culture as divorced from nature by pointing out how such conceptions can impact on the sustainability of local resources and could have contributed to the abandonment and the collapse of ancient cities such as Mapungubwe and the Great Zimbabwe. There is considerable evidence to suggest that both cities were abandoned due to climatic changes, environmental degradation and disease linked to the inhabitants' daily cultural practices (Garlake, 1982; McNaughton, 1987; Pikirayi, 2006).

The important point is that while the culture hut concept is a valuable curriculum innovation, with potential to contribute to successful reorientation of education in Zimbabwe towards ESD learning, its current use, where learning is not as holistic and integrative as it should be, needs careful review and reflexive expansion. Otherwise, there are likely to be many missed valuable ESD and social learning opportunities.

Continuing to limit the learning taking place around the culture village to learning about traditional artefacts, some of which are no longer as relevant or functional to the current context, is tantamount to freezing culture in space and time, dangerously neglecting the fact that culture evolves and is dynamic (Makhoba, 2009). Such framing of learning experiences is also arguably not interesting to learners. Asked why learners and other subject teachers are not showing interest in using the culture village at Primary School B, the head was quoted in Zazu (2013, p.188):

> Ahh ... it was not very appealing to them, they were saying ... some of these things are not really useful ... and I think to some of the teachers and kids these things do not make much sense.

While it is important within the ESD framework for learners to learn about their cultures, value systems and practices (Tilbury, 2011), presenting those cultures as things of the past and not making any effort to help learners relate the learning to their own immediate contexts and real-life challenges only serve to make learning disinteresting and not valuable. An example of more engaging and socioculturally situated learning (O'Donoghue, Lotz-Sisitka, Asafo Adjei, Kota, & Hansi, 2007) around the culture hut is where a complex scientific concept such as fermentation could be introduced to learners through reference to the *hodzeko* – a traditional clay pot (see Fig. 5.2) used in the past for souring milk and a common artefact in the culture hut. In terms of curriculum, in the Shona lesson, the teacher could use the *hodzeko* as a visual aid, which learners can touch and feel, to help them name and conceptualise such cultural artefacts and what they were used for. In the science lesson, the same *hodzeko* can then be used to introduce the concept of fermentation, looking at the shape and design of the pot, discussing how the milk was soured (fermented) in the past and finally making a connection to the science behind the current practices of fermentation. Such use of the culture village and the artefacts displayed is certainly more rewarding for learners and adds value to the investment made by schools to establish the village. It also helps learners to explore "the dialectic between tradition and innovation" (Tilbury, 2011, p. 104) and appreciate the epistemological dialogue between indigenous and western scientific knowledge (O'Donoghue & Neluvhalani, 2002; Zazu, 2008) as well as see the connection between learning about the past and the present contexts (Lupele & Lotz-Sisitka, 2012).

5.5 Conclusion: Future Engagement with the Culture Hut Concept as Curriculum Innovation

It is clear that the culture hut concept is a valuable curriculum innovation and that its use in the formal school system can go a long way to improving the relevance and quality of education in Zimbabwe. However, more educational benefits can be drawn from this curriculum innovation if the learning around it is redefined, or restructured, to work with a broader and inclusive construct of heritage as both cultural and natural (Makhoba, 2009; Shava & Zazu, 2012). Harnessing the dialectic and interconnectedness nature of heritage within the culture hut concept has considerable potential for reorienting education towards ESD learning (Zazu, 2013). It also has the potential to shape learners' attitudes to and practices in the natural environment (Daniel et al., 2012). This can result, as already argued, in the development of learners who are more prepared to take action for a sustainable future.

Capacity building or in-service training of teachers is needed to equip them with the necessary conceptual tools and skills required for the planning and delivery of a more engaging, socially inclusive and sustainability responsive ESD learning around the culture hut. As Zimbabwe strives to reorient its education system towards ESD, it can surely tap into the potential and opportunities that existing curriculum innovations, such as the culture hut concept, provide. Curriculum innovations such as the culture hut concept can provide opportunities to reorient learning to go beyond intergenerational transfer of culture and encompass aspects of environmental sustainability. This is, as pointed out in the opening section of this chapter, one of the issues that continues to motivate and shape my future engagement with research in ESD learning.

References

Cheru, F. (2002). *African renaissance: Roadmaps to the challenge of globalisation*. London: Zed Books.

Daniel, T. C., Muhar, A., Aznar, O., Boyd, J., & Tam, J. (2012). *Contributions of cultural services to the ecosystem services agenda*. Retrieved September 10, 2014, from www.pnas.org/cgi/doi/10.1073/pnas.1114773109.

Dumbrell, K. (2012). *Heritage resources management 1: Training module*. Grahamstown, South Africa: Environmental Learning Research Centre, Rhodes University.

Garlake, P. S. (1982). *Great Zimbabwe described and explained*. Harare, Zimbabwe: Zimbabwe Publishing House.

Government of Zimbabwe. (2004). *Cultural policy*. Harare, Zimbabwe: Ministry of Education, Sport and Culture, Harare, Zimbabwe.

Harrison, R. (2013). *Heritage: Critical approaches*. London: Routledge.

Hughes, J. D. (2009). *An environmental history of the world. Humankind's changing role in the community of life*. London: Routledge.

Lotz-Sisitka, H. (2010). *Conceptions of quality and 'learning as connection': Teaching for relevance*. Paper produced for the Second International Policy Dialogue Forum Providing Teachers for EFA: quality matters Conference. Amman, Jordan, July 2010.

Lowenthal, D. (2005). Natural and cultural heritage. *International Journal of Heritage Studies, 11*(1), 81–92.

Lupele, J., & Lotz-Sisitka, H. (2012). *Learning today for tomorrow: Education for sustainable development learning processes in sub-Saharan Africa*. Howick, New Zealand: SADC REEP.

Makhoba, K. L. (2009). Is the scope of heritage wide enough in the National Curriculum Statements (NCS)? *Journal of South African Heritage Resources Agency, 1*, 80–85.

Malegapuru, W. M. (1999). *African renaissance: The new struggle*. Sandton, South Africa: Mafube Publishing.

McKernan, J. (2008). *Curriculum and imagination: Process theory, pedagogy and action research*. London: Routledge.

McNaughton, D. L. (1987). Ancient civilization: The African connection. *Horizon Magazine (Khaleej Times)*, Dubai, UAE 6 February 1987.

Munjeri, D. (2004). Tangible and intangible heritage: From difference to convergence. *Museum International, 58*(1–2), 12–20.

Nyoni, T., & Nyoni, M. (2010). The 'culture hut' concept: A case of Danda and Chimedza schools in Zaka district. *Journal of Sustainable Development in Africa, 12*(1), 146–157.

O'Donoghue, R., Lotz-Sisitka, H., Asafo Adjei, R., Kota, L., & Hanisi, N. (2007). Exploring learning interactions arising in school-community contexts of socio-ecological risk. In A. Wals (Ed.), *Social learning towards a sustainable future* (pp. 435–448). Wageningen, Netherlands: Wageningen University Press.

O'Donoghue, R. & Neluvhalani, E. (2002). Indigenous knowledge and the school curriculum: A review of developing methods and methodological perspectives. In *Environmental Education, Ethics and Action in Southern Africa. EEASA monograph*. Cape Town. Human Sciences Research Council.

Obanya, P. (2005). *Culture-in-Education and Education-in-Culture*. Paper presented at the 5th Conference of Ministers of Culture, Nairobi, and African Union.

Perez, R. J., Lopez, J. M., & Ferres Listan, D. M. (2010). Heritage education: Exploring the conceptions of teachers and administrators from the perspective experimental and social science teaching. *Teaching and Teacher Education, 26*, 1319–1331.

Pikirayi, I. (2006). The demise of Great Zimbabwe, AD 1420–1550: An environmental re-appraisal. In A. Green & R. Leech (Eds.), *Cities in the world, 1550–2000: The society for post-medieval archaeology monograph* (3rd ed.). Leeds: Maney Publishing.

Shava, S., & Zazu, C. (2012). *Heritage education practices in South Africa. Training module*. Grahamstown, South Africa: Rhodes University, Environmental Learning Research Centre.

Smith, P., & Riley, A. (2009). *Cultural theory: An introduction*. London: Blackwell.

Tilbury, D. (2011). *Education for sustainable development: An expert review of processes and learning*. Paris: UNESCO.

UNESCO. (1972). *International convention on the protection world cultural and natural heritage*. Paris: UNESCO.

UNESCO. (2002). *World heritage in young hands. To know, cherish and act (An educational resource kit for teachers)*. Paris: UNESCO.

UNESCO. (2010). *Incorporating education for sustainable development into world heritage education: Perspectives, principles and values*. Bangkok, Thailand: UNESCO.

Zazu, C. (2008). *Exploring opportunities and challenges for achieving the integration of Indigenous Knowledge Systems into Environmental Education processes. A case study of the Sebakwe Environmental Education programme (SEEP) in Zimbabwe*. Unpublished masters thesis, Rhodes University, Education Department, Grahamstown.

Zazu, C. (2011). Heritage – A conceptually evolving and dissonant phenomenon: Implications for heritage management and education practices in post-colonial Southern Africa. *Southern African Journal of Environmental Education, 28*, 135–143.

Zazu, C. (2013). *Representation and use of indigenous heritage constructs: Implications for the quality and relevance of heritage education in post-colonial southern Africa*. Unpublished doctoral thesis. Grahamstown: Rhodes University, Education Department.

Personal Communications

Chataika, J. (2011, July 17). District Education Officer. Kwekwe, Zimbabwe.

Chauke, G. (2010, September 6). Director, Great Zimbabwe Monument. Masvingo, Zimbabwe.

Chapter 6
Integrating Afrocentric Approaches for Meaningful Learning of Science Concepts

Charles Chikunda and Kenneth Mlungisi Ngcoza

Economic and social development in any modern country relies heavily on a sound scientific and technological base. Essentially, science constitutes an area of any nation's education system where many of the skills that are needed to stimulate development are learned, such as securing good health, fighting diseases, protecting the environment, farming and developing agriculture and developing new industries and technologies and even building resilience to climate change. There is a need therefore for a country to harness the intellectual and scientific capacity of its young people. Ironically, however, science (especially physical sciences) is one of the least popular areas within the educational system of most developing countries. Research shows that students' and especially girls' low interest in science and their relatively negative attitudes are at least partially attributed to the way science is taught at school. The observation is that the science curriculum tends to emphasise only its academic, strongly intellectual and abstract character, and is presented in a decontextualised way, distanced from everyday life, resulting in a considerable mismatch between science-in-society and science-in-school, and hence rendering science as a school subject which appears irrelevant and therefore not useful in everyday life. This perception of school science has detrimental implications for the epistemological access to the discipline. In this chapter we use a case study to share our experiences of how the learning of science in schools could potentially be enhanced through the integration of Afrocentric-indigenous knowledge as a means

C. Chikunda (✉)
Association for Water and Rural Development, Rhodes University,
Grahamstown, South Africa
e-mail: charles@award.org.za; cchikunda2@gmail.com

K.M. Ngcoza
Senior Lecturer in Science Education, Faculty of Education, Education Department,
Rhodes University, Grahamstown, South Africa
e-mail: K.Ngcoza@ru.ac.za

H. Lotz-Sisitka et al. (eds.), *Schooling for Sustainable Development in Africa*,
Schooling for Sustainable Development 8, DOI 10.1007/978-3-319-45989-9_6

of strengthening epistemological access to complex and often abstract scientific concepts. We use ESD lenses to argue for science pedagogical practices that are creative, critical and empowering, as well as connected to real-life challenges and which engage with risk and uncertainty taking into consideration local and global cultures, practices and ideas.

In this chapter we share our experiences of how the learning of science in schools could potentially be enhanced through the integration of Afrocentric-indigenous knowledge. Using a case example, we argue for the integration of indigenous knowledge as a practice of equity and inclusion with a strong rationale for education for sustainable development (ESD). We believe that this integration is important as a means of broadening the ontology and strengthening epistemological access to complex and often abstract scientific concepts. Ontological assumptions are concerned with reality and how people view the world; epistemological assumptions entail how knowledge is constructed and communicated (Scotland, 2012). The case example discussed below foregrounds learners' everyday life experiences as playing an important role in the learning process. Furthermore, it illustrates that science is a human activity (Lemke, 2001), and thus there is a need to contextualise scientific knowledge so that it is accessible to learners.

Initially, we highlight the challenges faced in science education and emphasise the need to rethink pedagogy. We then use ESD lenses to argue for science pedagogical practices that are creative, critical and empowering. We show that such practices are connected to real-life challenges and engage with risk and uncertainty in the community.

6.1   The Need for Rethinking Pedagogy

Economic and social development in any country relies heavily on a sound scientific and technological base as well as the quality of its science and technology education. Essentially, science constitutes an area of any nation's education system where much of the knowledge and many of the skills that stimulate development are learned. These knowledge and skills are required for securing good health, fighting diseases, protecting the environment, farming and developing agriculture and developing new industries and technologies and even building resilience to climate change. There is a need, therefore, for any country to develop and nurture the intellectual and scientific capacity of its young people. Ironically, science subjects, especially physical sciences, are the least popular learning areas within the educational system of many countries as reflected in the Trends in International Mathematics and Science Study (TIMSS) (Howie, 2004; Reddy, 2005; Reddy et al., 2015). Howie (2004) reported that, since 1995, South African learners have performed poorly in TIMSS compared to learners from other countries such as Morocco and Malaysia. Focusing specifically on Africa, Clegg (2007) is of the opinion that economic and social development can be achieved by putting emphasis on science subjects at all levels of education.

Since the 1990s there has been a growing concern globally about the increasing lack of interest of young people in science, technology, engineering and mathematics (STEM subjects) from high school level and beyond. This disinclination emerged with several national reports identifying shortages of science graduates and declines in student interest in school science (Bøe, Henriksen, Lyons, & Schreiner, 2011). Chetcuti and Kioko (2012) reported on young people's experience of school science and the science curriculum, pointing out that some, especially girls, had the impression that science is a dry subject and only for the 'super brilliant' involving mainly the recall of factual knowledge rather than skills. Semela (2010) supported the idea, pointing out that students, in general, and girls, in particular, state that science as a school subject is irrelevant and therefore is not useful in everyday life. This concurs with earlier assertions by several researchers (e.g. Kalu, 2005; Mutasa & Willis, 1994) that students identify a considerable mismatch between science-in-society and science-in-school: school science is unattractive since it does not involve topics of interest; it does not provide students with opportunities for creative expression; and it is fairly alienated from society.

Furthermore, science curricula in almost all African schools have for many years disregarded indigenous knowledge (IK) or local knowledge (Breidlid, 2009; Cocks, Alexander, & Dold, 2012; Kibirige & van Rooyen, 2006; Le Grange, 2007; Shizha, 2007). Shava (2013) raised the concern that IK is unequally and unjustly represented in relation to formalised Western knowledge systems. He raises concerns with IK representation that include invalidation, devaluation, subjugation, marginalisation, primitivisation, exclusion and rejection. To this end, Kibirige and van Rooyen (2006) argued that the absence of IK in the science curricula has significant consequences for some learners because they experience conflict between their existing knowledge and the knowledge of the various science curricula which makes it difficult for learners to 'cross borders' from indigenous knowledge to scientific knowledge. Le Grange (2007) referred to this as 'cognitive conflict'. Yet, it is believed that if indigenous environmental knowledge is applied in the curriculum, meaning-making is enhanced (O'Donoghue, Lotz-Sisitka, Asafo-Adjei, Kota, & Hanisi, 2007). This suggests that there is a need to reimagine and rethink science pedagogy in schools.

The science education situation reported here is particularly dire for Africa which desperately needs scientists to tackle numerous problems such as hunger, diseases, environmental degradation as well as mitigating climate change-induced challenges. Enhancing school science would play a significant role in the social and economic development of the continent. In his study conducted in Namibia, Uushona (2013) explored teaching the concepts of fermentation and distillation through the inclusion of an indigenous practice of making *ombike*. Similarly, Shifafure (2014) investigated how science teachers mediate learning of scientific concepts associated with distillation through the indigenous practice of making *kashipembe*. In both cases studies, an enriching factor was the integration of contextual scientific knowledge in the classroom with community knowledge. Herein lies the importance of education for sustainable development (ESD).

6.2   ESD, Afrocentricity and IK

In this chapter we view ESD as a philosophy of education aiming to reorient thinking and practices in the whole education process, in response to the twenty-first-century complexities. The main question in ESD is 'What do we educate for?' (UNESCO, 2005). In responding to this overarching question, ESD qualifies to be a philosophy of education because it is concerned with cross-examining and transforming educational aspects such as ontology, epistemology, logic and ethics (Chikunda, 2016).

Ontology is concerned with what things are, that is, it is concerned with identifying, in the most general terms, the kinds of things that actually exist in an education set-up. Applying ESD lenses to science education in Africa, we are able to ask ontological questions such as:

• What is the nature of science education in a country?
• What can be said to exist in the science education system?
• What governs the properties of that system?
• What knowledge is valued in science education?

On the other hand, the epistemological concerns of ESD relate to the nature of knowledge itself and its possibility, scope and accessibility to the learner. More broadly, ESD epistemology interrogates, in an education system, the systematic ways we can determine when something is good or bad as well as issues of *equity* and *inclusivity*. In this regard, ESD as a philosophy of education provides sound rationality for respect and inclusion of local perspectives and IK to tackle local problems and think globally. With specific reference to teaching and learning of science in Africa, epistemological questions can include:

• How do we go about knowing science knowledge in a given education system?
• How do we go about accessing scientific concepts in a given education system?
• How do we separate 'useful' ideas from 'unviable' knowledge in science education?
• How do we facilitate cognitive access to scientific concepts in our education process?

Ethics, or the 'moral philosophy', of ESD is concerned primarily with the question of the best way for schooling or educating. Normative ethics for ESD are that children and adults alike are entitled to education that offers them considerable opportunities to live well in a context of increasing social and ecological risks. ESD therefore brings in broader moral questions in education that are normally not attended to in 'business as usual' education. Questions that relate to sustainability include, for example, interrogating the morality of girls or any group failing to access science because of the ontology and/or epistemology of school science.

The logic of ESD is mainly based on the principles of education that in a way are driven by the complexity in which we find ourselves in the twenty-first century. The instrumentalist economic drive has shaped school science education for centuries.

With ESD lenses the reasonableness of the school science curriculum will also be judged on whether the process of science education is contributing to sustainable livelihoods, for example, to poverty reduction, environmental protection, social justice and education for all.

From this description we can see that ESD urges science teachers to go beyond the traditional subject content mastery, to embrace values in their teaching. In an African context, embracing an Afrocentric thinking (Mukwambo, Ngcoza, & Chikunda, 2014) has the potential to fulfil some of these ESD dreams or ideals.

Afrocentric philosophy is a way of thinking from an African perspective, based on the principles of inclusivity, cultural specificity, critical awareness, committedness and political awareness (Mukwambo et al., 2014). African scholars (Asante, Kincheloe, Nyerere and others) have challenged Eurocentric perspectives and Eurocentric paradigm of knowledge construction as the only legitimate philosophy. Asante (1998), for example, argued that Afrocentricity as a philosophy is against the Western theory that dislocates and pushes Africans and their experience to the margins of human thought and experience.

Afrocentric-Indigenous Knowledge

Indigenous knowledge (IK) in an African context is born from Afrocentric philosophy. IK is conceptualised as a way of life of a specific group of people, defined by ancestral territories, cultural activities and historical locations (Van Wyk, 2014). Van Wyk further defined IK as knowledge and skills constructed by indigenous people with the purpose of advancing and sustaining their identity, culture and history for the next generation. The term indigenous denotes that the knowledge is local and specific to a people with common social and cultural connections. The knowledge is handed down from one generation to the next through symbols, rituals, practices, art, oral narratives, proverbs, performances, wise sayings and dances.

Adopting an Afrocentric paradigm will be helpful in pursuing simultaneously the goals of ESD and those of the science curriculum. Resonating with the ESD ideas put forward by Lotz-Sisitka and Lupele in Chap. 1 of this book, adopting an Afrocentric paradigm to science learning brings the importance of situated learning to the fore. This is an approach to science learning that focuses on systems in which individuals act as members of social groups and interact with material resources, situated in historical and cultural contexts. This means that in the curriculum process, African experiences of the learners would be placed at the centre, moving from the margins and empowering them by making them the subjects and not the objects of the learning encounter. Afrocentric educationists identify fundamental principles and these resonate well with ESD principles (see Table 6.1).

Ubuntu

Ubuntu is one of the Afrocentric philosophies found in different forms in many societies of Africa. Ubuntu is key to all African values, involving collective personhood and collective morality. Many African cultures across the continent carry the Ubuntu saying that translates to 'I am because we are'. Such adages thus recognise Ubuntu as a philosophy of humanism that links the individual to the collective. Other axioms across several African cultures, such as 'It takes a village to raise a child', emphasise the cultural connectedness of most African societies. Ubuntu manifests itself in attitudes and practices such as respect, empathy, accountability, responsibility, fairness, justice, compassion, unity, selflessness, compromise, love, caring, tolerance, forgiveness and many others. Nelson Mandela's humanism is a typical example of the qualities of Ubuntu. His famous saying "Education is the most powerful weapon, which you can use to change the world" suggests an education that has connections to sociocultural, social-ecological, contextual and historical dynamics of learners' life-worlds and experiences and communities' valued beings and doings while also giving attention to 'learning as mastery' (i.e. how to complete educational tasks successfully) and to 'learning as participation' (meaning: learners having a chance to voice their opinions and express their thoughts in an open, supportive learning environment).

Table 6.1 Fundamental principles underlying the Ubuntu approach to science education as pointers to ESD principles

Ubuntu principles and implications	Contribution to ESD (*Captures learning as democratic process* and *learning as connection in addition to learning as efficiency/mastery*) (*see Lupele and Lotz-Sisitka*, 2012)
Moral building: the teacher is vital in moral building, supporting and instilling good citizenship in his/her science teaching all the time. The teacher has an important duty to go beyond imparting science subject knowledge to caring for the learner and the community	Creation of science knowledge must be done in the context of school community. Schools become important spaces in which children not only gain subject knowledge but also learn about themselves and their culture and good personal and social conduct. Also involves learning to explore the dialectic between tradition and innovation, e.g. questioning harmful cultural practices. ESD learning links being to becoming (see Lupele and Lotz-Sisitka, Chap. 1 of this book). As children are given a chance to express their valued beings and doings and expand their knowledge, values and action competence through ESD learning, they can engage in a wider range of debates, think more critically and participate in the world differently

(continued)

Table 6.1 (continued)

Epistemological access: evaluation of whose knowledge is represented and how is it represented and applied is vital when considering epistemological access to science concepts	Mobilising indigenous and local knowledge in ESD learning processes is an act of *enabling epistemological access* to abstract forms of knowledge (O'Donoghue et al. 2007)
Pedagogies of empowerment and epistemological changes	ESD learning processes are not only concerned about conceptual access but empowerment and valuing as well. As Shumba, Kasembe, Mukundu and Muzenda (2008) questioned: does school science take note of local and indigenous forms of knowledge, counting it as being valid in the educational context, and is it 'brought into' processes of enabling epistemological access to new/more abstract science concepts in formal education settings?
Ontology of school science is concerned with what things are, that is, concerned with identifying, in the most general terms, the kinds of things that actually exist in the science education set-up. The how and what of school science learning	Principal ontological questions for ESD include: what knowledge is valued in school science curriculum in view of social and ecological risk? Good science teaching models good practice
	Learning to envision more positive and sustainable futures and learning to think and view things systemically. This requires complete departure from instrumentalist science teaching driven solely by economic development to embracing a 'caring science' that considers sustainability in development
Fairness and justice requires a learning process that is fair to all learners, which is applied with the well-being of all participants in mind including the community	Communities should have a stake in the education process of their children taking into consideration historical context in the learning process. In this context teachers are expected to learn to clarify their own values
Community involvement encourages teachers to accommodate learners of diverse learning needs in support of their own sense and beliefs about themselves and not someone else's expectations of them	The community needs to influence which of the many existing sustainability issues (e.g. biodiversity, climate change, equity, harmful practices, etc.) will be part of the curriculum
Commitment to the values of humanity – what is the logic of school science? This empowers learners, for example, to analyse science text and to critically uncover hidden assumptions in curriculum resources and processes. Learners and teachers need to be aware that knowledge is not value-free but is interwoven by its social and political contexts	Learning to be critical, learning to clarify one's own values, learning to envision more positive and sustainable futures and learning to think systemically in science education
	How is science contributing to sustainable livelihoods, for example, poverty reduction, environmental protection, social justice, education for all and many more?

The inclusion of Ubuntu in science learning can further contribute to science becoming more meaningful (Le Grange, 2012; Mukwambo et al., 2014). Ubuntu has the potential to promote 'caring science' in the school science curriculum. Interpreted through ESD lenses, the Ubuntu philosophy contributes to the ontology of ESD by broadening the 'what' question in school science curricula. For example, what is valued in science education in a context of socio-ecological risk? What is the logic to teach what we are teaching in view of societal challenges and needs? Do we consider social cohesion, value humanness, caring, etc. in our science curriculum? The elements of Ubuntu are also helpful to ESD to introspect the 'how' question, that is, the epistemology of school science. Simple questions such as is the science learning environment inclusive enough in terms of gender, race, ethnicity, culture and so forth? With these lenses (Ubuntu and ESD), inclusivity is broadened to go beyond physical access to school to include epistemological access and cognitive justice as well. According to Visvanathan (2005), there is 'a constitutional right for different systems of knowledge to exist as part of a curriculum' (p. 8).

Ubuntu values heighten the quality and relevance of school science education, broadening the ontology, epistemology and sense of the curriculum. In essence, the fundamental principles of an Ubuntu-informed science curriculum resonate well with ESD principles as shown in Table 6.1.

The rationale of IK in science is that science learning becomes effective when it is situated in time and place and hence has more meaning for learners. Reflecting on (or even just trying to investigate) indigenous ways of knowing enables learners to situate their learning within their cultural context and draw on their prior everyday knowledge and experiences (Kuhlane, 2011).

Case Study (Tooth Decay and Chemical Reactions)

Lwazi is an experienced and innovative physical sciences teacher in a rural school in Onangalo village in Namibia. Informed by the curriculum that acknowledges the centrality of learners in the learning process as well as valuing of their everyday life experiences, he always makes an effort to bridge the gap between everyday knowledge and scientific knowledge. He does this through elicitation and incorporation of his learners' prior everyday knowledge with a view to enhance meaning-making in science learning.

For instance, Lwazi recently taught the topic of chemical reactions to his grade 9 learners, and he explored using the tooth decay problem rampant in his village as a vantage point. He started by encouraging his learners to brainstorm causes of tooth decay in the village and how it could be avoided and treated. Thereafter, together with his learners, Lwazi designed an interview guide to get the perspectives of community members on the problem of tooth decay. In small groups, learners administered the guide, recording the information as agreed.

Information generated was presented in the following lesson. During the presentations, Lwazi was delighted to observe that learners had considerable

(continued)

knowledge about tooth decay from their homes and environment. He also found out that some of their knowledge was related to chemical reactions although they could not initially give proper scientific explanations. Some learners even used their home language to explain things so that others could understand their contributions. They had arguments within individual groups and these prompted further thinking. Furthermore, learners were keen to complement or ask for clarifications, something which is rare or non-existent in traditional science classrooms. They also freely talked about how their parents treat this malady in traditional ways.

Essentially, the main aim for recording community members' views was to complement what learners came up with in class in the brainstorming session. This helped him to understand the type of knowledge learners brought to his science classroom, that is, closing the gap between everyday knowledge and classroom science as proposed by Rennie (2011). The teaching approach, Lwazi was convinced, enhanced learner engagement, and learners demonstrated an understanding of what tooth decay is and explained different methods on how it could be treated.

In the following sessions, Lwazi and his learners decided to do a laboratory investigation on the cause of tooth decay that was rampant in the village. Their hypothesis pointed to local brews. They decided to focus on four locally brewed traditional beverages, namely, *Epwaka, Mundele, Otombo* and *Ondjidja*. Learners brought these from their homes. For ethical purposes, pig teeth instead of human teeth were used for the scientific investigations. Again these were locally sourced. Before the practical investigation, Lwazi asked his learners to predict what would happen when pig teeth were immersed into the different local traditional beverages and into the tap water which was used as a control. The investigation was done over a period of five days and the learners recorded their observations in worksheets. From the learners' investigation, they found that the tooth inserted in *Otombo* decayed the most. Lwazi then went on to consolidate the chemical reaction concepts related to tooth decay using chemical equations and ideal concepts as per curriculum requirements.

With reference to the whole community inquiry and practical investigation, learners commented that:

Learner 1: 'Yes, it was good because I see with my own eyes but not the teacher to tell me'.

Learner 2: 'Exciting to see the link of what we do at home and what happens in the laboratory'.

Learner 3: 'Yes, because we are going to tell our friends and parents not to drink the Otombo that cause tooth decay'.

Learner 4: 'Practical activity is good because we learn a lot of things on tooth decay and chemical reactions. Many practical activities. Scientific concepts emerged as a result of activities'.

The case study is an example of incorporating local knowledge into school science. As shown in the case study, the nexus of ESD and Afrocentric-indigenous knowledge in science learning opens up spaces for learners to gain traditional scientific knowledge and theories as well as acquire values and skills necessary for learning (Tilbury, 2011, p. 104) through:

- Asking questions
- Clarifying one's own values
- Thinking systemically
- Responding through applied learning
- Exploring the dialectic between tradition and innovation

This case study begins to show how an Afrocentric approach can enhance the quality and relevancy of school science education. The pedagogical privileges arising from this case resonate well with ESD. In essence, the pedagogical hybrid at the nexus of ESD and IK can be a response to some of the challenges bedevilling science education in Africa in that:

- Science knowledge will be taught and learned with responsibility on the part of both the teacher and learner to use that knowledge towards the community's well-being.
- IK is rooted in experiential learning; hence, science learning will be localised, contextual, relevant and therefore less abstract.
- IK connects school science knowledge to community knowledge and practices providing space to cross-examine culturally held beliefs and norms.
- IK transcends discipline boundaries. Dei (2002) gave an example of acquiring herbal knowledge, which involves plant biology, ecology, physiology and some familiarity with the historical, economic and social causes of illness, besides appropriate spiritual prayers and the rehearsal of sacred narratives.
- IK gives special place to knowledge rooted in oral traditions (such as storytelling, music and praise poetry). Van Wyk (2014) noted that these oral exchanges are never seen as locked up knowledge in history but they provide tangible realities that create a felt relationship with the past. This, in our view, has the potential to humanise science, making it possible to locate in people's history and practices.

In relation to this, Mukwambo et al. (2014) called for an 'Africanisation of the school science curricula'. Their point is that African communities already have beliefs and practices that can be incorporated in the classroom to enhance science teaching and learning. The point of departure for this argument rests in the fact that most curricula across the continent have shifted to focus on learner-centred approaches to align with the democratic dispensation. For example, the Namibian Ministry of Education (MoE) (2009) stated that:

> The starting point for teaching and learning is the fact that the learner brings to the school a wealth of knowledge and social experience gained continually from the family, the community, and through interaction with the environment. Learning in school must involve, build on, extend and challenge the learner's prior knowledge and experience. (Namibia. MoE, p. 4)

6.3 Reflections and Concluding Remarks

Adopting an Afrocentric approach in the school science curriculum in an ESD context is not without challenges. For instance, Kibirige and van Rooyen (2006) posited that the main challenge associated with inclusion of IK in the science classroom is the issue of multicultural classrooms in schools. As IK is dependent on the geographical location of people in a particular area, if one finds a situation (especially in towns), where there are different learners from different backgrounds, it becomes a challenge to bring into the classroom examples from the students' own contexts. There are many regional and cultural variations of indigenous knowledge, names and stories, which may lead to disagreement among learners. Rather than see this as a problem, we urge educators to use the opportunity to open up further deliberation about why such variations exist and what their implications are for contemporary science perspectives. Put differently, we need to embrace diversity as proposed by Zazu (2011) in his seminal work on what heritage entails in the Zimbabwean context (see also Chap. 5 of this book).

Kibirige and van Rooyen (2006) added that IK might have fallacies and should be properly scrutinised and corrected because knowledge that learners bring to classrooms is often comprised of both fact and fiction. Certain scientific concepts in terms of IK are associated with witchcraft, myths and superstitions. Care should therefore be exercised in facilitation of the content to the learners. Again, teachers should take advantage of such tensions to mediate science learning. Total snubbing of traditional beliefs and myths may lead learners to develop parallel conceptions. In this regard, teachers are thus challenged to know enough of the subject matter in order to be able to identify and avoid misconceptions and potentially harmful information embedded in IK. In addition to this challenge to teachers, IK as a field is not well documented. Agea, Lugangwa, Obua and Kambungu (2008) indicated that a limiting factor in enhancing indigenous knowledge lies in its lack of documentation.

References

Agea, J. G., Lugangwa, E., Obua, J., & Kambungu, R. (2008). Role of indigenous knowledge in enhancing household food security: A case study of Mukungwe, Masaka District, Central Uganda. *African Journal of Indigenous Knowledge Systems, 7*(1), 64–71.

Asante, M. K. (1998). *The Afrocentric idea in education* (2nd ed.). Philadelphia: Temple University Press.

Bøe, M. V., Henriksen, E. K., Lyons, T., & Schreiner, C. (2011). Participation in science and technology: Young people's achievement-related choices in late-modern societies. *Studies in Science Education, 47*(1), 37–72.

Breidlid, A. (2009). Culture, indigenous knowledge systems and sustainable development: A critical view of education in an African context. *International Journal of Educational Development, 29,* 140–148.

Chetcuti, D., & Kioko, B. (2012). Girls' attitudes towards science in Kenya. *International Journal of Science Education, 34*(10), 1571–1589.

Chikunda, C. (2016). Philosophies, theories and principles for ESD in teacher education. In *Education for sustainable development: Effective teaching and learning in teacher education institutions in Africa*. ESD UNESCO guide. Paris: UNESCO.

Clegg, A. (2007). *Girls into science: A training module*. Paris: UNESCO.

Cocks, M. C., Alexander, J., & Dold, T. (2012). *Inkcubeko Nendalo*: A bio-cultural diversity schools education project in South Africa and its implications for inclusive indigenous knowledge systems (IKS) and sustainability. *Journal of Education for Sustainability Development, 6*(2), 241–252.

Dei, G. J. S. (2002). Afrocentricity: A cornerstone of pedagogy. *Anthropology & Education Quarterly, 25*(1), 3–28.

Howie, S. J. (2004). A national assessment in mathematics within an international comparative assessment. *Perspectives in Education, 22*(2), 149–162.

Kalu, I. (2005). Classroom interaction in physics lessons, relative to students' sex. *African Journal of Research in Mathematics, Science and Technology, 9*(1), 55–66.

Kibirige, I., & van Rooyen, H. (2006). Enriching science teaching through the inclusion of indigenous knowledges. In J. de Beers & H. van Rooyen (Eds.), *Teaching science in the OBE classroom*. Braamfontein, South Africa: Macmillan.

Kuhlane, Z. (2011). *An investigation into the benefits of integrating learners' prior everyday knowledge and experiences during teaching and learning of acids and bases in Grade 7: A case study*. Unpublised master's thesis, Rhodes University, Grahamstown.

Le Grange, L. (2007). Integrating western and indigenous knowledge systems: The basis for effective science education in South Africa? *International Review of Education, 53*, 577–591.

Le Grange, L. (2012). Ubuntu, ukama, environment and moral education. *Journal of Moral Education, 41*(3), 329–340.

Lemke, J. L. (2001). Articulating communities: Sociocultural perspectives on science education. *Journal of Research in Science Teaching, 38*(3), 296–316.

Lupele, J., & Lotz-Sisitka, H. B. (2012). *Learning today for tomorrow: Sustainable development learning processes in Sub-Saharan Africa*. Howick, UK: SADC REEP.

Mukwambo, M., Ngcoza, K., & Chikunda, C. (2014). Africanisation, Ubuntu and IKS: A learner centred approach. In C. Okeke, M. van Wyk, & N. Phasa (Eds.), *Schooling, society and inclusive education: An African perspective* (pp. 65–80). Cape Town, South Africa: Oxford University Press.

Mutasa, N. G., & Willis, G. M. (1994). *Modern practice in education and science*. Gaborone, Botswana: Longman.

Namibia. Ministry of Education. (2009). *The National Curriculum for Basic Education*. Okahandja, Namibia: NIED.

O'Donoghue, R., Lotz-Sisitka, H., Asafo-Adjei, R., Kota, L., & Hanisi, N. (2007). Exploring learning interactions arising in school-in community contexts of socio-ecological risk. In A. Wals (Ed.), *Social learning towards a sustainable world* (pp. 435–447). Wageningen, The Netherlands: Wageningen.

Reddy, V. (2005). Cross-national achievement studies: Learning from South Africa's participation in the Trends in International Mathematics and Science Study (TIMSS). *Compare: A Journal of Comparative and International Education, 35*(1), 63–77.

Reddy, V., Zuze, T. L., Visser, M., Winnaar, L., Juan, A., Prinsloo, C. H., et al. (2015). *Beyond benchmarks: What twenty years of TIMSS data tell us about South African education*. Cape Town, South Africa: HSRC Press.

Rennie, L. J. (2011). Blurring the boundary between the classroom and the community: Challenges for teachers' professional knowledge. In D. Corrigan, J. Dillon, & R. Gunstone (Eds.), *The professional knowledge base of science teaching* (pp. 13–29). New York: Springer.

Scotland, J. (2012). Exploring the philosophical underpinnings of research: Relating ontology and epistemology to the methodology and methods of scientific, interpretive and critical research paradigms. *English Language Teaching, 5*(9), 9–16.

Semela, T. (2010). Who is joining physics and why? Factors influencing the choice of physics among Ethiopian university students. *International Journal of Environmental and Science Education, 5*, 319–340.

Shava, S. (2013). International handbook of research on environmental education. In R. B. Stevenson, M. Brody, J. Dillon, & A. E. J. Wals (Eds.), *The representation of indigenous knowledges* (pp. 384–393). London: Routledge.

Shifafure, A. M. (2014). *Understanding how grade 11 Physical Science teachers mediate learning of the topic distillation in the Kavango Region.* Unpublished master's thesis, Rhodes University, Grahamstown.

Shizha, E. (2007). Critical analysis of problems encountered in incorporating indigenous knowledge in science teaching by primary school teachers in Zimbabwe. *The Alberta Journal of Educational Research, 53*(3), 302–319.

Shumba, O., Kasembe, R., Mukundu, C., & Muzenda, C. (2008). Environmental sustainability and quality education: Perspectives from a community living in a context of poverty. *Southern African Journal of Environmental Education, 23*, 81–97.

Tilbury, D. (2011). *Education for sustainable development: An expert review of process and learning.* Paris: UNESCO.

UNESCO. (2005). *United Nations Education for Sustainable Development (2005–2014): Linkages between the global initiatives in education.* Paris: UNESCO.

Uushona, K. I. T. (2013). *An investigation into how Grade 9 learners make sense of the fermentation and distillation processes through exploring the indigenous practice of making the traditional alcoholic beverage called Ombike: A case study.* Unpublished master's thesis, Rhodes University, Grahamstown.

Van Wyk, M. (2014). Towards an Afrocentric-indigenous pedagogy. In C. Okeke, M. Van Wyk, & N. Phasha (Eds.), *Schooling, society and inclusive education: An Afrocentric perspective* (pp. 39–61). Cape Town, South Africa: Oxford University Press.

Visvanathan, S. (2005). Knowledge, justice and democracy. In M. Leach, L. Scoones, & B. Wynne (Eds.), *Science and citizens: Globalization and the challenge of engagement.* London: Zed Books, Open University.

Zazu, C. (2011). Heritage – A conceptually evolving and dissonant phenomenon: Implications for heritage management and education practices in post-colonial Southern Africa. *Southern African Journal of Environmental Education, 28*, 135–142.

Chapter 7
The Uptake of Education for Sustainable Development in Geography Curricula in South African Secondary Schools

Carolina Dube

Curriculum innovation that took place in the post-apartheid era in South Africa provided opportunities for integrating environmental concerns and sustainability issues in subjects such as geography and science at further education and training level (FET). However, evidence from my PhD study in geography education at FET level shows that there are major challenges in the implementation of education for sustainable development (ESD) in the school context. These relate, firstly, to teachers' lack of content knowledge (CK) and pedagogical content knowledge (PCK) and, secondly, to contextual issues and structural constraints within schools. In order to overcome some of the challenges, a standards-based teacher training model is recommended to ensure the mastery of appropriate competencies for the implementation of environmental education (EE) and ESD.

7.1 Policy Context in South Africa

According to Le Grange (2002), EE was first introduced into formal schooling in South Africa in 1997 through Curriculum 2005 (C2005) in the general education and training (GET) band. In the post-apartheid schooling system in South Africa, GET is the compulsory phase of schooling covering grade R to grade 9 (students aged approximately 7–15 years). In these grades, teaching and learning occurs through an integrated approach in eight learning areas. Environmental learning occurred through a cross-curricular approach in the learning areas where the environment theme was one of the phase organisers (Lotz-Sisitka & Raven, 2001). After the revision of C2005 (Chisholm, 2000), EE was integrated in the Revised National Curriculum Statement (RNCS) (Department of Education [DoE], 2002). Some of

C. Dube (✉)
Rhodes University, Grahamstown, South Africa
e-mail: carolina.dube@gmail.com

© Springer International Publishing Switzerland 2017
H. Lotz-Sisitka et al. (eds.), *Schooling for Sustainable Development in Africa*,
Schooling for Sustainable Development 8, DOI 10.1007/978-3-319-45989-9_7

the principles that underpin the curriculum are social justice, a healthy environment, human rights and inclusivity; these principles provided an opportunity to accommodate the shift from EE towards ESD. This is in line with the view that ESD is broader than EE (UNESCO, 2005).

The further education and training (FET) curriculum (DoE, 2003) for grades 10–12 (students aged approximately 16–18 years) was introduced in schools in 2006. Students take the National Senior Certificate school leaving examination at the end of the FET band. In contrast to the GET level, where environmental learning occurs through a cross-curricular approach, environmental concerns are infused in each subject in the FET band. A revised curriculum, the Curriculum and Assessment Policy Statement (CAPS) (Department of Basic Education [DBE], 2011), has been phased into schools between 2012 and 2014. Compared to the NCS (DoE, 2003), the CAPS curriculum (DBE, 2011) is more content based so as to improve the learners' knowledge and understanding of basic concepts in every subject or learning area.

In line with the 'knowledge turn' in the school curriculum, this chapter focusses on the importance of subject knowledge and pedagogical content knowledge of teachers for effective teaching of EE and ESD. In the discussions that follow, evidence is provided from my PhD study of geography teachers' experiences of implementing ESD at FET level in selected Western Cape secondary schools in South Africa.

7.2   Subject Knowledge and Pedagogical Content Knowledge

Shulman (1986, 1987) contended that his seminal teacher knowledge model could be a basis for education reform. The model comprises a number of elements that include subject matter or content knowledge (CK), pedagogical content knowledge (PCK) and curriculum knowledge. This chapter explores the importance of only two of the elements of the model, i.e. subject or content knowledge and pedagogical content knowledge. Researchers working on teaching and teacher education have acknowledged the importance of content knowledge (Abell, 2008; Ball, Thames, & Phelps, 2008; Corney & Reid, 2007). Some, for example Ball et al. (2008), argued that content knowledge is foundational to teacher competency. CK and PCK are crucial to the teachers' uptake of curriculum innovations such as the integration of ESD in the school context in light of the recent knowledge turn in the South African school curriculum.

7.3   The Contested Nature of ESD

The subject matter or content related to sustainable development and education for sustainable development is complex and contested (Bonnett, 1999; Corney & Reid, 2007; Fien & Tilbury, 2002). The interpretations of the meaning of sustainable development are 'value-laden' and vary in different societies and cultures. Fien and Tilbury proposed that:

> Many of these differences revolve around questions such as ... are we concerned with sustainability on ecological time-scales? And what kind of development do we want to sustain: social, cultural, political, spiritual and/or economic? (And are these separable?) What changes are required to achieve sustainability and how are they to be achieved? What are the implications for economic growth? Are there limits to economic growth in a sustainable society and, if so, what are they? Implicit within these questions are differing definitions and assumptions of "sustainability" and these, in turn, reflect both a variety of contesting ideologies and an ongoing political debate about the nature of sustainable futures. (p. 2)

Additionally, Fien and Tilbury (2002) identified two main perspectives of sustainable development. One tends to promote 'sustainable economic growth', while the other prioritises 'sustainable human development' through addressing 'issues of social equity and ecological limits'. According to UNESCO (2005), sustainable development aims to integrate economic and social development (including culture) with environmental protection. The World Commission on Environment and Development (WCED, 1987) in the Brundtland Report, *Our Common Future*, defined sustainable development as 'development that meets the needs of the present without compromising the ability of future generations to meet their own needs' (p. 45). As such, sustainable development is aimed at achieving both intergenerational and intra-generational equity. Intergenerational equity is concerned with the need to ensure that future generations enjoy the same quality of life as current generations, whereas intra-generational equity focuses on issues of poverty and the need to reduce social inequality within nations and between nations through better distribution of wealth (Hattingh, 2002).

There have been major international events that have helped to shape the nature and scope of EE and ESD including the implementation strategies. One of these, the United Nations Conference on Environment and Development, or the Rio Earth Summit held in 1992 in Rio de Janeiro, emphasised the link between environmental education and sustainable development (Irwin & Lotz-Sisitka, 2005). Agenda 21, a document produced at that conference, provides a programme of action by which governments and other parties can protect the environment. Wals (2012) in Dube (2014) traced the roots of ESD to Agenda 21. Chapter 36 (Para 2) of Agenda 21 calls upon nations to reorient 'education towards sustainable development' (UNCED, 1992).

Since then, there have been international debates on the nature and scope of education for sustainable development versus environmental education (see Jickling, 1992; McKeown & Hopkins, 2003; Robottom, 2007). In the context of South Africa, Lotz-Sisitka (2011) claimed that the EE subject matter has always been synonymous with that of ESD, resulting in much less contestation. Because of the country's unique history of oppression of the black population and human rights violations under colonial and apartheid rule, Lotz-Sisitka argued that the EE perspective has attempted to address issues of human rights and social justice in addition to those affecting the biophysical environment. In some countries, the biocentric perspective, a narrow view that mainly focuses on the protection of the biophysical component and downplays the human dimension of EE (i.e. social, economic and political), has been dominant (Andersen & Wennevold, 1997; McKeown & Hopkins, 2003).

The shift towards the ESD discourse would broaden the scope of EE in these countries. According to UNESCO (2006), 'ESD is fundamentally about values, with respect at the centre: respect for others, including those of present and future generations, for difference and diversity, for the environment, for the resources of the planet we inhabit' (p. 17).

ESD may thus be regarded as a 'process' that provides humanity with opportunities to engage in lifelong learning to address existing and emerging social, economic and environmental issues. The aim of this process and effort is to achieve the goal of sustainability at both local and global scales.

Additionally, in the context of education institutions, ESD implementation strategies include interdisciplinary and holistic approaches, inculcation of values, critical thinking and problem-solving pedagogy, multi-method approaches, participatory decision-making, applicability to learners' lives and contexts and being locally relevant (UNESCO, 2006). EE and ESD can be implemented effectively through the geography curriculum because the subject is interdisciplinary and deals with human-environment interactions. Drawing from Tilbury (1997), Dube (2012) showed that EE and ESD can greatly benefit from geography content and methodology.

Some geographical concepts such as sustainable development, exploitation, stewardship and responsibility, respect, protection, dependence and interdependence, co-operation, urbanisation and industrialisation, globality, complexity, citizenship and equity are 'integral' to the field of EE and ESD. Furthermore, approaches used in geography education such as problem-solving and enquiry-based learning, role play, simulations and fieldwork are similar to those recommended in the implementation of EE and ESD. It can be argued therefore that improved subject or content knowledge (CK) and knowledge of pedagogy (PCK) in geography education could enhance the teachers' competence to implement EE and ESD through the curriculum.

7.4   Research Context

The larger project on which this chapter is based followed a qualitative interpretivist and a multiple case study research design (Dube, 2012). The main aim of the research was to investigate how EE and ESD were being implemented through the geography curriculum. The research focused on questions related to: opportunities for teaching EE and ESD in the geography National Curriculum Statement, the geography teachers' perspectives on EE and ESD, the extent to which the teachers incorporate the teaching of EE and ESD in their geography lessons, pedagogical approaches used by the geography teachers and the barriers to teaching environmental concerns through the geography curriculum. The research context was FET level (grades 10–12) geography in five high schools in the Western Cape Province of South Africa. The sample, selected through purposeful sampling, comprised only public schools ranging from well-funded former whites-only schools to poorly funded township schools.

Two geography teachers were invited from each selected school to participate in the research project, making a total sample of ten teachers. The participants were given pseudonyms so as to observe ethical considerations of anonymity. The participants had all studied geography albeit at varying levels and trained to teach the subject at secondary school level. Eight of the ten participants pointed out that they were prepared to teach EE and ESD through the geography curriculum during their pre-service training. Only two stated that the pre-service training courses did not adequately prepare them to teach EE and ESD. Most of the teacher participants had a long teaching service (15–35 years). Their pre-service courses are likely to have included conservation education (see Dube, 2012) and not EE and ESD as it is currently conceptualised. Conservation education is based on a narrow view of the environment which excludes the human dimension (social, economic and political). The period during which the teacher participants were trained in South Africa was characterised by conservative attitudes towards both education (see Christian National Education, 1971; Flanagan, 1992; Nel & Binns, 1999) and the environment. Other studies have established that lack of training was found to be one of the main barriers likely to hinder the implementation of EE in the South African context by high school geography teachers (Ballantyne, Oelofse, & Winter, 1999) and primary school teachers (Reddy, 2000).

A number of research instruments, such as biographic questionnaires, semi-structured interviews, lesson observation and document analysis, were used to collect data. Data were analysed through thematic analysis which involved initial coding and categorisation into major themes. The research findings highlighted the difficulties experienced by teachers in implementing EE and ESD through the geography curriculum. These included lack of subject matter knowledge and pedagogical content knowledge related to teaching EE and ESD through the geography curriculum as well as contextual and structural constraints within the schools. In this chapter, I use some of the questionnaire and interview data from the research project to illustrate how lack of subject matter knowledge and pedagogical content knowledge act as a barrier to the effective implementation EE and ESD through the geography curriculum.

7.5 Geography Teachers' Knowledge of ESD

The research established that the geography teachers were experiencing conceptual difficulties regarding the meanings of sustainable development and education for sustainable development (Dube, 2012). The dominant perspective of sustainable development revealed by the geography teachers was the one promoted by WCED (1987) concerning intergenerational equity. Six of the ten geography teacher participants focussed only on intergenerational equity and excluded the issue of intragenerational equity as illustrated by interview extracts (see Box 7.1). The geography teachers' views are derived from the geography curriculum document (DoE, 2003) which adopts the WCED definition noted above.

> **Box 7.1: Participants' Views on the Meaning of Sustainable Development (Dube, 2012)**
>
> Maggie – We speak about the fact that their children [learners' children] should not be deprived of the resources [concern with future generations].
>
> Ian – Excessive extraction could result in some of these animals becoming extinct within a few years. I then told them that their kids [future generations] will not know what the rhinoceros, crayfish or abalone looked like. They will go to the museum to see specimens of the animals.
>
> Hilton – Environmental education and sustainable development should be integrated. You cannot separate the two entities. If you want sustainable development, you need to take care of the environment.
>
> Robert – You must look after the resource and make sure that there is something for the next person or the next generation.
>
> Oliver – So you have to take care of the environment as it is now and also how it will be tomorrow or 10 years from now.
>
> Godwin – People must be educated to sustain things. They must not just misuse them. They must conserve resources.
>
> Lloyd – Sustainable development talks about an ongoing process to secure life, to secure the future and that what we do today impacts tomorrow.
>
> Thomas – When we use our resources we must keep in mind that there are generations which are coming and so we must use them in such a way that we do not deplete the resources to enable future generations to benefit from them as well.

Additionally, the teachers' perspectives were biased towards the conservation of resources that emphasises the *environment* more than the other two pillars of the concept of sustainable development – the *economy* and *society* (see Box 7.1). The teachers had a narrow view of the concept of sustainable development that is synonymous with conservation education, mainly concerned with protection of resources provided by the biophysical environment. This perspective could be attributed to their training background when EE was more closely linked to conservation education. One of the teacher participants, Vena, had difficulties in understanding and articulating the meaning of sustainable development.

When asked about how they make sense of the concept of education for sustainable development, three of the participants understood the concept as having the same meaning as sustainable development, predominantly viewed by the teacher participants as being synonymous with conservation education as discussed above. This perspective is illustrated by Maggie's comment:

> I basically do the terms [sustainable development and education for sustainable development] with them [the learners] and ask them to be mindful of what they are doing in the environment because whatever they are doing in the environment has an impact on the sustainability of a resource.

Half of the teachers responded that they had no idea of the meaning of education for sustainable development. Only one teacher, Hilton, tried to communicate a more holistic view of the concept of education for sustainable development. He remarked:

> Education for sustainable development is necessary because the problem that we have with development is that you always have the risk when you develop that you exhaust and deplete your resources. It is necessary that we have environmental awareness about the way we use our resources so that we can modify our behaviour so that we can develop at a sustained rate; that we always have resources or alternatives at our disposal to keep on developing.

In the above statement, Hilton referred to the two pillars of the sustainable development concept. He was of the view that it was possible to maintain *economic growth* and adequately address the impacts of economic development on the *environment* at the same time. He then highlighted the importance of the third pillar – *society* – in the following comment:

> But the important thing that I want to stress about education for sustainable development is that it [development] should not only benefit the rich. Sustainable development can achieve the objective of getting a more even distribution of economic growth and economic wealth so that poverty in the process can be reduced. So the whole issue of poverty should also, in my understanding, be addressed in education for sustainable development.

Hilton's reflections on the meaning of education for sustainable development include the need to address both intergenerational and intra-generational equity (WCED, 1987) in his reference to sustained economic development resulting from better management of resources while tackling issues of poverty. However, there are tensions between 'sustainable economic growth' versus the 'sustainable human development' perspectives reflected in his views (see Fien & Tilbury, 2002). He was cognisant of the fact that one section of society often benefits from economic development programmes at the expense of the other.

The experience of teachers elsewhere indicates conceptual difficulties with subject matter related to sustainable development and education for sustainable development as revealed by the South African geography teacher participants in this study (Borg, Gericke, Höglund, & Bergman, 2014; Dube & Lubben, 2011; Munro & Reid, 2009; Taylor, Nathan, & Coll, 2003). According to Dube and Lubben (2011), more than half of the science teacher participants in a study in the context of Swaziland were initially unfamiliar with the concept and subject matter of ESD. Borg et al. (2014), in a study of secondary school teachers' conceptualisation of ESD in Sweden, found that social science teachers emphasised the social dimension, while science teachers were biased towards the ecological dimension. This was despite clear curriculum policy guidelines that ESD should be implemented through a holistic approach. Such conceptual issues highlight the problems resulting from a fragmented approach to implementing ESD in a discipline-based school curriculum.

The next section focuses on the teachers' interpretation and implementation of pedagogical approaches stipulated in the geography curriculum documents (DBE, 2011; DoE, 2003).

7.5.1 Pedagogical Content Knowledge Appropriate for ESD

According to Shulman (1987), pedagogical content knowledge refers to 'that special amalgam of content and pedagogy that is uniquely the province of teachers, their own special form of professional understanding' (p. 8). Lowery (2002) in Ball et al. (2008) stated that pedagogical content knowledge is 'that domain of teachers' knowledge that combines subject matter knowledge and knowledge of pedagogy' (p. 394). The geography policy documents at FET level suggest the use of the issue-based and the enquiry-based approaches including fieldwork activities. In the geography curriculum document (DoE, 2003), it is stated that 'the issues-based approach enables the geographer to focus on a specific issue in a natural, built or social environment in a locational (place or regional) context' and that the 'enquiry method provides learners with ways of thinking critically and creatively about the problems or issues they study' (p. 12). The enquiry-based approach is also well articulated in the Learning Programme Guidelines (DoE, 2008) as well as in the most recent iteration of the curriculum, the CAPS document (DBE, 2011, p. 10).

According to Roberts (2003), the enquiry process is guided by core questions; the 'kinds of questions that geographers ask' and Neighbour in Roberts claimed that the questions 'what, where, why, with what impact and what ought' (p. 37) guide most enquiry processes in geography education. The sequence of activities that make up the enquiry route for investigating environmental issues includes definition of the issue, description, analysis, explanation, evaluation, prediction, generalisation, decision-making, personal evaluation and judgement and personal response (Naish, Rawling, & Hart, 2002). Fieldwork activities may be incorporated as this is one of the ways in which information might be generated during the investigation process.

An analysis of the geography teacher participants' interview responses (see Dube, 2012) concerning their classroom experiences of the enquiry-based approach revealed that most follow the policy prescription that learners should be able to conduct an enquiry process on which they can be assessed. According to the geography curriculum document (DoE, 2003), the learners should be assessed on their competence to ask questions; acquire, organise and analyse information; and make judgements based on the information gathered (enquiry skills) (p. 14). It is suggested in the recently revised CAPS geography curriculum document that projects, among other tasks, can be used for assessment in the following statement (DBE, 2011, pp. 49–50): 'examples of formal assessments include … research tasks, fieldwork tasks, projects, oral presentations…'. The teachers' understanding of the enquiry-based approach is that it means that learners should be involved in research projects (see Box 7.2). These projects involve collecting factual information from libraries and the Internet, as mentioned by Johan and Thomas (see Box 7.2).

Some of the teachers had difficulties in using the enquiry-based approach as illustrated by Maggie who revealed that she was not comfortable with the approach (see Box 7.2). The teachers who remarked that the enquiry-based learning approach was time consuming, for example, Johan and Hilton (see Box 7.2), appeared to be more concerned with finishing the syllabus and then preparing learners for examina-

Box 7.2: Teacher Participants' Views on Enquiry-Based Learning (Dube, 2012)

Maggie – I do not give much attention to that [enquiry learning]. I use a method that I find comfortable for me to go through the syllabus. If I do not feel comfortable with it, I do not use it.

Ian – Yes that is research *[referring to the enquiry approach].* That is one of our aims for this year for the grade 10s, and for the grade 11s and 12s where they must do an assignment. They must do an enquiry and they must come and give feedback back to us.

Johan – The enquiry learning is only suitable for gifted learners. They are more likely to go out and find the information whereas the average learners are not likely to do it. Where can they get the information – libraries, the internet? Use of the method is restricted by time constraints.

Hilton – Some of the strategies like enquiry learning are very time-consuming and one has a lot to cover in the syllabus [in order to prepare the learners for the exam]. So … enquiry learning would be very difficult to implement due to the fact that one has to cover a huge amount of work.

Robert – They [learners] must go and do research, bring back their findings as a project or assignment. We mark them mostly on rubric … so that they know how their marks will be allocated.

Oliver – We are only using it [enquiry learning] in one term during the year when the learners do projects or research tasks on the environment, our cities or topics like pollution and so on.

Godwin – We make use of an assessment programme … we give projects following an assessment programme from the Department of Education. Each term there is one project per grade at FET.

Vena – The enquiry approach does not always mean the question and answer method. You [the teacher] can ask the learners to analyse information and pictures in a book. For example, in a lesson on population, I can ask questions based on the use of pictures or cartoons in a textbook.

Lloyd – As I understand it, it is more of the teacher posing a question to the learner and the learner does some research and then comes back and we discuss it in class.

Thomas – We do give projects, for example where they enquire about globalisation [Grade 11 syllabus], to get information from the internet and encourage them to go to libraries.

tions (Stevenson, 2007). The issue with time also reveals the structural constraints of implementing ESD in the school context (see Dube, 2012).

Only Hilton linked the enquiry-based approach to fieldwork which is suggested in the curriculum statement. He observed:

> We took the learners on an excursion and they had to go and collect information which they put into research reports afterwards.

Vena saw the possibility of using the enquiry-based approach in the classroom, whereas the majority of the participants associated the approach with activities outside the classroom such as doing research projects. The findings are corroborated by research in the Eastern Cape in South Africa which established that geography teachers did not fully understand the enquiry approach and that the approach was hardly used in the classroom (Wilmot & Dube, 2015a, 2015b). This resonates with the experiences of teachers in the English context where Roberts (2003) revealed that the enquiry-based approach 'is not well developed in the classroom' (p. 37).

Drawing from Boardman (1986), the enquiry route may be argued to be appropriate for implementing ESD through geography education because it creates opportunities for values education. Learners become not only aware of the values of decision-makers but can also analyse and clarify their own values. Furthermore, the application of the enquiry route is more likely to assist in the development of higher-order thinking skills in learners such as critical thinking as well as those required in evaluation, prediction, generalisation, decision-making, personal evaluation and judgement and personal response. The geography teacher participants' understanding and experiences in using the enquiry-based approach, as revealed in the above evidence, fell far short of the enquiry route prescribed in the geography curriculum (DBE, 2011, p. 10). The research projects assigned to learners involved collecting factual information resulting in possible development of only lower-order thinking skills. It is also not likely that the question of the need for values education would be adequately addressed by the teacher participants as ESD is ultimately about the development of values of respecting the environment.

7.6 Concluding Remarks

In this chapter, the role of practising teachers' knowledge of subject matter and PCK for effective implementation of ESD has been explored. Using some of the findings of a larger PhD study on the implementation of ESD in the context of geography education at FET level in South Africa (Dube, 2012), some of the challenges that the teachers face are illustrated. The geography teachers are experiencing conceptual difficulties regarding the subject matter knowledge pertaining to sustainable development and education for sustainable development. For most teacher participants, the subject matter of sustainable development is only concerned with intergenerational equity, and the concept is synonymous with conservation education. Most of the teacher participants lack a holistic view of sustainable development and education for sustainable development. Additionally, some of the teachers were yet to encounter the concept of ESD. The other challenge relates to the teachers' PCK for implementing ESD. The enquiry-based learning approach stipulated in the geography curriculum is not well understood. The geography teachers' interpretations and experiences of the enquiry-based approach fall far short of the enquiry route through which higher-order thinking skills such as critical thinking could be developed in learners. Better understanding and experiences of the subject matter of

ESD and PCK regarding the full enquiry route could improve the teaching and learning of values, the main goal of learning in ESD. The importance of the role of contextual factors and structural constraints is elaborated in Dube (2012).

As a way forward, a model has been proposed for addressing the issue of inadequate preparation of teachers for implementing ESD. The model, generated by European academics (Sleurs, 2008), proposes a competency-based approach that focuses on five dimensions that include knowledge, systems thinking, emotions, values and ethic and action. The European approach borrows ideas from Shulman's (1986) teacher knowledge model to explain the kind of knowledge that is appropriate in order to improve the capacity of teachers to implement ESD. As discussed in this chapter, aspects such as subject matter or content knowledge (CK) and pedagogical content knowledge (PCK) are emphasised. Although knowledge of curriculum is not mentioned in the European competency-based model, it needs to be emphasised too in the knowledge dimension because it is in the curriculum that sustainability issues are incorporated. The teachers should be able to not only identify the opportunities for ESD in the curriculum but also to adequately interpret and use the strategies that are stipulated in the curriculum such as enquiry-based learning.

References

Abell, S. K. (2008). Twenty years later: Does pedagogical content knowledge remain a useful idea? *International Journal of Science Education, 30*(10), 1405–1416. doi:10.1080/09500690802187041.

Andersen, H. P., & Wennevold, S. (1997). Environmental education in Norway – some problems seen from the geographer's point of view. *International Research in Geographical and Environmental Education, 6*(2), 157–160. doi:10.1080/10382046.1997.9965041.

Ball, D. L., Thames, M. H., & Phelps, G. (2008). Content knowledge for teaching: What makes it special? *Journal of Teacher Education, 59*(5), 389–407. doi:10.1177/0022487108324554.

Ballantyne, R., Oelofse, C., & Winter, K. (1999). Geography educators' perceptions of teaching environmental education in South African schools. *South African Geographical Journal (Special Issue), 81*(2), 86–90.

Boardman, D. (1986). Geography in the secondary school curriculum. In D. Boardman (Ed.), *Handbook for geography teachers* (pp. 9–26). Sheffield, UK: The Geographical Association.

Bonnett, M. (1999). Education for sustainable development: A coherent philosophy for environmental education? *Cambridge Journal of Education, 29*(3), 313–324. doi:10.1080/0305764990290302.

Borg, C., Gericke, N., Höglund, H.-O., & Bergman, E. (2014). Subject- and experience-bound differences in teachers' conceptual understanding of sustainable development. *Environmental Education Research, 20*(4), 526–551. doi:10.1080/13504622.2013.833584.

Chisholm, L. (2000). *A South African curriculum for the twenty-first century: Report of the review committee on Curriculum 2005*. Pretoria, South Africa: Department of National Education.

Corney, G., & Reid, A. (2007). Student teachers' learning about subject matter and pedagogy in education for sustainable development. *Environmental Education Research, 13*(1), 33–54. doi:10.1080/13504620601122632.

Department of Basic Education. (2011). *The Curriculum and Assessment Statement (CAPS) for geography in the further education and training band*. Pretoria, South Africa: Department of Basic Education.

Department of Education. (2002). *Revised national curriculum statement grades R-9: Overview*. Pretoria, South Africa: Government Printer.

Department of Education. (2003). *National curriculum statement grades 10–12 (general): Geography*. Pretoria, South Africa: Government Printer.

Department of Education (DoE). (2008). *National curriculum statement grades 10–12 (general): Learning programme guidelines geography*. Pretoria, South Africa: Department of Education.

Dube, C. (2012). Implementing education for sustainable development: The role of geography in South African secondary schools, Unpublished doctoral thesis, University of Stellenbosch, Stellenbosch. Available at http://scholar.sun.ac.za/handle/10019.1/71683.

Dube, C. (2014). Environmental concerns in the geography curriculum: Perceptions of South African high school teachers. *Southern African Journal of Environmental Education, 30*, 130–146.

Dube, T., & Lubben, F. (2011). Swazi teachers' views on the use of cultural knowledge for integrating education for sustainable development into science teaching. *African Journal of Research in MST Education (Special Issue), 15*(3), 68–83. doi:10.1080/10288457.2011.10740 719.

Christian National Education. (1971). *Standard encyclopedia of Southern Africa*. Cape Town, South Africa: Nasau.

Fien, J., & Tilbury, D. (2002). The global challenge of sustainability. In D. Tilbury, R. Stevenson, J. Fien, & D. Schreuder (Eds.), *Education and sustainability: Responding to the global challenge* (pp. 1–12). Cambridge: IUCN.

Flanagan, W. (1992). Pedagogical discourse, teacher education programmes and social transformation in South Africa. *International Journal of Education Development, 12*(1), 27–35.

Hattingh, J. (2002). On the imperative of sustainable development: A philosophical and ethical appraisal. In J. Hattingh, H. Lotz-Sisitka, & R. B. O'Donoghue (Eds.), *Environmental education, ethics and action in Southern Africa* (pp. 5–15). Pretoria, South Africa: Human Sciences Research Council.

Irwin, P., & Lotz-Sisitka, H. (2005). A history of environmental education in South Africa. In C. Loubser (Ed.), *Environmental education: Some South African perspectives* (pp. 35–54). Pretoria, South Africa: Van Schaik.

Jickling, B. (1992). Why I don't want my children to be educated for sustainable development. *Journal of Environmental Education, 23*(4), 5–8. doi:10.1080/00958964.1992.9942801.

Le Grange, L. (2002). Towards a "language of probability" for environmental education in South Africa. *South African Journal of Education, 22*(2), 83–87.

Lotz-Sisitka, H. (2011). National case study – teacher professional development with an education for sustainable development focus in South Africa: Development of a network, curriculum framework and resources for teacher education. *Southern African Journal of Environmental Education, 28*, 30–65.

Lotz-Sisitka, H., & Raven, G. (2001). *Active learning in OBE: Environmental learning in South African schools* (Research Report of the National Environmental Educational Programme – GET Pilot Research Report). Pretoria, South Africa: Department of Education.

McKeown, R., & Hopkins, C. (2003). EE ≠ ESD: Defusing the worry. *Environmental Education Research, 9*(1), 117–128. doi:10.1080/13504620303469.

Munro, R. K., & Reid, A. (2009). Beginning Scottish geography teachers' perceptions of education for sustainable development. In *Developing critical perspectives on education for sustainable development/global citizenship in initial teacher education*. UK Teacher Education Sustainable Development/Global Citizenship (UK TE ESD/GC) (pp. 57–72). ISBN 978-0- 946786-56-5.

Naish, M., Rawling, E., & Hart, C. (2002). The enquiry approach to teaching and learning geography. In M. Smith (Ed.), *Teaching geography in secondary school: A reader* (pp. 63–69). London: Routledge Falmer.

Nel, E., & Binns, T. (1999). Changing the geography of apartheid education in South Africa. *Geography, 84*(2), 119–128.

Reddy, C. P. S. (2000). Issue-based curriculum development as a professional development process in environmental education: A case study of primary school teachers in the grassy park area. In N. Jenkin, L. Le Grange, H. Lotz, K. Mabunda, K. Madisakwane, T. Makou, … I. Robottom (Eds.) *Educating for socio-ecological change: Case studies of changing practice in South African tertiary institutions* (pp. 22–32). Grahamstown, South Africa: Australia-South Africa Institutional Links Programme.

Roberts, M. (2003). *Learning through enquiry: Making sense of geography in the Key Stage 3 classroom.* Sheffield, UK: Geographical Association.

Robottom, I. (2007). Re-badged environmental education: Is ESD more than just a slogan? *Southern African Journal of Environmental Education, 24,* 91–96.

Shulman, L. S. (1986). Those who understand: Knowledge growth in teaching. *Educational Researcher, 15*(2), 4–14.

Shulman, L. S. (1987). Knowledge and teaching: Foundations of the new reform. *Harvard Educational Review, 57*(1), 1–22.

Sleurs, W. (Ed.). (2008). *Competencies for ESD (Education for Sustainable Development) teachers. A framework to integrate ESD in the curriculum of teacher training institutes.* Comenius 2.1 project 118277–CP–1–2004–BE–Comenius–C 2.1.

Stevenson, R. B. (2007). Schooling and environmental/sustainability education: From discourses of policy and practice to discourses of professional learning. *Environmental Education Research, 13*(2), 265–285.

Taylor, N., Nathan, S., & Coll, R. K. (2003). Education for sustainability in regional New South Wales, Australia: An exploratory study of some teachers' perceptions. *International Research in Geographical and Environmental Education, 12*(4), 291–311. doi:10.1080/10382040308667543.

Tilbury, D. (1997). Environmental education and development education: Teaching geography for a sustainable world. In D. Tilbury & M. Williams (Eds.), *Teaching and learning geography* (pp. 105–116). London: Routledge.

UNESCO. (2005). *United Nations Decade of Education for Sustainable Development: International Implementation Scheme.* Paris: UNESCO. Retrieved April 1, 2013, from http://unesdoc.unesco.org/images/0014/001486/148654E.pdf

UNESCO. (2006). *Framework for the UN DESD International Implementation Scheme [Online].* Paris: UNESCO. Available from http://unesdoc.unesco.org/images/0014/001486/148650E.pdf. Accessed on 27 May 2009.

United Nations Conference on Environment and Development (UNCED). (1992). *Earth summit'92, chapter 36.* Conches, Switzerland: UNCED.

Wilmot, D., & Dube, C. (2015a). Opening a window onto school Geography in selected public secondary schools in the Eastern Cape Province. *South African Geographical Journal.* doi:10.1080/03736245.2015.1028989.

Wilmot, D., & Dube, C. (2015b). School Geography in South Africa after two decades of democracy: Teachers' experiences of curriculum change. *Geography, 100*(2), 94–101.

World Commission Environment and Development (WCED). (1987). *Our common future.* Oxford: Oxford University Press.

Chapter 8
Education for Sustainable Development in the Namibian Biology Curriculum

Sirkka Tshiningayamwe

As a response to the United Nations Decade of Education for Sustainable Development, Namibia has incorporated Education for Sustainable Development (ESD) into its education policy. There has thus been a growing recognition of the significance of ESD across the school curriculum. However, of the various subjects taught in secondary schools, science subjects (especially biology) are often perceived as subjects that can make a significant contribution to ESD. Drawing on research[1] that was conducted at three schools in Windhoek, this chapter will comment critically on the uptake of ESD in the Biology Namibian Senior Secondary Certificate (NSSC) curriculum. The chapter reflects on the influence of learner-centred education in the Namibian curriculum, the challenges and successes of ESD integration in the biology curriculum and how ESD has potential to strengthen and expand policies on learner-centred education, helping to translate them into practices.

The United Nations' launch of the Decade of ESD in 2005 focused international attention to the concept of ESD. The United Nations Decade of ESD's overall goal was to 'integrate the principles, values and practices of sustainable development into all aspects of education and learning' (Republic of Namibia, 2009–2014). Namibia is signatory to numerous international agreements on sustainable development such as the Convention for International Trade in Endangered Species (CITES), Convention on Biological Diversity (CBD), the International Framework Convention on Climate Change (IFCCC) followed by the Kyoto protocol, the Convention on Wetlands of International Importance (RAMSAR), SADC Policy and Strategy for Environmental and Sustainable Development, among others (Fröhlich, 2006). These conventions and international treaties have shaped the

---

[1] The research has been written up as a master's thesis (Tshiningayamwe 2011).

S. Tshiningayamwe (✉)
Environmental Learning Research Centre, Rhodes University,
Grahamstown, South Africa
e-mail: sirkka.ts@gmail.com

© Springer International Publishing Switzerland 2017
H. Lotz-Sisitka et al. (eds.), *Schooling for Sustainable Development in Africa*,
Schooling for Sustainable Development 8, DOI 10.1007/978-3-319-45989-9_8

introduction of ESD in Namibia, by giving prominence to the promotion of environ-mental literacy in its constitution, development plans and educational policies. ESD seeks to orientate education in order to foster values and attitudes that will help achieve a world where people peacefully coexist; where rights are respected; where there is less suffering, hunger and poverty; and where the natural environment and its biodiversity are conserved (UNESCO, 2005). In response to this, 'Education 2030 will ensure that all individuals acquire a solid foundation of knowledge, develop creative and critical thinking and collaborative skills, and build curiosity, courage and resilience' (UNESCO, 2015, p. 2).

In Namibia, ESD is integrated in both the formal and non-formal education pro-grammes. ESD is integrated across the formal school curriculum, but of the various subjects taught in schools, there is a perception that biology is one of the subjects that broadly covers environmental topics and can make a significant contribution to ESD. Thus, all teachers and particularly biology teachers need to be knowledgeable about ESD content, pedagogies and assessment to be able to teach and facilitate ESD in the curriculum. This is in line with Shulman (2004) who noted that teachers need to master two types of knowledge: content knowledge and knowledge of cur-ricular development. Writing from Australia, Ferreira, Ryan and Tilbury (2007) indicated that teacher education is supposed to prepare teachers and ensure that they are fully ready to implement ESD when they start their teaching profession. However, they recognised that ESD in initial teacher training programmes has been neglected and does not prepare teachers fully to be able to implement ESD. This is because, as in most countries, in Namibia (Kanyimba, Hamunyela, & Kasanda, 2014; Tshiningayamwe, 2011), most teacher educators have not been trained to sup-port teachers to teach environmental knowledge. Thus, despite the importance of ESD, the lack of teachers' environmental knowledge results in a mismatch between ESD theories and practice (Tshiningayamwe, 2011). This chapter thus comments critically on the uptake of ESD in the biology curriculum in Namibia. It reflects on the influence of learner-centred education in the Namibian curriculum, the chal-lenges and successes of ESD integration in the biology curriculum and how ESD has the potential to strengthen and expand policies on learner-centred education, helping to translate them into practices.

8.1 Namibian State of the Environment and ESD Uptake in the Curriculum

Namibia is the driest country south of the sub-Sahara with no perennial rivers except at the borders (Namibia. Ministry of Environment and Tourism [MET], 2002). Namibia experiences unpredictable annual rainfall except in the northern parts of the country. Its environment is semiarid and fragile, yet the majority of its popula-tion is dependent on the land (Fröhlich, 2006). Sustainable development issues in Namibia are complex and interlinked (ibid.). The sustainable development issues can be classified as social, economic and environmental. Societal issues include human health and welfare, cultural diversity, human rights abuse, erosion of cultural values and morals, gender inequality, poverty and inequality and concerns for poor governance. The economic challenges revolve around unemployment, rural to urban migration, corruption, limited human resources and capacity, increasing competition for shared resources, population growth and settlement patterns. The environmental issues include limited water resources, unsustainable natural resource management, loss of biodiversity, drought and climate change (Republic of Namibia, 2009–2014). These sustainable development issues needed to be addressed through education to create an environmentally literate and responsible nation, whose citizens understand the consequences of the past and who have the competence to take the necessary action to live sustainably for the benefit of the present and future generations. This influenced the integration of ESD in the formal schooling system (Republic of Namibia, 2009–2014).

The implications of inclusion of ESD in formal education as a distinct separate subject or infusion or integration of ESD across the curriculum have been carefully studied (Lotz-Sisitka, 2002). The interdisciplinary character and complexity of ESD have led to ESD being placed as an orientation in the curriculum within different disciplines and subjects (ibid.). Environment and sustainability issues are cross-cutting and require multidisciplinary responses in the mainstream curriculum. However, curriculum documents favour the combination of science and ESD, stating that teachers should aim 'to understand the importance of the endangered natural environment and how the problems related with it can be solved' (Namibia. Ministry of Education [MoE], 2009, p. 23). All subjects have a special relationship to the environment; for example, biodiversity makes up an important part of the biology curriculum in Namibia (ibid.). Teachers are, however, mostly unaware of the actual purpose of these components in the curriculum, as they appear in the curriculum without any special distinction being made between them and other components (Haindongo, 2013). Thus, teachers mostly teach the ESD components in the same way as other components in the curriculum, using traditional methods such as lecturing (Haindongo, 2013). To implement ESD, teachers require not only curriculum policy changes, strategies and plans; they also need a deeper understanding of what is required and what and how to implement in particular contexts (Shulman, 2004; UNESCO, 2012). The major challenge of ESD in Namibia is thus to meet the training needs of teachers with the view to effect profound changes in their ways of thinking, attitudes and behaviours for sustainable development. This has an effect on teachers' practices as they seek to implement ESD pedagogies in their teaching. The University of Namibia (UNAM) is responsible for training teachers who are expected to implement ESD. However, UNAM does not prepare teachers adequately for ESD (Kanyimba et al., 2014). Therefore, little attention is paid to ESD in the school curriculum, and teachers are not able to convince the learners about the importance of ESD (Tshiningayamwe, 2011; Haindongo, 2013). A number of teachers may currently be perceived as uncommitted to ESD because they are not implementing ESD due to their lack of capacity.

In Namibia, the infusion of ESD across the curriculum approach has been supported by three major ESD development projects: the Enviroteach project, Life Sciences project (1991–2000) and the Supporting Environmental Education in Namibia (SEEN) project (established in 2001). Through the Life Sciences project, the in-service teacher education programme was designed to develop subject knowledge and teachers' pedagogical knowledge, through enhancing their methodological and didactic skills to enable them to adopt a learner-centred approach (Kristensen & Andersen, 2001). The Life Sciences project supported learner-centred education and was devoted to encouraging learners to solve problems, to promote critical thinking, practical work and self-confidence (ibid.). By the time the project was complete, life science was a fully developed subject among the natural sciences subjects (such as biology), and teachers are trained to teach it on an ongoing basis (ibid.). The natural sciences subjects are offered in different phases of the curriculum. The basic aims of the natural sciences syllabi are the same for all learners (Namibia. Ministry of Education, 2009). Relevant to ESD, the natural sciences subjects have common environmental topics integrated in them, but learning objectives are at different levels (ibid.). The cross-curricular issues serve as a direct link of the natural sciences syllabi to other subjects for the Namibian school curricula. Life science has inspired other subjects, thus becoming one of the front runners of the educational reform in Namibia supporting the goals of the Ministry of Education's policy document *Towards Education for All*, the cornerstones of which are access, equity, quality and democracy (Bones, 1994).

The Enviroteach project spearheaded and strengthened a cross-curricular approach to ESD in Namibia as the Life Sciences project only worked at the level of one subject (Enviroteach, 1998). The Enviroteach project explored options for effective incorporation of relevant environmental information into formal education (ibid.). ESD is seen as not only contributing to an environmentally literate society but also to the education reform process by promoting cross-curricular and learner-centred approaches to teaching and learning (UNESCO, 2005). The Enviroteach project had three major components: curriculum development and production of learning support materials, supply of teaching equipment to schools and institutions and in-service training and pre-service teachers' education (ibid.). By the time the project ended in 2000, ESD was established as a cross-curricular learner-centred concern in the Namibian curriculum. Further impetus was, however, needed to train teachers and curriculum advisors on this new cross-curriculum issue, and the SEEN project was established as a follow-up project to the Life Sciences and the Enviroteach projects. Issues related to the environment, such as global warming, greenhouse effect, environmental degradation and impacts of floods on human life as well as problems of feeding the world's population, were extensively addressed across the curriculum through SEEN (SEEN, 2005). These were addressed under the six broad themes: natural resources and their management, poverty and inequality, society and governance, development and the environment, health and the environment and globalisation (ibid.). The SEEN project also dealt with pedagogical issues and provided guidelines on issues-based approaches to ESD.

ESD stimulates learners to achieve sustainability and encourages them to respect and live within the limits of nature and to evolve social, production, technological and economic systems that are creative, innovative, equitable and sustainable (SEEN, 2005). It is important to promote education that builds capacity to engage critically with contemporary development discourses and practices that nurtures and strengthens dialogues and advocacy skills (ibid.). The aims and objectives of ESD in Namibian curricula are to:

- Encourage and promote in learners a holistic understanding of the dynamic interdependence of all living things and their environment.
- Promote a sense of responsibility towards restoring and maintaining ecological balances through the sustainable management of natural resources.
- Encourage an involvement in practical activities to preserve and sustain the natural environment (SEEN, 2005).

8.2 ESD Within the NSSC Biology Curriculum

As part of ESD, the biology syllabus aims at increasing the learners' knowledge that can help them understand the physical and biological world (SEEN, 2005). This includes understanding how people use the natural environment to satisfy human needs. The application of scientific knowledge and attitudes, in the framework of sustainable use of resources, is of relevance for the individual, the family and the society as a whole (ibid.; Namibia. MoE, 2009). Process and manipulative skills are essential to understand the value and limitations of natural scientific knowledge and methods and their application to daily life (Namibia. MoE, 2009). In order to meet these biology aims, environmental topics have been integrated within the biology curriculum (ibid.). These aims enable learners to acquire and develop competencies in ESD. These competencies are meant to be achieved through effective implementation of certain biology syllabus learning objectives (ibid.).

However, the current biology syllabus adheres to international standards linked to accreditation and certification and tends to focus more on universal science knowledge, especially at higher grades (Namibia. MoE, 2009). As a result, it does not influence the way Namibians think about the environment locally. Most of the conventions and knowledge related to the local environment are accommodated in the Junior Secondary syllabi, which is not a sufficient knowledge for learners. For quality and relevance, the biology syllabus can be improved by responding to more local issues than global issues, by strengthening participation into practice, which might result in fully promoting socioecological resilience. Sustainable practices transpire through teaching of concepts like conservation, which has a broader scope of interpretation and could be tailor-made for the particular setting, e.g. Erongo region would look more in depth at the issue of the lagoons and how the communities could use them sustainably, while the Northern regions could explore the sustainability of the Oshanas and the perennial rivers. Educational approaches must look into strengthening situated learning while promoting capabilities (Hogan, 2008). Environmental thoughts and environmental ethics are dynamic; thus individual and group practices must also be dynamic, and hence the recommendation from the Ahmedabad declaration of 2007 for a paradigm shift in our practices and the need for transformation in ESD (UNESCO, 2008).

8.3 Pedagogical Approaches for ESD in the NSSC Curriculum

The pedagogy of ESD encompasses progressive constructivist education approaches: critical thinking, participation, contextualised learning, use of local materials, problem solving, community engagement, action-oriented, socially critical and student-led (UNEP, 2006). ESD pedagogies also encourage reflection and questioning. According to UNESCO ESD is based on the premise of 'learning by doing' (ibid.). Each cultural or societal group will choose to address ESD in the context of its own aspirations for sustainable development. Thus, there can be no 'one size fits all' approach to ESD. In Namibia, teaching methods have shifted from early positivist approaches, where ESD was about transferring information and raising awareness towards participatory methods based on social constructivism. More new methods are emerging which are ontologically situated, for example, the inquiry-based methods. In the Namibian context, curriculum builds on learner-centred education in accordance with the 'Towards Education for All' policy (Namibia. Ministry of Education and Culture [MEC], 1993).

Central to learner-centred education is the assumption that knowledge is created by individuals through a process of making sense that involves establishing a link between prior knowledge and new information. The process happens both individually and socially by constructing, deconstructing and reconstructing knowledge (Kristensen & Andersen, 2001). Learner-centred education encourages learner participation and involvement in pedagogy. Learner-centred education sees a learner as an active, inquisitive human being, eager to learn, to investigate and to make sense of his or her surrounding world (Namibia. MEC, 1993). Thus it takes into account that learners are individuals with their own needs. Teachers thus need to give attention to individual differences through differentiation of teaching methods and instructional materials and the way teaching is organised. This is often a challenge for most Namibian biology teachers (Tshiningayamwe, 2011). With regard to ESD, learner-centred education promotes critical thinking and problem-solving skills in learners. The implementation of ESD promotes progressive constructivist pedagogy, integration of disciplines and use of everyday knowledge related to disciplinary knowledge and structure (Kanyimba, 2002). Teachers need to guide the learners in acquiring new knowledge and skills. This does not mean that the teacher is the source of knowledge, but rather the facilitator of the learning processes of the learners. Active approaches to teaching and learning are highly recommended for ESD

(UNESCO, 2012). This implies that for successful ESD, learners should be active participants in knowledge construction rather than receivers of others' knowledge (Wood, 2007).

According to UNESCO (2012), successful implementation is shaped by pedagogies as much as it is by content. Methods which can be used to implement ESD include investigation and problem solving, demonstrations, cooperative group work and experimental methods (Dreyer & Loubser, 2005; O'Donoghue, 2015). These include teaching strategies such as group work, project work, eliciting prior knowledge, excursions, drama and role plays (Namibia. MoE, 2009). These strategies provide opportunities for addressing other goals of education such as democracy. With a clear understanding of the why and how learner-centred education is to be implemented, many of the education goals would be achieved. A learner-centred curriculum is a holistic and integrated curriculum (Dahlstrom, 1995). Home and school curricula should be interrelated in order to make sense for the learners (Hogan, 2008). Integrating environment and sustainability issues into the daily lives of learners, schools and communities will help bridge the gap between schools and community (ibid.). Therefore, in designing the curriculum, the learner must be considered first and should be placed at the centre of the design. Learner-centred education and related pedagogical approaches which are currently offered in Namibia are, however, being challenged for not meeting the educational goals of social justice and denying learners' access to powerful knowledge systems (Hoabes, 2004). There is thus a need to strengthen teachers' knowledge of learner-centred education and of how to plan and implement it in Namibia.

8.4 Challenges and Successes of ESD in the NSSC Biology Curriculum

Despite the Namibian ministerial policy documents that support the incorporation of ESD within the biology curriculum, ESD continues to suffer from constraints that hinder its effective incorporation into the curriculum (Tshiningayamwe, 2011). ESD aims at preparing individuals to be responsive to a rapidly changing technological world, to understand contemporary world problems and to provide the skills needed to play an effective role in the improvement and maintenance of the environment (SEEN, 2005). The major challenges of meeting the ESD goals are to translate these goals into practice (Tshiningayamwe). The goals of ESD are to develop and adopt a set of interventions that will eliminate environmental problems (UNESCO, 2012). This can be achieved by establishing sustainable systems and producing knowledgeable, competent and innovative learners and graduates who will solve environmental problems (UNESCO, 2015).

Among many, some of the challenges of ESD implementation are the provision of human resources and ensuring that the policies in place are implemented by all biology teachers (Namibia. MET, 2006; Tshiningayamwe, 2011). Effective implementation of the policies and curriculum involves financial implications and the training of teachers and others to coordinate and regulate the implementation process (Namibia. MET, 2006). The vastness of Namibia as a country poses many problems. This affects the review of ESD in the curriculum negatively and affects curriculum policy implementation. Many policies and laws have been created to restrict certain human activities that degrade the environment, but they are not adequately reinforced; thus some biology teachers' attitudes and behaviour towards the teaching of the environmental concepts and content knowledge in the curriculum are still negative (Tshiningayamwe, 2011). Because of the ineffectiveness of some of the biology teachers' practices, tensions arise among individuals and organisations, even among curriculum developers (ibid.). Other factors that hamper the delivery of environment and sustainability knowledge in the biology curriculum in Namibia include higher learner teacher ratios, lack of qualified teachers, lack of education facilities in schools, inadequate lesson preparations, poor school management and administration and lack of motivation among teachers (Enviroteach, 1998; Hoabes, 2004; Tshiningayamwe, 2011). With regard to ESD implementation, some of the problems experienced during the implementation phase of the Enviroteach project included:

- Resistance to change on the part of the teachers, school management, learners and parents
- Lack of confidence and experience on the part of the teacher
- Lack of support from school management who in many instances do not understand new methodology being promulgated through educational reform
- Lack of relevant, appropriate and user-friendly resources (Enviroteach, 1998, p.16)

In Namibia, learner-centred education (where learners determine the direction of the learning experience) is not implemented (Hoabes, 2004; Tshiningayamwe, 2011). Therefore, a large number of learners experience a poor match between their everyday world at home and their school world. Teachers mainly interpret learner-centred education in terms of changes in methods only (Hoabes, 2004). Some biology teachers appear to have little understanding of the learner-centred approach; thus, their teaching practices have not been fully in line with the learner-centred policy framework (Tshiningayamwe, 2011). There is thus a need to support teachers continuously in the form of in-service training to help them understand learner-centred education better and to help them understand how the methods they use reflect learner-centred education. There is a need for teacher education and training in ESD, appointment of ESD coordinators in schools, ESD policies for schools, interdisciplinary collaboration and fieldwork as an instructional technique and strengthened capacity of Namibian Environmental Education Network (NEEN). These have the potential to increase the success of ESD implementation in Namibian schools.

Most tertiary lecturers in Namibia, including those who prepare teachers for teaching, recognise the importance of integrating ESD in all subjects (Kanyimba et al., 2014). However, there are dispositional, situational and institutional barriers that hinder the implementation of ESD in Namibian higher education institutions (Dawe, Jucker, & Martin, 2005). These include a lack of expertise, limited institutional drive and commitment and perceived irrelevancy by some academic staff (ibid.; Kanyimba et al., 2014). The introduction of the Eco-Schools initiative in Namibia is helping to enhance learner-centred education in relation to diverse needs and allows for learner-initiated contributions (Haingura, 2009). Integrating an Eco-Schools framework across the curriculum provides opportunities to enhance learner-centred education and to strengthen school-community interactions and enables active involvement of learners in decision-making (Haingura, 2009). For successful implementation of ESD in the biology curriculum, there is a need to enhance teachers' ESD knowledge, pedagogies and assessment approaches, improve attitudes towards ESD, align ESD theories to practice and reorient curriculum documents and other learning support materials used for ESD (Tshiningayamwe, 2011).

However, it is clear that there is some enabling education reform in Namibia including an environmental focus in the biology curriculum (Namibia. MoE, 2009). Most teachers are qualified to teach the subjects they teach (Hoabes, 2004; Tshiningayamwe, 2011). Teachers receive support through workshops that acquaint them with subject content and methodology both at national, regional and cluster levels, but the support is sporadic or insufficient for ESD implementation (Hoabes, 2004). Thus teachers do need support to implement an ESD focus, for example, using learners in planning lessons, using visual aids and materials in ways that involve learners and using local environment when teaching environmental topics to contextualise this within learners' prior knowledge and experience (ibid.; Tshiningayamwe, 2011).

8.5 The Potential of ESD to Strengthen Policy on Learner-Centred Education

The United Nations Decade of ESD has supported the role of ESD in contributing to educational relevance and quality, which are issues of concern in the biology curriculum. SEEN has introduced curriculum guidelines for environmental learning in Namibia (SEEN, 2005), which support the incorporation of teaching and learning and school-based educational approaches. Principles in the guidelines supporting ESD in Namibia are:

- Transforming relationships and improving partnerships
- Practising and modelling what is 'preached'
- A focus of change to improve the future
- Learning across the subject boundaries
- A focus of exploring values
- An emphasis on critical thinking and reflection
- Valuing and encouraging active participation (SEEN, 2005)

These curriculum guidelines support the United Nations Decade of ESD by incorporating some of the best practices of teaching and learning and school-based educational approaches trialled by SEEN pilot schools (SEEN, 2005). These imply that ESD can play a pivotal role in supporting the transformation in Namibia to 'good' learner-centred education. ESD calls for a rethink and reform of current practice in formal education. It requires that stakeholders in education constantly improve their capacity and skills through the retraining of biology teachers using appropriate courses, methodologies and materials. It implies the complete integration of sustainability issues within the lives of the schools and their communities. It complements Namibia's aim to provide education for all and achieve its Millennium Development Goals by promoting relevant values, processes and behaviours.

8.6 Conclusion

This chapter has highlighted the fact that ESD implementation in Namibia is complicated and is taught using approaches based solely on the transmission of knowledge that are not suited for it. The chapter further highlighted the major challenge of ESD, that is, to meet the training needs of educators with the view to effect profound changes in their ways of thinking, attitudes and behaviours for sustainable development. There is, thus, a need for teachers' professional development to ensure effective implementation of ESD, particularly in the Namibian biology curriculum. In addition to curriculum policies on ESD and strategic plans, teachers need a deeper understanding of what is required and what and how to implement ESD in particular contexts. Teachers also need to be supported on different pedagogical approaches to ESD. These have the potential to strengthen learner-centred education and can help achieve education for all. There is however a need to explore the different models that are effective for teachers' professional development. This has the potential to inform professional development activities that will lead to transformation of teachers' classroom practices as they relate to the implementation of environmental topics in the curricula.

References

Bones, B. (1994). *Getting started. A guide to bringing environmental education into your classroom*. Washington, DC: The National Environmental Education and Training Foundation.

Dahlstrom, L. (1995). Teacher education for independent Namibia: From the liberation struggle to a national agenda. *Journal of Education for Teaching, 21*, 273–288.

Dawe, G., Jucker, R., & Martin, S. (2005). *Sustainable development in higher education: Current practice and future developments a report for the higher education academy*. Heslington, NY: Higher Education Academy.

Dreyer, J., & Loubser, C. (2005). Curriculum development, teaching and learning for the environment. In C. P. Loubser (Ed.), *Environmental education: Some South African perspectives* (pp. 127–153). Cape Town, South Africa: Van Schaik.

Enviroteach. (1995). *Investigating of opportunities for the implementation of Enviroteach programmes in Namibia colleges of education*. Windhoek, Namibia: DRFN.

Enviroteach. (1998). *The pilot phase, reflections on exploring mechanisms for integrating learner-centred cross-curricular, activities based environmental education into basic education in Namibia*. Windhoek, Namibia: DRFN.

Ferreira, J., Ryan, L., & Tilbury, D. (2007). Mainstreaming education for sustainable development in initial teacher education in Australia: A review of existing professional development model. *Journal of Education for Teaching: International Research and Pedagogy, 3*(2), 225–239.

Fröhlich, G. (2006). *Namibian environmental education certificate. Module 1. Study guide*. Windhoek, Namibia: Centre for open and lifelong learning, Polytechnic of Namibia.

Haindongo, N. (2013). *Environmental Education in Namibia: A case study of the Biology teachers*. Unpublished master's thesis, University of Stellenbosch, Stellenbosch.

Haingura, R. (2009). *Enhancing learner centred education through the Eco-school framework: Case studies of Eco-schools practice in South Africa and Namibia*. Unpublished master's thesis, Rhodes University, Grahamstown.

Hoabes, R. (2004). *Investigating teaching strategies used by teachers to foster environmental learning in the Namibian Life Science curriculum*. Unpublished master's thesis, Rhodes University, Grahamstown.

Hogan, R. (2008). Contextualizing formal education for improved relevance: A case from the Rufiji wetlands, Tanzania. *Southern African Journal of Environmental Education, 25*, 44–56.

Kanyimba, A.T. (2002). *Towards the incorporation of Environmental Education in the Namibian Secondary School Curriculum*. Unpublished master's thesis, UNISA, Pretoria.

Kanyimba, A., Hamunyela, M., & Kasanda, C. D. (2014). Barriers to the implementation of Education for Sustainable Development in Namibia's higher education institutions. *Creative Education, 5*, 242–252.

Kristensen, J. O., & Andersen, H. M. (2001). *Life science project 1991-2000. Educational reform in Namibia: Changing the way we teach, learn and live*. Windhoek, Namibia: Capital Press.

Lotz-Sisitka, H. (2002). Curriculum patterning in environmental education: A review of developments in formal education in South Africa. In J. Hattingh, H. Lotz-Sisitka, & R. O'Donoghue (Eds.), *Environmental education, ethics and action in Southern Africa* (pp. 97–120). Pretoria, South Africa: Environmental Education Association of Southern Africa/Human Sciences Research Council.

Namibia. Ministry of Education. (2009). *The national curriculum for basic education*. Okahandja, Namibia: NIED.

Namibia. Ministry of Education and Culture. (1993). *Toward education for all: A developmental brief for education, culture and training*. Windhoek, Namibia: Gamsberg Macmillan.

Namibia. Ministry of Environment and Tourism. (2002). *Atlas of Namibia*. Windhoek, Namibia: Directorate of Environmental Affairs.

Namibia. Ministry of Environment and Tourism. (2006). *Vital signs of Namibia 2004. An integrated state of the environment report*. Windhoek, Namibia: Directorate of Environmental Affairs.

O'Donoghue, R. (2015). *Environment and sustainability education: Quality teaching in relation to human conduct and the common good. BEd Honours environmental education elective*. Grahamstown, UK: Environmental Learning and Research Centre, Rhodes University.

Republic of Namibia. (2009–2014). *Education for Sustainable Development strategy 2009–2014*. Namibia: Windhoek.

Shulman, L. (2004). *Teaching as community property. Essays on higher education*. San Francisco: Jossey Bass Wiley.

Support Environmental Education in Namibia (SEEN). (2005). *Education for sustainable development in Namibia: The experience of SEEN: past responses and future directions.* Windhoek, Namibia: SEEN.

Tshiningayamwe, S. A. N. (2011). *Implementation of Environmental Learning in the NSSC Biology curriculum Component: A case study of Namibia.* Unpublished master's thesis, Rhodes University, Grahamstown.

UNEP. (2006). *Education for sustainable development innovations – programmes for universities in Africa.* Howick, New Zealand: Share-Net.

UNESCO. (2005). *Education for all: The quality imperative. EFA Global monitoring report.* Paris: UNESCO.

UNESCO. (2008). *Ahmedabad Declaration 2007: A call to action education for life: Life through education* (Report by deputy general on the follow-up to decisions and resolutions adopted by the executive board and the general conference at the previous sessions). Paris: UNESCO.

UNESCO. (2012). *The education for sustainable development sourcebook. Education for sustainable development in action learning and training tools* (Vol. 4). Paris: UNESCO.

UNESCO. (2015). *Framework for Action Education 2030: Towards inclusive and equitable quality education and lifelong learning for all (DRAFT).* Paris: UNESCO.

Wood, D. R. (2007). Professional learning communities: Teachers, knowledge, and knowing. *Theory Into Practice, 46*(4), 281–290.

Chapter 9
Developing Problem-Based Learning Approaches to Water Education in Mauritius

Ravhee Bholah

This chapter reflects on how to develop a problem-based learning (PBL) approach as a strategy to strengthen the quality and relevance of water education in Mauritius. Problem-based learning is situated in real-life situations and is usually designed to engage students in a problem-solving activity that they can relate to and that can be seen as meaningful while emphasising the authentic feature of learning in context. This chapter therefore explains how a problem-based learning approach was developed in one secondary school context and describes possibilities for transforming classrooms into active learning environments with a dynamic interplay of questioning, explaining and designing investigations, communicating ideas, collaborating and reflecting. Assessment is critically discussed for both the processes and the products of the work and the roles of different key players, especially educators, within the school community in helping learner capabilities to deal with water-related issues. The chapter also highlights how problem-based learning eventually helps to develop an appropriate water-related environmental education strategy that can foster appropriate water management at the school.

From the perspective of a small island state, it is notable that over two thirds of the earth's surface is covered by water and less than a third is taken by land; yet, water is one of the most critically scarce resources that is under threat around the world. Water is unevenly distributed through Africa. Central Africa alone accounts for 48 % of the continent's total internal waters, while Northern Africa accounts for only 15 %. The small island of Mauritius is vulnerable to the intensity and frequency of natural and environmental disasters in Africa and their increasing impact. The country is also feeling the effects of climate change on its water resources and water availability. For instance, it is presently experiencing unprecedented drought and has recorded a decline of about 50 mm annual rainfall over the last 50 years.

R. Bholah (✉)
Science Education Department, School of Science and Mathematics, Mauritius Institute of Education, Réduit, Mauritius
e-mail: r.bholah@mieonline.org

© Springer International Publishing Switzerland 2017
H. Lotz-Sisitka et al. (eds.), *Schooling for Sustainable Development in Africa*, Schooling for Sustainable Development 8, DOI 10.1007/978-3-319-45989-9_9

Global water use figures estimate that people are already using about 50 % of accessible freshwater, and this may increase to 75 % by 2025. This has become a major concern in many countries and water resources management is becoming increasingly crucial in such situations. In Mauritius, the heavy consumption of and demand for water for domestic, agriculture and industrial purposes are placing stress on available water resources, and their management has become an increasing challenge, made worse by climate change and related consequences. According to the National Environmental Action Plan of Mauritius, if effective water resources management is not introduced, the demand would surpass the supply in coming decades (Ministry of Education, 2005). This clearly shows the need for assessment, management, development and conservation of water resources and education about water resources.

Water education is needed to facilitate changes in behaviour and personal attitudes among water consumers and help to promote a better understanding of the environment in a water context. The literature shows that water-related environmental education strategies have been used to manage water resources in different countries. One of the curriculum strategies that has been widely employed is **problem-based learning (PBL)**. This is based on real-world situations where learning results from the process of working towards the understanding of, or resolution of, a problem. It is aimed at engaging learners in a problem-solving activity that they can relate to and see as meaningful, while emphasising the authentic features of learning in context. PBL is a systematic method that engages learners in learning knowledge and skills through an extended inquiry process structured around complex, authentic questions and carefully designed products and tasks (Markham, Larmer, & Ravitz, 2003). This method is intended to enable learners to understand the concept of a driving question so that they feel able to contribute meaningfully to a group brainstorming exercise and demonstrate independent critical thinking, particularly when designing an investigation and/or in the analysis and interpreting of data.

In our study, a PBL approach was chosen to help the students to identify any water-related problems in their school context and ultimately develop appropriate strategies to manage the situation.

In PBL the role of the classroom teacher is mostly as a facilitator of discussion among students, leading to self-directed learning that culminates in meaningful comprehension. In PBL, learners are encouraged to pursue their own problem solutions by clarifying a problem, posing necessary questions, researching these questions and producing a product that displays their thinking. These activities are generally conducted in collaborative learning groups that attempt to solve the same problem in different ways, and they often arrive at different answers. In this chapter we explore how a water education project was implemented in one context.

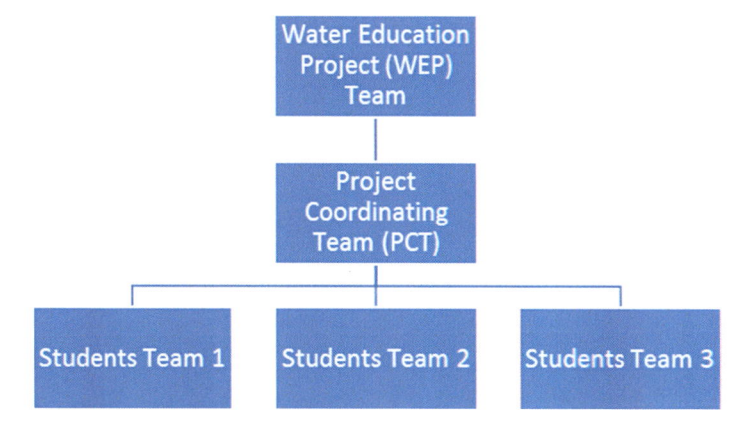

Fig. 9.1 Teams of the water education project

9.1   Qualitative Research Design and Methodology

An interpretive research methodology involving a qualitative case study approach was employed to engage and dialogue with the school community members (school manager, teachers from a number of subject disciplines, non-teaching staff and students). This enabled the water education project (WEP) team members to develop an understanding of how education could help learners acquire agency through developing and expanding their capabilities (Walker, 2005).

Purposive sampling was used to select one school from the central part of the country in order to portray, analyse and interpret the uniqueness of the school community members and the school situation. The research design is action-oriented as the school community, especially the form III students (13–14 years old) engaged in an inquiry-based approach examining issues around sustainable use of water as well as developing appropriate strategies to deal with emerging problems. Other members of the school community were also encouraged to provide their support and help the students to complete various stages of PBL.

The WEP was systematically planned and implemented at the school. Initially, a project coordination team (PCT) was set up at the school level, and it comprised nine teachers (from different subject areas) who were working with the form III students (Fig. 9.1).

The PBL approach involved the following steps (see Fig. 9.2 that follows):

- Identification of the problem/environmental issue (phase 1)
- Planning and investigation of the problem (phases 2 and 3)
- Data collection and data analysis, interpretation of data and discussion (phase 4)
- Presentation of data and recommendations (phase 5)

The project coordination team carried out a brainstorming session on 'environmental issues at school' among 24 students in a form III class. A wide range of

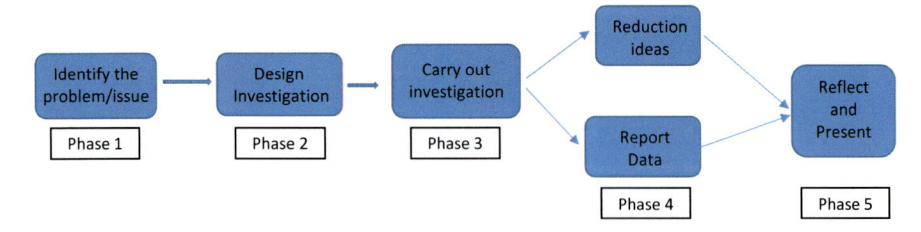

Fig. 9.2 Steps involved in the PBL

environmental issues were identified and noted on a whiteboard but the most common issue was water. The class was then split into three groups of eight students. Each student team was asked to develop an action plan to investigate the water situation at the school. During the project each student team had regular meetings to discuss their project work and was assigned to a set of three teachers of the project coordination team. These educators acted as facilitators.

Educators agreed to meet their respective students' team regularly (once or twice a week) and to communicate and discuss the progress of the study in the project coordination team meeting. Meetings were usually convened during school recess time or free periods to avoid any disturbance to the normal school activities. This timeline also ensured appropriate and timely coverage of environment or water-related concepts in the class as per the national curriculum framework for secondary education so that the learners could better apply knowledge gained in various project activities. Thus the project was not isolated from the curriculum and any other teaching and learning transactions occurring within the school context.

All the steps and major discussions of the PBL activities were recorded by the student teams in their learning logs. Data obtained during the investigation were analysed, interpreted and discussed in each team. Each team then proposed strategies to better manage water at school which were later presented in the class. Together with the help of members of the project coordination team, appropriate post-PBL tasks were carried out to manage water at the school. The project coordination team also compiled a short report of the implementation of the school project on water education (Table 9.1).

Formal and informal meetings were designed and used by the WEP team to gather information and triangulate data. Some data were collected from the students' learning logs. This provided depth of understanding from two different perspectives. It is believed that understanding this case could lead to better understanding, perhaps better theorising, about a still larger collection of cases (Stake, in Denzin & Lincoln, 2000).

Data obtained from the study were grouped, organised, synthesised, critically analysed and interpreted from different perspectives. This was important for better analysis of the data which would eventually inform the development of appropriate sustainable water resources management and help to minimise environmental impact as well as safeguard health.

Table 9.1 Stages of PBL implementation, students' activities, and tools used

Stage	What students did	Tools used
1. Identifying the problem	Walk-through investigation at school	Observation checklist
	Formal and informal interviews	Mind map
2. Designing investigation	Identified resources and tasks to solve their problem	Learning log
3. Carrying out the scientific inquiry (investigation)	Collected data through questionnaire survey	Questionnaire, learning log Interview schedule
	Conducted field studies, questionnaire survey, interviews	
4. Putting the information together (reduction of data)	Completed learning logs	Learning log
	Recorded what they have learnt; reviewed consolidated, information gathered	
5. Presenting findings, evaluation and self-evaluation (reflection and presentation of data)	Oral presentation using PowerPoint	
	Completed self-reflection forms	

9.2 Findings

All students and members of the project coordination team participated in the various stages of the PBL. The most common environmental issues identified during the brainstorming session of the PBL and students' presentation at the school were water-related problems. Student teams were able to develop 'unstructured' walk-through investigation and identify sources of water in the school environment and local surroundings where they investigated both water quality and quantity (see Table 9.2). Learners could also meet and discuss water-related problems regularly. They tried to interpret the results in connection with school population density and water supply.

The most common teaching method used in the process was the traditional method of information exchange. Active teaching and learning strategies, such as storytelling, games and role play, were seldom used. The learning activities or tasks were usually assessed at the end of the class.

Each student team proposed a number of recommendations. The most feasible recommendations were deliberated and were implemented with the approval of the school management. These included a sensitisation programme for students including peer counselling on water management, demonstrations of 'water wastage' activities and a seminar for students and members of the local community. Materials, including a flyer, were developed. It was also found that the PBL approach helped the educators to plan their work in a very systematic way so that appropriate water-related concepts were covered at different times during the school year. These were also integrated in different subjects and addressed in a holistic way.

Table 9.2 Summary of water-related problems obtained from walk-through investigation

Water-related problems
No water supplies during the day
No water to carry out experiments in the science lab
Not enough water to meet the school population
Water leakage problems in the toilet
Younger students break the toilets and some of them play with water and there is wastage
Some students do not close the tap
Some take much time to wash their hands or to refresh after the recess time
Water utility bills for several months mentioned showed a low supply of water
Water crisis in the island has worsen the water availability problems at school

Based on the findings, several activities on water education were organised, and other members of the school community, parents, non-governmental organisations and the local community were invited to participate. The students were found to have effectively conveyed appropriate water education-related messages to their friends, parents and others in the wider community.

9.3 Discussion

Findings from this study show that learners were effectively engaged in identifying environmental/water-related problems in their school context and eventually developed appropriate strategies to manage the situation. This is in accordance with Vandenbosch's view (2007) that a school is an institution that provides opportunities to help learners develop creatively and emotionally while acquiring skills, knowledge, values and attitudes necessary for responsible and productive citizenship.

The teaching and learning resource materials currently in use in the selected school were found to be unsatisfactory for supporting water education. This project enabled the different educators to revisit their school programme and develop more appropriate lesson plans and teaching resources that could support effective integrated water management skills. In addition, the student teams developed various resources (e.g. posters, flyers and poems) during the project, and these served to motivate and sensitise their peers about the importance of managing water resources at the school.

9.3.1 Curricular and Extracurricular Activities

The Mauritius education system is examination-driven. There were few opportunities for co-curricular activities and these were usually carried out during an activity period of 30–40 min, insufficient for the project as a whole. The interest generated by the project had learners and educators using their free time to work on the water concerns. This work has provided an opportunity for learners to engage educators from different subject areas as well as to discuss and design cross-curriculum aspects and uses of the project.

9.3.2 Teaching and Learning Methods and Assessments

Observation from this study showed that students encountered and learned water concepts via the project. The brainstorming stage served to focus students on environmental issues especially water. The developing activity engaged students and boosted cooperative learning skills and critical and decision-making skills. The learners felt like the owners of the project because they had planned, implemented and evaluated a project that had real-world application beyond the classroom. The process emphasised water learning activities that are interdisciplinary, long term and student-centred. It was also concluded that the capacity building session on water education and the regular interactions at informal meetings helped the learners to identify water-related problem(s) in their context and enabled them to develop appropriate strategies to manage the situation.

Moreover, this approach allowed educators to situate learning processes in the school. Curriculum processes could be linked to the local contextual realities based on the local issues or concerns that were identified through developing environmental learning interactions. Benzie, Mavers, Somekh and Cisneros-Cohernour (2005) argued in favour of situated learning by suggesting that the context and activities through which learning takes place are an integral part of what is learned and the environment in which the learner engages in learning is learned.

It was also found that most assessment methods used at the school were summative, and these were not effective for promoting integrated water management actions. The project created a school climate that allowed educators in the project coordination team to work together and develop learning activities that required critical thinking and also encouraged assessment for learning at every stage. The participation of the educators with different experience, knowledge, skills, values, attitudes and expertise added value to the learning processes in the case study school.

9.3.3 School Community and Community Learning

Studies show that there are three key learning environments for students: the school, the home and the wider community (Taylor & Mulhall, 1993). The PBL method, being an active approach across these, strengthened the links, and the experiences gained in each helped to bring together and integrated the learning process. School management supported the water management strategies of the project and contributed by exploring possibilities to connect the school to the community. These school-community linkages are important. Moreover, it is noted that the capacity building of educators during the workshop on water education empowered them to work with students and develop learning activities that address water management issues as well as develop learner capabilities to deal with these issues. This type of interaction encouraged dialogue with the school community and other relevant stakeholders including parents.

9.3.4 Agents at School, Home and Community

A PBL approach aims at promoting action learning tailored to meet the needs of the community and to address the environmental issues in the school and their surrounding community. This type of action-oriented and problem-solving pedagogy enabled the learners to address local issues especially water in this study. Findings show that they addressed water management issues as well as developing learner capabilities to deal with these issues. For instance, some comments recorded during post-PBL were:

> When I went back home, I told my mother how much water we are wasting, now I avoid to waste water
> At school we told our friends to stop wasting water

9.4 Conclusion

The project-based learning framework explored in this research consists of identifying and prioritising the environmental and sustainability issues of concern, exploration of options and potential actions to address the issues, integrating issues and potential actions into school environmental policies and real action taking to address the issues in the context through a range of activities and practices.

Project-based learning has been shown to have the potential to be an effective teaching approach for the twenty-first century that can be used to facilitate education for a sustainable future. The processes reviewed clearly show that a PBL approach can involve a 'whole brain approach' which enables learners to develop knowledge and different skills (e.g. critical negotiation and decision-making) and to develop appropriate attitudes towards water preservation. The students also developed appropriate capabilities that were effectively used to promote water education in school, home and community environments.

References

Benzie, D., Mavers, D., Somekh, B., & Cisneros-Cohernour, E. J. (2005). Communities of practice. In B. Somekh & C. Lewin (Eds.), *Research methods in social sciences*. London: Sage.

Denzin, N., & Lincoln, Y. S. (Eds.). (2000). *Handbook of qualitative research*. Thousand Oaks, CA: Sage.

Markham, T., Larmer, J., & Ravitz, J. (2003). Introduction to project based learning. In *Buck Institute for Education project based learning handbook* (2nd ed., pp. 3–12). Oakland, CA: Wilsted and Taylor.

Ministry of Education. (2005). *Mauritius: Staking out the future*. Government of Mauritius Press.

Taylor, P., & Mulhall, A. (1993). Linking learning environments through agricultural experience – Enhancing the learning process in rural primary schools. *International Journal of Educational Development, 21*, 135–148.

Vandenbosch, T. (2007). *Contextualizing learning in primary and secondary schools using natural resources*. Paper presented at World Environmental Education Congress, Durban, July 2–6.

Walker, M. (2005). Amartya Sen's capability approach and education. *Educational Action Research, 3*(1), 103–107.

Chapter 10
Issues-Based Enquiry: An Enabling Pedagogy for ESD in Teacher Education and School Geography

Di Wilmot

This chapter addresses the need for innovations in geography teacher education programmes in a developing world context. More specifically, it responds to the need for practical 'how to' examples for ESD integration into school geography by describing a pedagogical experiment that was piloted with in-service Namibian teachers and education development officers (EDOs) enrolled for a Bachelor of Education (Honours) degree in 2014. The theoretical constructs underpinning the experiment's design and pedagogical approach as well as the teacher professional development model are described. This is followed by a description and justification of the methodology used to answer the research question: 'How can issues-based enquiry enable the integration of ESD at the micro level of the classroom?' The findings of the experiment provide evidence of how issues-based enquiry, underpinned by active learning and constructivist epistemology and a model of teacher professional development located in reflexive practice, enabled the teachers to acquire foundational knowledge and pedagogical content knowledge for effective integration of ESD into school geography. This chapter may offer other teacher educators some guidelines on how to develop teacher capacity to integrate ESD into their own programmes.

We live in a rapidly changing global environment in which there are significant socioecological challenges that need addressing for a sustainable future. The international geography community recognises the contribution school geography ought to play in helping young people to acquire knowledge, skills and values necessary for engaging critically with complex environment and development issues, both immediately and in their future lives as adults in a global society (Haubrich, Reinfried, & Schleicher, 2007; Lee & Williams, 2006). The *2007 Lucerne Declaration on Geography Education for Sustainable Development* acknowledges that ESD is culturally defined and a contentious issue because it is subject to

D. Wilmot (✉)
Faculty of Education, Rhodes University, Grahamstown, South Africa
e-mail: d.wilmot@ru.ac.za

© Springer International Publishing Switzerland 2017   129
H. Lotz-Sisitka et al. (eds.), *Schooling for Sustainable Development in Africa*,
Schooling for Sustainable Development 8, DOI 10.1007/978-3-319-45989-9_10

different interpretations according to the needs of different nations, groups and societies (Haubrich et al., 2007). The declaration calls for school geography which inter alia:

- Enables young people to acquire knowledge and understanding of the 'human-earth ecosystem' and a 'systems thinking approach', that is, an ability to think ecologically and holistically in order to understand how nature, society and individuals are interconnected
- Develops skills and capacities in learners, including critical thinking, an ability to communicate and argue effectively as well as cooperation and conflict skills and an ability to challenge injustice and inequalities
- Focuses on major issues in the contemporary world, identifying and evaluating solutions and alternatives and taking action
- Promotes values and attitudes including empathy, value and respect for diversity; a sense of identity, the belief that people can make a difference (agency); commitment to social justice and equity; and stewardship.

More recently, the International Geographical Union (IGU) Commission on Geographical Education (2014, p. 1) reaffirmed that 'geography is the study of the Earth, its natural and physical environments, human activities, the interrelationships and interactions of these and their effects, from local to global scale'. Systems thinking, a defining attribute of geography, is seen as necessary for understanding and addressing the problem of global environmental change (International Social Science Council (ISSC), 2013).

Like many other developing countries in Africa, South Africa and Namibia face sustainable development challenges, notably water scarcity, climate change mitigation and adaptation and loss of biodiversity. These are exacerbated by the legacy of a divided and unequal past which is still impacting on the present. Building capacity for adaptation and sustainable livelihoods and lifestyles is seen as a critical challenge, especially for those in rural areas (38 % of South Africans and 62 % of Namibians live in rural areas) where people are most vulnerable to social, environmental and economic risks (Lotz-Sisitka, 2011).

In South Africa and Namibia, where the pedagogical experiment described in this chapter was conceptualised and implemented, good progress has been made in incorporating key principles for ESD into the national curricula (Namibia. Ministry of Education, National Institute for Educational development (MoE), 2009a; South Africa. Department of Basic Education (DBE), 2011; South Africa. Department of Education (DoE), 2003). In a developing world context, school geography is seen as having an important role to play in enabling societal transformation. The South African Geography curriculum aims to develop 'informed, critical and responsible citizens who are able to participate constructively in a culturally diverse and changing society and contribute to a just and democratic society' (DoE, 2003, p. 4). It emphasises the need to foster agency and self-efficacy, choice and participation in society through informed decision-making. Similarly, Namibian school geography should ensure that learners develop an understanding of the political, social, economic and biophysical dimensions of the world so that they can operate effectively and responsibly in their society. Geography must provide learners with 'an understanding of the risks and challenges in their world that need to be addressed in order to improve the quality of their lives and health of their environment' (Namibia. MoE, 2009b, p. 3). In spite of good progress having been made in integrating ESD into school geography at the level of policy, the enactment of ESD by teachers at the micro level of the classroom is an ongoing and as yet unresolved challenge. Teachers' understanding of sustainable development is poor, and they have little capacity for integrating these issues into their teaching (Dube, 2012; Lotz-Sisitka, 2011). Furthermore, the knowledge framework of the South African curriculum:

> tends to be limited by content on problems and issues for raising awareness, but fails to develop deeper conceptual depth and understanding of environment and sustainability, as issues-based knowledge dominates. (Lotz-Sisitka, 2011, p. 6)

Large-scale national studies undertaken by the sustainability pathways sector in South Africa point to the need for teachers' knowledge of environmental and sustainable development content and pedagogical content knowledge (how to teach) to be improved (Lotz-Sisitka, 2011, p. 10). In Namibia, teachers struggle to implement learner-centred pedagogy, the epistemology underpinning Namibian education reform, because they lack experiential knowledge of the pedagogical approach and because officials cling to traditional teacher-centred pedagogic structures and arrangements (Nyambe & Wilmot, 2012). Self-demeaning feelings, a lack of confidence and inadequate professional support for in-service Namibian teachers militate against meaningful enactment of the curriculum (Nyambe & Wilmot, 2014).

10.1 Purpose of the Research

This chapter describes a pedagogical experiment that responds to the need for practical 'how to' examples of how to develop teachers' foundational knowledge and pedagogical content knowledge through an active learning, experiential, issues-based enquiry approach. The experiment took place with ten teachers and two education development officers (EDOs) enrolled for a Geography Education course that formed part of a Bachelor of Education (Honours) degree. The group included two primary and eight secondary school teachers with between 4 and 10 years of teaching experience. The EDOs, employed by the Namibian Ministry of Education's National Institute for Education Development (NIED), support in-service teachers through workshops and school visits. It is hoped that the explanation of what was done and why it was done will offer other geography teacher educators some guidelines on how to use issues-based enquiry as a pedagogical strategy for integrating ESD into their own programmes.

The theoretical constructs, underpinning the experiment's design and pedagogical approach, and teacher professional development model are described. This is followed by an explanation of the methodology used to answer the research question:

> How can issues-based enquiry enable the integration of ESD in school geography at the micro-level of the classroom?

In order to answer this question, I describe how an active learning approach, which models theory in practice, enabled and supported in-service teachers' acquisition of foundational knowledge and pedagogical content knowledge to reorient school geography towards ESD.

10.2 Theoretical Perspectives Informing the Design of the Pedagogical Experiment

There are a range of interlinked socioecological, socio-economic and sociopolitical problems and challenges that affect development and quality of life in Namibia. Kraft (2014) cautioned that, since achieving independence in 1990 and in spite of many notable political, social and economic achievements, Namibia still faces important development challenges. Four key challenges, inherited by Namibia at independence and identified in Namibia's Fourth National Development Plan (NDP) 2012/3–2016/7, are low economic growth, a high rate of poverty, inequitable distribution of wealth and income and high unemployment (National Planning Commission, 2012, as cited in Kraft, p. 19). The Fourth NDP also identifies the country's unhealthy, shallow economic structure and resource base, the latter being overdependent on primary commodities, as a development challenge. The NDP calls for education to play a critical role in addressing these challenges so that economic and social life is enhanced, inequality and social justice is achieved, and natural resources are utilised in sustainable ways.

The experiment took place in Okahandja, a town 110 km north of Windhoek, the capital city of Namibia. Okahandja, a nodal town located at the junction of transport links to northern and western Namibia, has many tourists passing through it. It is an important centre for woodcarvers who, operating out of two informal markets located alongside arterial roads, sell their carvings to tourists passing through the town. Wood, harvested from indigenous forests in the north of Namibia, is transported to Okahandja where it is carved into a variety of wooden products, including carved animals, bowls and full-size carvings of people. The woodcarving industry is classified as a small and medium enterprise (SME). In Namibia, as in other developing world contexts in Africa, SMEs are an important part of the economy, providing employment and a source of income to approximately one third of the Namibian workforce (Ogbokor & Ngeendepi, n.d). In spite of this, the woodcarving industry also encourages deforestation and unsustainable natural resource utilisation.

The pedagogical experiment responds to the need for school geography to 'address the challenges and risks Namibians face if they do not care for and manage their natural resources, and challenges and risks to health caused by pollution, poor sanitation and waste' (Namibia. MoE, 2009b, p. 5). According to the teachers who participated in the experiment, the dominant teaching approach is teacher centred and transmissive with little, if any, engagement with local issues and challenges. By modelling an issues-based enquiry and then getting teachers to design and implement an enquiry in their classrooms, the experiment provided experiential learning opportunities and built teacher capacity and agency for implementing ESD through active learning in geography.

The experiment consisted of two phases: the first modelled an issues-based enquiry in the local environment for the teachers. Working under the guidance of and in collaboration with a more experienced learner (the university tutor), the teachers planned and undertook an issues-based enquiry framed by the questions: *What are the challenges facing the woodcarvers? And how do these link to broader environment and sustainability issues associated with SME in Namibia?* The enquiry involved a review of literature and gathering data in the field through observations and interviews. Phase 2 consolidated and applied the learning acquired in Phase 1. It involved the teachers planning, implementing, analysing and reflecting on an issues-based enquiry in their professional context. This phase took place at different schools in Namibia and at two teacher professional development workshops run by the two EDOs.

The time allocated to the experiment was constrained by the Bachelor of Education (Honours) model of delivery which consists of five teaching blocks of 5 days each which are evenly spread between March and November. Phase 1 of the experiment took place in April 2014 (the second block), with Phase 2 (the planning, implementation, analysis, report writing and reflection of a school-based enquiry) taking place between May and October.

Table 10.1 summarises the two phases of the pedagogical experiment. It shows the enquiry process, the questions that framed the different activities that the teachers participated in and how evidence of learning was gathered.

Table 10.1 shows how the pedagogical experiment adopted a *question-led (enquiry) framework* based on the work done by contemporary geography teacher educators (Davidson & Catling, 2000; Martin, 2006; Roberts, 2003, 2013) and applied in a Southern African context (Wilmot, 2000). The experiment also drew on the work of Environmental Education educators (O'Donoghue & Fox, 2009; Rosenberg, 2009) and curriculum policy (South Africa. DBE, 2011; Namibia. MoE, 2009a, b).

The experiment was *contextually relevant* to the lived experiences of the teachers. By focusing on a local issue (woodcarvers) that the teachers were familiar with, they were able to draw from, and build on, the diversity of experiences in the group. The experiment included activities which supported individual and social learning. These included working with a partner, working cooperatively in a group and participating in plenary discussions. This helped to create spaces in which the teachers could articulate and share their ideas with others.

Table 10.1 shows how the experiment was *change oriented*. It moved beyond awareness raising which, according to Lotz-Sisitka (2011), has dominated issues-based teaching and sought instead to develop conceptual understanding of how a local issue is linked to broader concepts of development and sustainability. The teachers' foundational knowledge was developed through a literature review of selected documents relevant to the issue. These included a research paper (Ogbokor & Ngeendepi, n.d) and a discussion paper of the Namibia Economic Society

Table 10.1 Summary of the pedagogical experiment

Phase	Enquiry questions and pedagogical activities	Evidence of learning
Phase 1	What is the issue? What do we know and do we not know about the issue? [Picture reading activity]	Ask questions and participate in class discussion
Plan, implement and evaluate an issues-based enquiry in the local environment: the case of the woodcarvers in Okahandja	What do we not yet know and need to know in order to answer the question: what are the challenges facing the woodcarvers? [Generating questions using Wood's (n.d) snowball technique]	Formulate questions individually, in pairs and as a group
	How can we find out more about the issue? Where can we find more information? Who can we contact and how? [Document analysis]	Analyse document and participate in discussion on information accessed from different sources
	How can we find out more about the issue? Where can we find more information? Who can we contact and how? [Document analysis and design an interview and observation schedule]	Design an interview schedule
	What do we need to investigate the issue and answer the question? [Undertake cooperative groupwork in the field, observing and interviewing the woodcarvers]	Work collaboratively in group. Gather data through observations and interviews. Make field notes and audio recordings and take photographs. Participate in a plenary discussion
	What has changed? What is changing? How fast is it changing? Why has it changed? What are the possible causes and consequences of the issue? Who can tell us about change? Who/what is affected by the change? How do they feel about change? Who decides about change? [Record, analyse and interpret the data gathered]	Report and discuss the findings. Evaluate what has been done and what emerged
		Identify the actions needed to solve the problem and improve or change the situation. Take appropriate action
	What can we do about it? How can we do something about it? What are the possible solutions? What alternatives do we have? What resolutions are possible? What sort of action needs to be taken? By whom?	
Phase 2	Analyse the curriculum and identify an appropriate environmental and development issue to investigate with a selected class	

(continued)

Table 10.1 (continued)

Phase	Enquiry questions and pedagogical activities	Evidence of learning
Design, implement and evaluate an issues-based enquiry at the micro level of the classroom	Plan an issues-based enquiry	Develop an enquiry plan
	Present plan to peers, receive feedback, refine the enquiry before implementing	Written feedback from peers and evidence of how this has been used to develop the enquiry
	Implement the issues-based enquiry at the micro level of the classroom	
	Analyse and evaluate what worked well, what did not and how the enquiry should be revised before being implemented again	Report on the implementation
	Present findings to peers and tutors	Seminar presentation

(Gaomab, 2004). These documents, downloaded from the Internet, modelled for the teachers the importance of accessing information and using resources to deepen their content knowledge. This is important in a developing world context where resources for teaching are limited and where there is often an over-reliance on a single textbook as a source of information. The literature helped the teachers to think relationally and holistically about natural resource utilisation and challenged their common-sense understanding of small businesses like the woodcarvers. This was evident from the questions asked, and comments made, by the teachers in class discussions. Furthermore, in their written report on the woodcarvers, the teachers provided evidence of understanding how socioecological processes and interactions have impacts and consequences. Woodcarving, an important SME in Namibia, impacts on the biophysical environment (natural resources) and has consequences for the economy (generation of money through tourism) and society (jobs and livelihoods).

In order to develop the teachers' pedagogical content knowledge (how 'to do' issues-based enquiry in their geography classrooms), the experiment was purposively designed to model *learner-centred pedagogy* and its underpinning constructivist epistemology. Modelling theory in practice and using experiential learning responds to the problem of teachers not having experience of implementing learner-centred pedagogy (Nyambe & Wilmot, 2012). It enabled the teachers to experience, in a practical and concrete way, constructivist theory and a participatory, interactional pedagogy. The latter consisted of both negotiated and framed styles of teaching and learning (Roberts, 1996 in Kent, 2006, p. 65) congruent with the mixed pedagogical approach which Nyambe and Wilmot (2014) contend was appropriate in Namibia where teachers struggle to make the leap from a traditional, teacher-centred pedagogy to a democratic, participatory, learner-centred pedagogy.

Table 10.1 illustrates a model of teacher professional development underpinned by constructivist epistemology and located in reflexive practice (Wilmot, 2009). It shows how different teaching activities were used to engage the teachers physically and mentally in constructing knowledge and making sense of the local environment and how the teachers were actively engaged in activities that elicited, built on, challenged, modified and transformed their ideas and practices. Activities included picture analysis, document analysis, cooperative groupwork, fieldwork, practical work (designing an interview schedule) and making observations, participating in a plenary session, seminar presentations and writing a research report.

The purpose of Phase 2 was to consolidate and apply the knowledge acquired in Phase 1 at the micro level of the classroom. Table 10.1 shows how in Phase 2 teachers had to enact theory in practice using a type of action research process that involved planning an issues-based enquiry, sharing the plan with and receiving feedback from peers, refining the plan, implementing the enquiry, analysing and reflecting on the implementation and reporting on what was done and achieved both verbally and in writing. All ten teachers and the two EDOs successfully completed Phase 2. They designed and implemented issues-based enquiries which mirrored the pedagogical approach of the experiment. This included using photographs and artefacts as stimuli to engage learners, generating questions using the snowball technique, accessing and using additional sources of information (some teachers used their school's computer lab to access Internet resources, while others provided reference books for learners), undertaking fieldwork, engaging learners in finding solutions to issues and taking action. The teachers' written reports and seminar presentations provided rich evidence of exciting shifts in pedagogy and a reorientation of geography towards ESD.

10.3   Lessons Learned

The pedagogical experiment, underpinned by constructivist epistemology and modelling learner-centred pedagogy, seems to have been very successful in making teachers aware of the need for a more active, relevant and engaged way of teaching school geography. It provided them with a lived experience of how this can be achieved in practice using issues-based enquiry. The experiment was very useful in equipping the teachers to do this in practice.

The experiment engaged the teachers in critical and creative thinking. It made them aware of inequalities and injustices and offered practical solutions to the challenges faced by the woodcarvers. If we want teachers to think creatively and critically, be resourceful and innovative, we need to model this in teacher education programmes. We need to be the change we expect our teachers to be. Similarly, we need to model the values and attitudes that school geography should develop, including a sense of fairness and justice, a respect for diversity and empathy and a concern for the environment and commitment to sustainable development. Issues-based enquiry is a good vehicle for this.

If we want teachers to become change agents and reorient school geography towards ESD, we need to adopt a model of teacher professional development located in reflexive practice (Wilmot, 2009). As demonstrated in the above case study, this approach enabled and supported the teachers' acquisition of foundational knowledge and pedagogical content knowledge (issues-based enquiry) for effective integration of ESD into geography lessons.

10.4   Conclusion

This chapter has addressed the need for empirical studies which offer geography teacher educators some suggestions on how ESD can be integrated into school geography and teacher education programmes. Using the case of a Namibian teacher education pedagogical experiment, this chapter has provided evidence of how issues-based enquiry characterised by active and experiential learning and underpinned by constructivist epistemology is an effective vehicle for integrating ESD into school geography and teacher education programmes. It has described how a model of teacher professional development located in reflexive practice enabled and supported teachers to acquire knowledge and skills to re-implement ESD in their geography lessons through issues-based enquiry. This experiment provides evidence of the power and efficacy of experiential learning, modelling theory in practice and learning through apprenticeship (where the preliminary application of theory was done in the teacher education classroom) for developing teacher resourcefulness, agency and change-oriented practices.

References

Davidson, G., & Catling, S. (2000). Towards a question-led curriculum. In C. Fisher & T. Binns (Eds.), *Issues in geography teaching* (pp. 5–14). London: Routledge.

Dube, C. (2012). *Implementing education for sustainable development: The role of geography in South African secondary schools*. Unpublished doctoral dissertation, University of Stellenbosch, Stellenbosch.

Gaomab, M. (2004). *Challenges facing SMEs. Namibia Economic Society*. Downloaded from http://www.sarpn.org/documents/d0001004

Haubrich, H., Reinfried, S., Schleicher Y. (2007). The Lucerne declaration on geographical education for sustainable development. In S. Reinfried et al. (Eds.), *Geographical views on education for Sustainable Development*. Proceedings, Lucerne Symposium Switzerland, July 29–31, 2007, Geographiedidaktische Forschungen, Vol. 42, Hochschulverband für Geographie und ihre Didaktik, Nürnberg, 243–250.

International Geographical Union Commission (IGU) on Geographical Education. (2014). Draft Executive Statement on the International Declaration on Research in Geography Education, August 2014. Downloaded from http://www.igu-cge.org/Krakow2014-results.htm

International Social Science Council (ISSC). (2013). *World social science report: Changing global environments. Summary*. Paris: UNESCO.

Kent, A. (2006). Changing learning and teaching. In J. Lidstone & M. Williams (Eds.), *Geographical education in a changing world: Past experience, current trends and future challenges* (pp. 23–37). Dordrecht, The Netherlands: Springer.

Kraft, R. (2014). The Namibian socio-economic landscape. In M. Schafer, D. Samson, & B. Brown (Eds.), *Namibia counts*. Grahamstown, South Africa: Rhodes University.

Lee, J. C., & Williams, M. (2006). Geography, environment, sustainability, culture and education. In Z. Li & M. Williams (Eds.), *Environmental and geographical education for sustainability: Cultural contexts*. New York: Nova Science.

Lotz-Sisitka, H. (2011). *National case study: Teacher professional development within an education for sustainable development focus in South Africa: Development of a network, curriculum framework and resources for teacher education*. Paper presented at the African (ADEA) Triennale on Education and Training in Africa, Ouagadougou, Burkina Faso, 7–12 February 2011. Downloaded from http://www.adeanet.org/triennale/Triennalestudies/subtheme1/1_3_09_GIZ_en.pdf

Martin, F. (2006). *Teaching geography in primary schools*. Cambridge, UK: Chris Kingston.

Namibia. Ministry of Education, National Institute for Educational development (NIED). (2009a). *The national curriculum for basic education 2010*. Okahandja, Namibia: NIED.

Namibia. Ministry of Education, National Institute for Educational development (NIED). (2009b). *Junior secondary geography syllabus grades 8–10*. Okahandja, Namibia: NIED.

Nyambe, J., & Wilmot, D. (2012). New pedagogy, old pedagogic structures: A fork-tongued discourse in Namibian teacher education reform. *Journal of Education, 55*, 55–83.

Nyambe, J., & Wilmot, D. (2014). Post-apartheid pedagogic reform in Namibia: The route to learner-centred pedagogy and its implementation. In M. Schafer, D. Samson, & B. Brown (Eds.), *Namibia counts*. Grahamstown, South Africa: Rhodes University.

O'Donoghue, R., & Fox, H. (2009). *Handprint resources books: Action towards sustainability*. Howick, South Africa: Share-Net.

Ogbokor, C. A., & Ngeendepi, E. J. (no date). *Investigating the challenges faced by SMEs in Namibia*. Research Paper. Department of Economics, Polytechnic of Namibia. Windhoek.

Roberts, M. (2003). *Learning through enquiry*. Sheffield, UK: The Geographical Association.

Roberts, M. (2013). *Geography through enquiry*. Sheffield, UK: The Geographical Association.

Rosenberg, E. (2009). *Teacher education workbook for environment and sustainability education in Southern Africa*. Howick, South Africa: Share-Net.

South Africa. Department of Basic Education (DBE). (2011). *Curriculum and Assessment Policy Statement (CAPS) geography*. Pretoria, South Africa: Department of Education.

South Africa. Department of Education (DoE). (2003). *National curriculum statement grades 10–12 (General): Geography*. Pretoria, South Africa: Department of Education.

Wilmot, D. (2000). Making issues-based enquiry a reality in South African classrooms through co-operative fieldwork. In A. Kent & S. Jackson (Eds.), *Geography and environmental education: International perspectives* (pp. 126–131). London: Institute of Education, University of London.

Wilmot, D. (2009). A critical review of a school-based intervention in grade 9 human and social sciences at two South African schools. *Journal of Educational Studies, 8*(3), 94–111.

Wood, P. (no date). *The importance of questioning in developing an enquiring classroom*. Downloaded from www.geography.org.uk /gtip/thinkpieces/questioning/

Part III
Integrated Approaches to Education for Sustainable Development in Schools

Chapter 11
Integrating School Community Concerns in Framing ESD and Educational Quality

Raviro Chineka and Cecilia Mukundu

This chapter focuses on bridging the educational quality gaps for vulnerable groups when dealing with ESD concerns valued by both teachers and parents. The chapter draws from a qualitative case study that sought to develop teachers' capabilities and agency to integrate HIV/AIDS education in different subjects across the secondary school curriculum in one secondary school in Zimbabwe. The research site was chosen due to its predisposition as a rich source of data for HIV/AIDS and ESD interactions. Data was gathered from 50 teachers through interactive workshops, interviews, focus groups and open-ended questionnaires. Findings from the study suggest the curriculum could work as a mediating device making connections between the school and its community. The study highlights the role of teachers in mobilising opportunities for learning in diverse classrooms where HIV/AIDS pose unique challenges for particular learners. The study concludes that given the right training and mindsets, teachers become agents of change, and classrooms become therapeutic laboratories where learners' concerns are valued. Consequently, quality of ESD learning improves as the community of teachers and learners are empowered to live differently. The study recommends reorientation of teacher education to embrace skills for teaching in contexts of risk, vulnerability and uncertainty. Legislators may review and enact policy changes so HIV/AIDS teaching takes a whole-school approach. More so, conceptions of quality education and teacher competency ought to be broadened beyond learner pass rates to include aspects such as creativity and making the world a better place.

R. Chineka (✉)
School of Education Faculty of Arts and Social Sciences, University of Technology, Sydney, NSW, Australia

Faculty of Education, University of Zimbabwe, Harare, Zimbabwe
e-mail: rchineka@gmail.com

C. Mukundu
Faculty of Education, University of Zimbabwe, Harare, Zimbabwe
e-mail: ckmumundu@gmail.com

© Springer International Publishing Switzerland 2017
H. Lotz-Sisitka et al. (eds.), *Schooling for Sustainable Development in Africa*, Schooling for Sustainable Development 8, DOI 10.1007/978-3-319-45989-9_11

This chapter provides an overview of how a school community can work together to address sustainability issues of mutual concern. Using HIV/AIDS as an example, the chapter gives a summary from a study (see Chineka, Mukundu, & Nyamukunda, 2013, for a full report) that sought to design curriculum interventions that improve responsiveness to sustainability issues and explored how these issues can be addressed to improve quality of education and hence quality of life. The research sought to investigate the interactions between education for sustainable development (ESD) and educational quality and relevance in contexts of risk and vulnerability related to HIV and AIDS. Basing on the premise that ESD enables capabilities (what people value doing), action competence (their ability to learn how to act and do new valuable things) and agency (taking action), it became imperative to question what, in an educational setting, constrains people's abilities to turn their existing capabilities into action.

Further to that the question of linking capabilities to the quality and especially relevance of education becomes critical. Naturally people would value what they think is relevant to them. Our research was motivated by the desire to locate opportunities for improving the quality and relevance of education in a school and its community. The project tapped into existing structures and addressed capacity gaps within one Zimbabwean school community to enable them to improve HIV/AIDS learning. This participatory action research was conducted within the context of the ESD thinking which stresses the value of education in promoting sustainable development. We were particularly guided by Chap. 36 of Agenda 21, which recognises the importance of providing education, thus empowering and leading people to adopting more sustainable lifestyles as they are empowered to make more informed choices.

Traditionally, the priority and essence of HIV/AIDS education in Zimbabwe was to promote awareness. Our research transcended this aim as we endeavoured to change routine and everyday practices and enabled participants to assume new roles as participants in a social learning process. The immediate objective was to enhance teachers' capacity for creating child-friendly classroom environments for children living in contexts of risk and vulnerability related to HIV/AIDS.

This chapter will argue that HIV/AIDS is a reality facing Zimbabwe. It has caused untold human suffering (Juma, 2001; Kelly & Bain, 2003) and has impacted negatively on the provision of quality education in Zimbabwe (Chineka et al., 2013; Nyamukapa, Foster, & Gregson, 2003). However, it is possible to curb the epidemic through education (Juma, 2001; Kelly & Bain, 2003). People are willing to change their practices because the impacts of HIV/AIDS are real and being felt by teachers, students and the community at large. HIV/AIDS is a challenge that knows no boundaries and this makes it a sustainability issue of mutual concern. However, the old type of education will not work. We need a new type of education that is holistic, participatory, collaborative, flexible and relevant to the students' needs and those of society. To put the chapter into context, we will briefly discuss the nature and evolution of HIV/AIDS education, the structure of secondary school education and secondary teacher education in Zimbabwe.

11.1 Contextual Background

In the wake of increased environmental and sustainability challenges, education institutions have been challenged to take stock of society's needs and develop curricula that are responsive to the students' needs and encourage positive social transformation (Juma, 2001; Kasembe & Moonga, 2009). One such sustainability issue confronting Zimbabwe is HIV/AIDS. With the HIV/AIDS prevalence rate at 14.3 % (United Nations Children's Fund [UNICEF], 2011), Zimbabwe is rated among the worst affected nations in southern Africa and the world.

A typical challenge for Zimbabwe is the increasing number of children orphaned as a direct effect of HIV/AIDS. Such orphans may face great trauma, stigma and sometimes the burden of caring for sick parents or relatives (Nyamukapa et al., 2008). These children eventually end up in school where teachers are not formally trained to handle the 'socio-economic, socio-ecological and/or psychosocial aspects of HIV/AIDS in the classroom' (Chineka et al., 2013, p. 204). Our research project was therefore conceived to provide a pragmatic solution to this apparent gap. Our research endeavoured to enhance teachers' capacities to integrate HIV/AIDS education effectively in the secondary school curriculum. The interventionist research aimed to enhance agency and capabilities to break the HIV/AIDS risk cycle and bring about positive change in the quality of education and hence the quality of life of youths living in contexts of risk and vulnerability related to HIV and AIDS.

11.1.1 HIV/AIDS Education in Zimbabwe

It is important to note that we will limit this discussion about the provision of HIV/ AIDS education to the secondary school level of schooling. We chose to work with a secondary school because this particular age group of learners is at the greatest risk of HIV infection (Pembrey, 2009; Ross, Dick, & Ferguson, 2006).

Zimbabwe faces a multitude of environmental and sustainability issues, but HIV/ AIDS deserves special attention because of its impacts on educational quality. As the AIDS pandemic wreaked havoc on the nation, the government launched the School AIDS Programme in 1994 to integrate HIV/AIDS into the national education system. Syllabi and policy guidelines were developed to guide HIV/AIDS education. The focus was primarily on awareness creation to help students understand the basics of HIV/AIDS. This approach however falls short of the demands of HIV/ AIDS education, particularly where the objective is to acquire attitudes and skills on how to deal with the challenges posed by the pandemic. HIV/AIDS presents unique challenges for education, hence the need to devise alternatives to existing forms of education (Chineka et al., 2013).

For education to be regarded as a viable tool in the fight against HIV/AIDS, it must transcend awareness creation and go further to help children build resilience to subvert the negative impacts of HIV/AIDS on the quality of their education and their life (UNICEF, 2006). Teaching should be tailor-made to develop reflexivity leading to responsible decision-making (SADC Regional Environmental Education Programme [SADC REEP], 2012). This demands a change in pedagogical tools to embrace participatory methodologies that encourage reflexivity leading to transformative learning. It has been argued that critical reflection by all individuals influences the participatory process and the ethics and process of participation. Consequently, participants become more effective agents of social change (Chineka et al., 2013; Janks, 2014).

11.1.2   Structure of the Secondary School Curriculum

The curriculum in Zimbabwean secondary schools is subject based with subjects like mathematics and English being core and compulsory. Naturally, such subjects are valued and prioritised on the school timetable. In addition, the curriculum is examination driven. Consequently, non-examinable subjects are generally neglected and regarded as inferior. By the time this chapter was concluded, HIV/AIDS was not a stand-alone subject but integrated into other subjects. Its teaching is left to the discretion of the school and the teacher. We traced some of the challenges bedevilling HIV/AIDS education within the country's teacher education system.

11.1.3   Teacher Education and ESD in Zimbabwe

Teacher education in Zimbabwe is provided in 14 teacher training colleges and 14 universities. In response to global trends in curriculum reform, the Zimbabwean teacher education curriculum has developed continuously to accommodate contemporary issues arising in human society and affecting the quality of education. In the context of HIV/AIDS, a programme worth mentioning in advancing ESD in teacher education in Zimbabwe is the Secondary Teacher Training Environmental Education Programme (St^{2}eep). St^{2}eep's emphasis has been on advancing environmental education and mainstreaming of ESD in all subjects in the curriculum. Many of the teachers who received training under this programme were deemed to have the capacity to deal with sustainability issues in the curriculum and in their environment. However, not every teacher in Zimbabwe has had the opportunity to participate in the St^{2}eep. Consequently, a significant number of secondary school teachers are not skilled to handle HIV/AIDS education and its associated challenges. Our research project was a step towards closing this gap.

11.1.4   Loopholes in Teacher Education

As we deliberated on the possible causes of the neglect of HIV/AIDS education, one important issue teachers raised was lack of capacity which goes back to the way the Zimbabwean teachers were trained. Although the school's curriculum and the teacher education curriculum are designed separately, an intricate relationship exists between the two. In as much as HIV/AIDS is taught in teacher training institutions, there is no programme where trainee teachers graduate with HIV/AIDS as a major. In addition, there are scores of teachers who graduated from college before HIV/AIDS became a topical issue. In addition, professional development for teachers after initial qualification is limited due to the country's economic circumstances. Taking cognisance of these and other factors, we were challenged to design a programme to assist teachers to integrate HIV/AIDS education effectively in the different subjects of the formal school curriculum.

11.2   The Research Context

The study site was a high school within a growth point outside Harare, Zimbabwe's capital city. Growth point communities are generally characterised by high levels of unemployment and extreme poverty. Consequently, many women turn to sex for money exacerbating the spread of sexually transmitted infections including HIV/AIDS (Chineka et al., 2013). When we made our initial entry into the school, the teachers indicated that HIV/AIDS was rife within the school and its community. On average, between 10 and 15 students per class were orphaned. They cited the influence of the growth point as a driver behind the high prevalence of HIV/AIDS within the school:

> The influence of the growth point on the school is very significant … commercial sex, normally in the form of brothels is common even among adolescent girls. One of our former students for instance has been recruiting girls from the school to partake in commercial sex work. (Teacher in Charge of the Child Protection Services Committee)

Current school children have been born into a world where AIDS is a harsh, unavoidable reality. The children in the school community found themselves in an environment where HIV/AIDS naturally flourished posing serious threats to the provision of quality education. However, we recognised the school had the capacity to break the risk cycle. School education can have a powerful preventive effect that has been described as a 'social vaccine' (Pembrey, 2009). Schools can strengthen capabilities and agency for improved risk assessment and perception, thus enabling better risk management to vulnerabilities created by HIV/AIDS.

11.3 Methodology

The research took a qualitative case study approach in which one school was sampled as an information-rich site by virtue of its predisposition to risk and vulnerability to HIV/AIDS. We worked with a group of 50 teachers, in an action research process that spanned a period of 2 years. Our strategy was action research because it has been used in situations where the goal is to transform. Action research is also upheld for introducing innovations in teaching as it empowers the teachers with the dual role of theory developer and user (Riding, Fowell, & Levy, 1995). Teachers take ownership of the innovations making the implementation much easier. Since the goal of our research was positive social transformation, action research was an appropriate approach to adopt. Through interactive workshops, semi-structured interviews, focus group discussions and open-ended questionnaires, we created opportunities for dialogue, where the teachers identified what constrained them from effectively implementing HIV/AIDS education within the school. Training was designed to suit the specific needs raised by the teachers themselves.

The research team explored a number of issues including but not limited to:

- What are the risks and vulnerabilities within the school community and how can they be addressed? What structures exist and how can they be tapped into to improve provision of quality education within this community? Where are the gaps and how best should we address them?
- How can school-community relations be strengthened so that both learn from each other, i.e. creating social interaction and operating as a community of practice with the ultimate goal of improving quality of life within the community?
- How can teachers in the school work harmoniously together to address a common cause (HIV/AIDS) and how can all these efforts be coordinated to improve the provision of quality and relevant education to disadvantaged children within this school?

We created platforms for dialogue among teachers and between teachers and their students which eventually led to a dialogue between teachers and parents. Our emphasis was on encouraging people to take responsibility for action directed at addressing the impacts of HIV/AIDS. We used case scenarios and stories as well as discussions to help participants realise that HIV/AIDS is everyone's concern and everyone must be seen to be playing a role in reducing its impact on education and humanity. We endeavoured to dispel the notion that HIV/AIDS is a specialty subject area which should be left to a few teachers to handle, the 'none of my business approach'.

The research team explored teachers' capacity and knowledge gaps in relationship to and attitudes towards HIV/AIDS, building on existing knowledge and support structures within the school to improve HIV/AIDS learning in the school. The essence was to identify training needs of teachers to enhance capacity for effective teaching/learning of HIV/AIDS. This was done collaboratively in ways that made the teachers feel that they were important participants in the whole process.

We adopted a participatory approach marked with dialogue and negotiation in an open action research process. Our project's ultimate goal was to improve HIV/AIDS learning through working with teachers and students to identify challenges, risks and vulnerabilities related to HIV/AIDS that students face and devise curriculum interventions to mitigate such challenges.

11.4 Theoretical Framework

The research was heavily influenced by Sen's (1990) capabilities approach which argued that people will do what they value to do but may fail to do so due to typical constraints. The research team worked with a group of teachers in ways that allowed them to identify capacity gaps that constrained them from doing what they valued. We created an environment where teachers reflected on their attitudes and practices and identified opportunities for improving their practice. Teachers were then supported with training in areas they had identified to develop capabilities that enabled them to mitigate the impacts of HIV and AIDS. Training offered included, but was not limited to, how to teach for the development of life skills, how to mainstream HIV and AIDS in various subjects, dealing with abuse and trauma, child rights and child participation and how to strengthen school-community linkages.

Since the essence of the study was to improve the quality and relevance of education of vulnerable youth, we synthesised ideas from a number of authors (Harvey & Green, 2006; Kissack & Meyer, 1996; Lotz-Sisitka, 2011; Nikel & Lowe, 2010) to provide a theoretical lens through which to analyse whether the intervention had made any contribution in advancing the quality of education in the study school. Quality in education is a highly contested concept, and a universal definition of the concept has not yet been agreed upon. Several conceptions of the term have emerged, but, for our purposes, we believed that quality education should be responsive to students' needs. Therefore, the aspect of relevance was a key defining feature for our study. We also thought that quality education must be empowering and transformative. Education of a good quality is supposed to help people to make more informed decisions and adopt more sustainable lifestyles. We further believed that good quality education must be inclusive and democratic. It should fully embrace and add value to the lives of disadvantaged children including those infected and affected by HIV/AIDS (Chineka et al., 2013).

11.5 Findings and Discussion

We engaged in practical activities that involved collaboration and teamwork to demonstrate that it is possible to integrate people affected by HIV/AIDS in classroom activities without discrimination. We can change the face of HIV/AIDS if we are prepared to change the way we perceive our roles and the way we perceive others. It is about embracing change (adaptation) and a willingness to effect that change (agency). These changed roles have serious implications for teachers and the teaching methodologies they adopt. Below are highlights of the major findings.

11.5.1 Teaching in a Changing Context: A Pedagogical Shift in Paradigm

'Education for Sustainable Development (ESD) is synonymous with quality education but requires far-reaching changes to the way education functions in society' (Ofei-Manu & Didham, 2014: 2). There are increased calls for new orientations to education foregrounded in capability and agency (Jensen & Schnack, 1997; SADC REEP, 2012). This particular study highlighted that we are living in a dynamic society which makes it mandatory for teachers to embrace changed roles and changed modes of operation. An important pedagogical dimension in the context of HIV/AIDS is to embrace the fact that we do not 'teach our subjects' but we 'teach our students'. Our research was tailored to reorient methods of delivery to embrace skills for teaching in changing contexts characterised by risk, vulnerability and uncertainty, to embrace life skills education and the techniques for creating child-friendly schools, to create healthy environments for both physical and psychosocial support within the school system and to open dialogue on a sensitive issue (HIV/AIDS) and encourage interaction between the school and its community (Chineka et al., 2013).

11.5.2 Working Collaboratively to Improve Educational Quality Through Changed Attitudes to HIV/AIDS

Teachers claimed that the intervention changed the way they viewed their students and the way they perceived their roles. They realised their own potential and capacity to counter the realities of HIV/AIDS, an effort which draws on the aspect of being able to meet various challenges, developed in quality education. This was articulated by one teacher in the following way:

> As a teacher I have a great role; I need to give guidance and counselling to all affected pupils ... I have learnt that pupils carry different burdens and as teachers there is need to be cognisant of these ... I have been looking at people with HIV and labelling them promiscuous. I now have a more positive attitude and I have learnt not to discriminate people with HIV... I was empowered by the workshop to create healthy relationships with students so that they can share their problems.

Teachers became more open to discussion, more approachable and more tolerant to their learners, thus creating a healthy environment for psychosocial support. If quality of education is measured from the dimension of inclusivity, it can then be argued that the intervention added value to the quality of education of the disadvantaged children. Despite enabling teachers to be more responsive to the learners' needs, the study also cultivated a culture of dialogue among teachers that encouraged them to work collaboratively. By its very nature, HIV/AIDS transcends disciplinary boundaries which calls for collaboration among teachers.

11.5.3 Policy and Provision of Educational Quality in the Context of ESD

The practice of labelling some subjects as core and some as noncore creates the impression that certain subjects are more important than others. In this case, where school authorities and systems adhered to traditional modes of education geared for examinations, HIV/AIDS suffered neglect. In this particular school, it did not even appear on the timetable. The existing obsession with academic subjects being core is detrimental to the country and students' needs. In fact, if we are to improve quality of education in a broader sense, it is imperative to strike a balance between the importance accorded to academic and skills development-oriented subjects, including HIV/AIDS (Chineka et al., 2013). Towards the end, the teachers had developed an understanding that integrating HIV/AIDS in all subjects in the school curriculum, particularly the core and examinable subjects, was possible and brought relevance to what students learn.

11.5.4 Reorienting Teacher Education to Enable Provision of Quality Education

It has been argued no education system can rise above the level of its teachers (Darlington-Hammond, 2000; Hock, 2007; Kasembe, 2011; Stronge, Ward, Tucker, & Hindman, 2007). Teacher training is therefore critical to successful implementation of quality HIV/AIDS education. Agency for teaching HIV/AIDS is sometimes constrained by lack of skills as hinted by Sen's capabilities and functioning approach. Teacher education thus needs reorientation to promote experiential, participatory and collaborative learning platforms which are ideal to achieve interactive learning and enhance quality of learning. We must develop a new kind of teacher who is creative, adaptive and sensitive to student needs and teaches for diversity.

11.5.5 Learning as Connection and Educational Quality

We observed that learning becomes meaningful and empowering if it is based on things that can be directly applied in real-life contexts (learning as connection). The teachers demonstrated a natural willingness to take part in the project because it involved things they valued and things they could easily connect with that could be directly and immediately applied to solve real-life problems within the community in which they lived and worked. Engaging in ESD initiatives like this intervention enabled teachers to explore and realise their capabilities, action competence and agency. Teachers acknowledged the awareness that developed during the workshops as a result of participating and deliberating on HIV/AIDS-related issues that bring conflicts in the classroom and impact negatively on the quality of education.

11.5.6 Stigma, Culture and the Need for Improved School-Community Linkages

Some teachers expressed sentiments of stigma associated with the teaching of HIV/AIDS:

> We can effectively teach about HIV/AIDS without including it on the timetable – *zvinobva zvatibetsera nokuti tanga tichipiwa mazita kuti vaye ve AIDS as if ndiwe unenge watorwara ne AIDS yacho* (It will be good for us because by virtue of being allocated to teach HIV/AIDS, we have been labelled the 'AIDS ones' as if we are suffering from the disease).

In addition, teachers face sociocultural restrictions regarding the content they teach. They operate in a community where issues of sexuality are regarded as taboo. Topics like condom use, for instance, have stirred up much debate as people fail to agree on whether it is morally acceptable to teach teenagers about premarital sex and condom use. Certain sections of Zimbabwean society regard this kind of education as inappropriate and believe it encourages children to engage in premarital sex. In view of these concerns, teachers have called for closer collaboration between the school and its community. They also argued for a whole-school approach as a mediating strategy to provision of quality education in the context of ESD where the issues under consideration are generally complex and contested.

Since our study was a case study localised in a particular school facing unique challenges, we will not offer any recommendations. However, we will highlight the lessons we drew from the study, which we believe might be useful for schools in similar contexts.

11.6 Lessons Learned

HIV/AIDS can be integrated and be taught in any subject of the curriculum for as long as the teachers are prepared to embrace changed roles and are equipped with the requisite skills to integrate it into their teaching. The challenge for teacher education is to design programmes which transform the role of the teacher from the expert and giver of knowledge to that of guide who facilitates learning. Related to this, and in view of increased environmental complexities, we need a new breed of teachers, who can take the extra role of caregiver, nurse, counsellor and innovator (Chineka et al., 2013). Like all sustainability issues, HIV/AIDS is a complex and contested issue that calls for a pedagogical shift in paradigm if it is to be successfully addressed through education.

Government policy pertaining to which subjects are core calls for review as it constrains the teaching and learning of HIV/AIDS. There is a need to strike a balance between the importance accorded to academic and skills development-oriented subject including HIV/AIDS. More so, effective HIV/AIDS teaching should transcend the disciplinary divide and incorporate community concerns. School-community linkages thus should be strengthened for the mutual benefit of both.

11.7   Conclusion

In this chapter, we have sought to provide insights into how integration of HIV/AIDS in the secondary school curriculum of one school in Zimbabwe strengthened the ESD concept of learning as connection. We conclude that, despite several challenges, classrooms provide platforms for dialogue and can become therapeutic laboratories where teachers and learners engage in dialogue, learners' concerns are valued and action competence towards mitigating the impacts of HIV/AIDS is developed. Consequently, the quality of ESD will improve leading to improved choices and the adoption of more sustainable lifestyles. By its very nature, HIV/AIDS knows no boundaries, either you are affected or you are infected. This is what makes it a sustainability concern of mutual interest. This chapter has demonstrated that, despite all the complexities surrounding the teaching of the subject in Zimbabwe, HIV/AIDS as an ESD issue is a reality that deserves special attention. Its integration in the formal curriculum has potential to improve quality of education and hence quality of life of youths living in contexts of risk and vulnerability related to HIV/AIDS.

References

Chineka, R., Mukundu, C. K., & Nyamukunda, M. (2013). Exploring opportunities for mainstreaming HIV/AIDS in the secondary-school curriculum as an ESD strategy. *Southern African Journal of Environmental Education Research, 29*, 203–215.

Darlington-Hammond, L. (2000). Teacher quality and student achievement: A review of state policy evidence. *Education Policy Analysis Archives, 8*(1), 1–44.

Harvey, L., & Green, D. (2006). Defining quality. *Assessment and Evaluation in Higher Education, 18*(1), 9–34.

Hock, R. H. (2007). Examining the relationship between teacher quality as an organizational property of schools and student achievement and growth rates. *Educational Administration Quarterly, 43*(4), 399–432.

Janks, M. (2014). Redesign - from critical literacy to social action. In H. Janks, K. Dixon, A. Ferreira, S. Granville, & D. Newfield (Eds.), *Doing critical literacy: Texts and activities for teachers* (pp. 145–158). New York: Routledge.

Jensen, B., & Schnack, K. (1997). The action competence approach in environmental education. *Environmental Education Research, 3*(2), 163–178.

Juma, M. N. (2001). *Coping with HIV/AIDS education: Case studies of Kenya and Tanzania.* London: Commomwealth Secretariat.

Kasembe, R. (2011). Teaching science through the science technology and society lens in Zimbabwean high schools: Opportunities and constraints. *Zimbabwe Journal of Educational Research, 23*(2), 314–348.

Kasembe, R., & Moonga, M. S. (2009). Opportunities for universities in southern Africa to participate in the UN-DESD. *Environmental Education Bulletin, 33&34,* 22–23.

Kelly, M. J., & Bain, B. (2003). *Education and HIV/AIDS in the Caribbean.* Kingston, CA: Ian Randle Publishers.

Kissack, M., & Meyer, S. (1996). *Vantages of a new horizon: Knowledge, judgement and the concept of quality in the 1995 white paper. In pursuit of equality.* Cape Town, South Africa: Kenton Education Association/Juta.

Lotz-Sisitka, H. (2011). *Education for sustainable development research network: Research programme and network progress report.* SADC-REEP research programme and network, Rhodes University, South Africa (unpublished).

Nikel, J., & Lowe, J. (2010). Talking of fabric: A multi-dimensional model of quality in education. *Compare: A Journal of Comparative International Education, 40*(5), 589–605.

Nyamukapa, C. A., Foster, G., & Gregson, S. (2003). Orphan's household circumstances and access to education in a maturing HIV epidemic in eastern Zimbabwe. *Journal of Social Development in Africa, 18*(2), 7–32.

Nyamukapa, C. A., Gregson, S., Lopman, B., Saito, S., Walts, H. J., & Monash, R. (2008). HIV – associated orphanhood and children's psychosocial distress: Theoretical framework tested with data from Zimbabwe. *American Journal of Public Health, 98*(1), 133–141.

Ofei-Manu, P, & Didham, R. J, (2014). *Quality education for sustainable development a priority in achieving sustainability and well-being.* Policy brief number 28 March 2014. Kamiyanyaguchi: Institute for Global Environmental Strategies (IGES). Retrieved January 5, 2014 from www.iges.or.jp

Pembrey, G. (2009). *HIV/AIDS and schools.* AVERT. Retrieved September 20, 2013 from http://www.avert.org.html

Riding, P., Fowell, S., Levy, P. (1995). An action research approach to curriculum development. *Information Research, 1*(1).

Ross, D. A., Dick, B., Ferguson, J. (2006). (Eds.). *Preventing HIV/AIDS in young people: A systematic review of the evidence from developing countries.* Geneva: WHO Technical Report Series Number 938.

Sen, A. K. (1990). Development as capability expansion. In K. Griffin & J. Knight (Eds.), *Human development and the international development strategy for the 1990s* (pp. 41–58). London: Macmillan.

Southern African Development Community Regional Environmental Education Programme (SADC REEP). (2012). *Learning together for a sustainable future: 15 years of Swedish-SADC co-operation.* Howick, New Zealand: SADC REEP.

Stronge, J. H., Ward, T. J., Tucker, P. P., & Hindman, J. L. (2007). What is the relationship between teacher quality and student achievement: An exploratory study. *Journal of Personnel Evaluation Education, 20,* 165–184.

United Nations Children's Fund (UNICEF). (2006). *Which skills are life skills?* UNICEF.

United Nations Children's Fund (UNICEF). (2011). *A situational analysis on the status of women and children's rights in Zimbabwe, 2005–2010: A call for reducing disparities and improving equity.* Harare, Zimbabwe: Government of Zimbabwe.

Chapter 12
Integrating Learners' Voices into School Environmental Management Practices Through Dialogue

Nthalivi Silo

Drawing on an example from a primary school in Botswana, this chapter sheds light on how goals for Education for Sustainable Development (ESD) initiatives for ensuring a more sustainable future in social and environmental terms can be met by engaging learners in participatory learning processes that are meaningful, that are purposeful and that broaden their action competence through dialogue. In Botswana society which is largely authoritarian in the teaching and learning processes, if opportunities for dialogue exist between teachers and learners, positive changes for a healthier environment can be created in schools. This requires scaffolding as part of the teacher's role in supporting the learning process. Data were largely generated from focus group interviews with learners and observations of learner activities. These two methods were complemented by informal interviews with teachers and other actors in the school and show-and-tell explanations of their activities by learners themselves. What is required to develop sustained action competence are critical reflections on teaching strategies as well as support for teachers to identify ways of engaging children's views on issues in the school in open, dialogical ways. Such teaching strategies should deepen teachers' understandings of learners' potential, demonstrating how dialogue and scaffolding are part of a teacher's role in supporting learning.

This chapter draws on a study focused on dialogue between teachers and learners, which is seen to be an important dynamic for transformative Education for Sustainable Development (ESD) in Botswana (Silo, 2011). Botswana society and culture are largely authoritarian, hence the need for such a focus in transformative ESD (Maundeni, 2002; Silo, 2011; Tabulawa, 1997). To respond to this ESD focus, the investigation into how dialogue and action competence can be developed is

N. Silo (✉)
Faculty of Education, University of Botswana, Gaborone, Botswana
e-mail: nthasilo@gmail.com

© Springer International Publishing Switzerland 2017   153
H. Lotz-Sisitka et al. (eds.), *Schooling for Sustainable Development in Africa*,
Schooling for Sustainable Development 8, DOI 10.1007/978-3-319-45989-9_12

examined in the context of environmental management activities in Botswana schools where it has been found that dialogue is a crucial missing tool for learners to participate effectively in school environmental management activities (Silo, 2011). A focus on dialogue offers an opportunity for ESD learning processes that can explore opportunities to engage learners in environmental management activities that contribute to their ability to act and effect change as well as to develop their action competence (Carlsson & Jensen, 2006; Jensen & Schnack, 2006). This should be a departure from an approach that simply involves learners in environmental management activities through normalised strategies (Ketlhoilwe, 2007) under the prescription and authority of teachers (Ketlhoilwe, 2007; Silo, 2011). It then follows that any associated knowledge and insight that learners acquire during action in these activities should, in essence, bear some element of action competence being action oriented in a way that takes account of learners' voices and thoughts (Carlsson & Jensen, 2006; Jensen, 2004; Jensen & Schnack, 2006).

This chapter critically explores how learners can be engaged through dialogue and considers the importance of dialogue in scaffolding learners' participation in school environmental management activities. It is important to recognise that dialogue as a mediating tool for engaging learners entails both verbal and non-verbal interactions through guided support in children's participation for any activity they are engaged in (Rogoff, 1995). The importance of guided participation and scaffolding as a process for action competence development is also explored. To achieve this, the chapter draws on a case study in Botswana where the researcher, with the support of teachers, attempted to engage learners in environmental management activities by providing support through dialogue to a selected group of learners.

12.1 Background and Context

At national policy level, the 1994 Revised National Policy on Education (RNPE) was the first major attempt in Botswana to place a focus on environmental education in the school curriculum by making specific reference to a strong drive towards recognising and affirming the role of learner participation in the care and preservation of the environment. This is evident in its recommendation that, in so doing, this should be coupled with the training of teachers 'in the methodologies, at both pre-service and in-service levels, for environmental education to ensure that learning results in attitudinal changes and learner participation' (Botswana Government, 1994, p. 26).

Consequently, schools have been charged with the responsibility of producing environmentally responsible learners who will be able to handle the demands of an ever increasing pressure of environmental challenges in their society. Ketlhoilwe (2007), in his analysis of the policy goals, argued that the text was based on a linear casual logic that assumes that:

... environmental education would lead to individual attitudinal and social change and the development of desirable behaviour. The text does not take into account unexpected factors that may emanate from economic status or personal or social (communal/cultural) attitudes and experiences that may constrain the achievement of the policy objectives (p. 197).

What Ketlhoilwe is alluding to is that the policy is not clear on the social and situated nature of learning processes and the systems of education and training which both enable and constrain such learning, particularly change-oriented learning in a context where democratic learning environments and transformation should be high on the agenda.

In its 2007 review, the National Environmental Education Strategy and Action Plan (NEESAP) (Botswana Government, 2007), which was tasked with the responsibility of translating the 1994 RNPE into action and of ensuring that environmental education was infused into the national curriculum, responded to the SADC Regional Environmental Education Programme's (REEP) initiatives to meet the ESD objectives of integrating sustainability practices into aspects of education and learning (Lotz-Sisitka, 2006). NEESAP tasked the Ministry of Education (Botswana Government, 2002) through its Department of Curriculum and Evaluation (DC&E) with the responsibility of designing curriculum guidelines for schools to infuse environmental education. The guidelines contain the national goals of environmental education as recommended in the RNPE, and one of the goals relates to the development of 'critical thinking, problem solving ability, individual initiative, interpersonal and inquiry skills to make informed decisions when dealing with environmental issues and willingness to participate in environmental protection and conservation' (Botswana Government, p. 1).

According to Ketlhoilwe (2007), this understanding is consistent with Stevenson's (2007) thinking. Learners, according to Stevenson, should be exposed to processes of inquiry, critique and reflection to develop and defend their own social inquiry and moral deliberations because engaging them in a rational process of social inquiry and moral deliberation would enable them to pursue actions they deem appropriate and justifiable for achieving environmental sustainability (ibid.). This would support learners in formulating a moral code concerning environmental issues and in developing a willingness to act on their personal values by participating actively in environmental education (ibid.). Hence, one of the specific objectives of environmental education in the guideline document is for learners to:

- Be able to work towards resolving environmental problems as well as *actively participating* in the care and conservation of the environment (Botswana Government, 2002, p. 2, my emphasis).

This is consistent with the ESD objectives which call for learning processes which have to draw on learners' social and cultural capital as propounded by Tilbury (2011) and Lupele & Lotz-Sisitka (2012). The UN 2005 World Summit Outcome document referred to the 'interdependent and mutually reinforcing pillars' of sustainable development: social development, economic development and environmental protection. The three pillars are intertwined in ESD, and none can be neglected in any teaching and learning process if education is to be relevant and meaningful to learners so as to lead to a change in attitudes, behaviours and values to ensure a more sustainable future in social and environmental terms (Lupele & Lotz-Sisitka, 2012). For Botswana, the hope was that this would be achieved through the participation of learners in environmental education practices as recommended by the NEESAP document.

In Botswana, a number of approaches have been taken up within the learner-centred initiatives that call for the learner to be a key participating agent in the learning process if learning for change for a sustainable future is to be achieved (Silo, 2011). But for Botswana where the education is mediated in a largely authoritarian society and culture (Maundeni, 2002; Silo, 2011; Tabulawa, 1997), children have been largely excluded in genuine participation (Hart, 1997). In a study (Silo, 2011) on learner participation in Botswana schools, findings revealed that due to culturally and historically formed views of environmental education, teachers have consistently wanted children to pick up litter, and this has been their primary environmental education concern. Learners, on the other hand, identified sanitation issues in the school toilets as their primary environmental management concern. Teachers had not considered this as an environmental education concern. A solution to this tension is therefore to open up a dialogue between teachers and learners and amongst the learners themselves.

12.2 The Concept of Dialogue in Learner Participation

Karlsson (2001, p. 212) defined dialogue as an exchange of opinions, information, ideas and meanings, in which the purpose is learning; it is more than simply utterances. He saw the discourse of dialogue as 'a spontaneous movement between asking and answering questions about issues that pertain to each other's lives' with the aim of establishing, maintaining or developing social contact. According to him, a dialogue should take the form of engagement that takes interlocutors (Cheyne & Tarulli, 1999) beyond mere discussion characterised by unidirectional verbal exchange. Learners should rather decide collaboratively how things are or should be. This should respond to the dialogue that appears to be missing in Botswana schools' environment teaching and learning process activities where teachers prescribe rules and ascribe roles to learners without democratic consultation (Silo, 2011).

According to Silo (2011) and Ketlhoilwe (2007), the participation of learners in environmental learning activities falls far short even of this definition as it is mainly characterised by unidirectional instructions from teachers to children. The way teachers engage with children in teaching and learning in environmental activities is merely 'technical'; teachers use their authority in mediating participation of learners by telling them what they should and should not be doing (Ketlhoilwe, 2007; Silo, 2011). Critical dialogue between teachers and children has seemingly been stifled by the traditional cultural value systems (Silo, 2011; Maundeni, 2002; Tabulawa, 1997) that have pervaded pedagogical practices in Botswana schools. This has led to narrow forms of learner participation. As a result of the lack of dialogue in their interactions in the environmental management activities, there is a resultant parallel social interaction and a mismatched shared objective for the activities between teachers and learners (Silo, 2011).

Genuine dialogue should be concerned with an exchange of ideas that allows for mediated forms of participation of learners. Such a process should also embrace the learners' needs and diverse views as well as the views of the teachers (Hart, 1997; Jensen & Schnack, 2006; Simovska, 2008) which is a characteristic feature of democratic pedagogy (Mogensen & Schnack, 2010; Schnack, 2008). Dialogue in the participation of learners should be looked upon as a 'locus of democratic process' (Karlsson, 2001; Lansdown, 2001) that allows an open line of communication between learners and teachers.

12.3 The Role of Dialogue in Engaging and Scaffolding Participation of Learners

Participation of learners in environmental management activities is a critical concept of environmental and ESD learning processes that are located within social learning and its role in the learner's development of action competence (Jensen, 2002, 2004; Jensen & Nielsen, 2003; Jensen & Schnack, 2006). Development of action competence relies upon enhancing the learners' cognitive structures and requires the kind of instruction that Palinscar (1986) described 'as a joint venture between students and teachers [who] share responsibility for learning and refining [these cognitive], strategies' (p. 73). Simovska's (2008) work provided a three-pronged framework that highlights the quality of learner participation between learners and their teachers. This work identified (i) the focus of learning activities in which learners are participating, (ii) the expected outcomes of the activities and (iii) the target of change for their participation. Her framework drew on the democratic approach to participation that is key to action competence development and socio-cultural approaches to learning which highlight the importance of collaborative participatory approaches to learning in activities in the learners' social contexts. Palinscar (1986) also highlighted the importance of dialogue and reflexivity in participatory approaches necessary for the transformative process needed to respond to social-ecological issues in learners' communities and society at large.

As part of environmental education and an ESD learning process which is mandatory in formal education in Botswana, schools should be able to explore opportunities to engage learners in environmental learning activities that contribute to learners' ability to act and effect change in their environments (Carlsson & Jensen, 2006; Jensen & Schnack, 2006).

In the research study reported in this chapter, an action competence model was used to guide learner engagement as it provided an appropriate framework for a dialogical, democratic way of engaging learners (Jensen, 1997, 2004; Jensen & Schnack, 2006). The model was adapted and used, encouraging learners to identify

Table 12.1 The action competence development cycle

Component of action competence cycle	Area of focus
(A) Selection of environmental management themes (issues, problems and concerns)	What are our issues of concern/problems?
	What are the causes of the problem?
	What influences are we exposed to and why?
	Why is this important to us?
	What is its significance to us/others? Now/in the future?
	What influence do lifestyle and living conditions have?
	How were things before and why have they changed?
(B) Vision building	What alternatives are imaginable?
	What alternatives do we prefer and why?
	How can we contribute to these alternatives?
(C) Activities (action and change)	What changes will bring us closer to the visions?
(Direct and indirect actions)	Changes within ourselves? In the classroom/school? In the community?
	What action possibilities exist for realising these changes?
	What barriers might prevent the undertaking of these actions?
	What barriers might prevent actions from resulting in change?
	What actions will *we* initiate?
(D) Evaluation	How will we evaluate those actions?
	What comes out of this evaluation?

and select problems or issues of concern, envision possible solutions and then, based on these, take action to address these problems through selected activities and then reflect on and evaluate their new activities in an action competence development cycle (see Table 12.1). This provided a methodology for guided participation through dialogue.

While it might appear that the first three components of the action competence development cycle (see Table 12.1), i.e. selection of themes and identification of problems, envisioning and action taking, were all undertaken in a linear way, practically these components formed an integral part with the fourth component of the action competence development cycle, reflecting and evaluating change, that occurred throughout the study and even after.

Using the action competence framework shown in Table 12.1, a case study of a primary school is discussed. The aim was to explore opportunities for engaging learners to identify issues and concerns relating to environmental problems in their school, envision what their aspirations were for a healthy environment and take action by generating solutions to the identified problems. The study focussed on a case analysis of one primary school in a peri-urban area, characterised by informal settlements which emerged as a result of illegal occupation of land, leading to overcrowding and poor sanitation conditions.

12.4 The Research Process

The case study involved a small group of eight primary school learners ranging from age 11 to 12. The selection was done by the teachers, based on mixed parameters, which included academic aptitude, character disposition such as shyness and outspoken individuals, introverts and extroverts and social background (Hennessy & Heary, 2006). Five girls and three boys were selected and consent from learners, parents and school authorities was given (Masson, 2000; Yin, 2009).

The data were largely generated from focus group interviews with learners and observations of learner activities. These two methods were complemented by interviews with teachers and other actors in the school where this was deemed necessary, learners' activities and work, learners' notes, as well as show-and-tell explanations by learners themselves. Focus group interviews with learners were chosen instead of individual interviews as these provided children with the safety and support of their peers, an environment within which there was a power balance, a crucial factor for optimal participation (Green & Hogan, 2006). Focus group interviews with children are also important because, according to Levine and Zimmerman (1996, cited by Hennessy & Heary, 2006), focus group interviews acknowledge children 'participants as experts. Thus a child participating in a focus group should not feel that he or she is being questioned by an adult but rather that he or she is sharing experiences with a group of peers' (p. 239).

Observations of learners busy with activities that they had initiated allowed the researcher to examine the type of activities chosen as well as the ones they would have liked to initiate but were discouraged from doing so (Tudge & Hogan, 2006). Observational methods also allowed for examination of some key aspects of what learners did to start activities, how they engaged in those activities and how the school community responded. Engaging and guiding learners in dialogue took the form of using questions to engage learners in the action competence development process (reflected in Table 12.1).

12.5 Learner Activities

As part of the action competence model, the benefits of taking concrete action during the participation process were stressed to learners. Learners were encouraged to develop solutions to the problems that they identified. The extent to which their activities involved some degree of action was analysed according to the children's attempts to take the initiative and to take the action whether direct or indirect (Jensen, 2002; Jensen & Schnack, 2006). They initiated and participated in a wide range of different types of activities through various actions. Some were direct concrete actions and some indirect: all actions formed a vital step in the action competence development process. These are summarised in Table 12.2.

12.6   The School's Action Competence Development Cycle

This section presents detailed empirical results and evidence of children's voices through dialogue in the action competence that developed (or lack of it) in the environmental management activities initiated by learners in the school, mainly guided by me with the help of their teachers. The main objective of the section is to present and analyse activities carried out by learners using the different components of the model of action competence outlined above. The criteria for presentation and analysis of evidence of action competence described dynamics of learners' activities that:

- Considered environmental management problems and proposed actions which were chosen jointly by the participating learners
- Selected environmental management problems which provided a scope for a local solution
- Involved concrete actions (direct or indirect) on the part of learners as integrated elements of the participation process
- Involved the development of new human relationships, i.e. social capital, in the community as a consequence of the activities
- Involved the strengthening of insight, commitment and visions, i.e. action competence, on the part of the participating learners
- Shared dialogue between participants and/teachers, including a common understanding of the processes and aims of activities (Jensen, 2004)

Table 12.2 below provides an outline of the selected environmental management issues identified, as well as the solutions and activities envisioned by learners in the school.

12.7   Learners' Identification and Selection of Problems
and Their Visions

Table 12.2 highlights the action competence cycle for the school. The first step that learners were guided through involved developing a critical starting point and discussing issues and problems that were of major concern in the school environmental management activities. They were encouraged to discuss these in order to identify those that needed attention, stating the causes of the problem, how it affected them

Table 12.2 The action competence cycle in the school

Component of action competence cycle	Area of focus
(A) Environmental management concerns	Issues, problems and concerns
Poor toilet sanitation	Inadequate and poor maintenance of toilets; poor usage of toilets by children; the use of area outside toilets for sanitary needs; litter pick-ups around toilet area without gloves; lack of proper sanitary facilities in girls' toilets; shortage of toilet paper
Poor maintenance of school infrastructure	Locked up non-functional toilets; old school buildings with peeling paint which makes the school ugly while other schools are well maintained; broken windows, locks and doors largely caused by children but not repaired; overcrowded small classrooms which make cleaning difficult; shortage of cleaning material resources; lack of plants and flowers which makes the school look ugly; broken fence and gate allowing goats into the school
Poor litter management	Littering caused by children buying from vendors outside school gate
	Lack of bins in classrooms and inadequate litter tanks for the school
	Council's failure to regularly collect and dispose of litter
Poor maintenance of grass	Uncleared grass which provides a hide-out for children to use for sanitary needs
	Provides breeding environment for mosquitoes
Lack of teacher support	Teachers not listening to children when they need help nor taking their welfare and concerns seriously
(B) Vision building	To be allowed to contribute in making their school a clean, sanitary healthy and aesthetic environment, e.g. writing to a builders shops requesting paint
Children's visions	To open a school tuck shop so that children are not allowed to buy from vendors outside the school gate
	To create a culture in which they could freely approach and communicate with their teachers and school head to discuss their issues and problems
	To be allowed to clean toilets and clear grass as they used to do previously
	To ask teachers to support, respect and listen to them when they ask for help
(C) Activities *Action and change* (direct and indirect actions)	Meeting with their teachers and school head to discuss their issues and problems
	Addressing other children in class and at assembly
	Toilets, classroom doors, windows, locks fixed
	Tuck shop opened but operated by teachers
	Grass partially cleared
(D) Evaluation	Need to be partners with their teachers in evaluating activities
	Actions still largely undertaken without learner consultation
	Development of children's social interaction skills; conflict resolution and consultation

and others and its significance. According to Jensen (1997, p. 422), the major first component of action competence development in learners 'is concerned with pupils acquiring a coherent knowledge of the problem of concern to them – a knowledge about the nature and scope of the problem, how it arose, who it affects and the range of possibilities that exist for solving it'.

Their visions were mainly centred on creating a school culture with balanced and responsible interaction between them and their teachers and other learners in order to collectively address environmental management issues that affected them. They also considered the contribution that they could make towards the activities in the school. The main issues of concern and problems that were selected and their envisioned changes and alternatives were captured under the themes discussed in the following section.

The following extracts illustrate how this process was opened up:

Extract 1 – Researcher (R) asking learners to open up the discussion on learners' environmental management issues in the school:

R: Do you like your school?
L: Nooo!!! [chorus]
R: What is it that you don't like about your school?
L1: I don't like my school because as you can see it doesn't look nice. I'm worried about toilets, the toilets are so dirty.... ...
R: Can you tell me more about that? Are you saying the council truck doesn't come at all?
L: It comes, but sometimes!! [A couple of them responding at the same time] (FLI)

In all the problems identified, they had to ultimately agree with each other, and to do this I had to keep supporting their ideas and settle their arguments by making reference, for example, to their photographs or drawings as illustrated in the next extract:

Extract 2 – Researcher supporting and moderating argument on prioritising problems with learners:

R: If you were to put those issues in order of importance which ones would you say are the major problems and why?
L5: Leaking taps...
L1: No, dirty toilets, leaking taps are better
L3: I think littering everywhere...
R: Who litters everywhere?
All: Children!!
L5: But we are always made to pick up litter every morning, but at the end of the day there will still be litter everywhere. We follow a litter pick-up rota.
R: So littering is your main concern?
L6: No!!! Dirty toilets please what do you think?
L1: Because we must have a healthy environment
L5: Leaking taps because we cannot live without water.
L1: We can!
L5: Yes, we can't
R: Why don't we look at the photos and see which ones you took pictures of most?
L7: Toilets   L1: Toilets!!   L3: Toilets!! (FLI)
All: Toilets!!!
L: And there is only one lady who cleans them and she doesn't manage cleaning them all the time, so they are Aaaaagggg!!!

12.7.1 Sanitation as a Health and an Environmental Issue

It was clear that learners in this school were concerned about the environmental health condition of their school with particular emphasis on toilet sanitation (see Table 12.2). They largely blamed this poor state of sanitation to lack of care and concern for their welfare by their teachers whom they felt did not take their plight seriously. In this school, they noted that poor environmental management was due to inadequate labour, with only one cleaner for a school as large as theirs. All this pointed towards the children's appreciation of the aesthetic value and health of the environment. Through dialogue, learners in the school could directly or indirectly have contributed towards the solutions to the problems. Some examples of the ways learners identified their concerns and how they were supported through dialogue are illustrated in the extracts below:

Extract 3 – Learners' problems and concerns:

R: Of the tasks that you do, which one do you find difficult?

L4: The difficult one is where we pick up litter without gloves around toilets and sometimes you find soiled toilet papers which has been used and you are told to pick it up and we complain that it can cause you disease and it's a hazard to your health we are told 'what do you know, just pick it up?' And we are not given gloves.

L3: The lady who cleans toilets is also overburdened because there is a lot of work in those toilets. When girls use toilets, they leave without flushing them and they are not nice to look at.

R: So in other words what you are trying to say is children also don't use toilets properly?

L1: What I am trying to say is that toilet paper is not placed in the toilets for us and so the children will then use the toilets carelessly.

L1: They [older girls] are showing kids horrible things [soiled sanitary pads]. Because after that, they [little children] touch and then go and eat food without washing hands, because they don't know what these things are. They should see what to do about this (FL).

The emphasis on sanitation revealed that the children had knowledge of, concern about and were fully aware of environmentally related health issues, their causes and how they affect them. They were able to link these to environmental issues affirming that health and environmental issues cannot be separated.

12.7.2 Teacher Support as an Enabling/Constraining Factor in Creating Dialogue in Action Competence Development

Because learners were neither afforded opportunities for engagement with their teachers nor were their needs and views ever discussed with them, they developed assumptions about teacher actions. This concern was addressed with their teachers and their desire to contribute positively to the environmental welfare of their schools. It was this tension that I encouraged learners to focus on in their envisioned attempts to address their school's environmental management problems. Extract 4 is an example of how this was supported:

Extract 4 – Tensions in learners' perception of teachers as mediating tools:

L2: This issue, really Mrs Silo, we can improve by … if teachers can be serious about these issues, it would be easy to address them.

R: So how do you think you should approach them to make it easier and bring in teachers and have them take these issues seriously?

L6: They can first take us seriously and listen to us and also be involved in the contribution towards what we want to do, say each teacher paying P20. (AFL)

L6: I mean imagine for us girls, only two toilets for so many of us!! How many are we? I guess more than 500 because I know we are more than boys (AFL).

R: Let's hope that after we have compiled a list of all these problems you will go and talk to them, and they will listen, you will approach them, won't you?

L3: We are afraid of them, they won't listen.

R: Remember I have talked to them about this project, so they know that you want to come up with solutions to the problems in the school, they will listen. Because they are also interested in what you are doing.

L: Aaah!

R: You don't think so?

L2: Maybe when you talk to them yourself they will start to listen to us.

R: We will try and see (FL).

For learners, understanding and being involved in the identification and selection of problems, being allowed to come up with their visions and suggestions for solutions to the problems, was not merely an interactive exercise, but rather an exercise for learning how to interact positively with their teachers and each other in the whole school. This was the starting point of a decision-making process and development of responsibility that would contribute towards their action competence development. Jensen (1997) contended that 'the fact that they have been given the opportunity to develop, discuss and share their visions with others – or perhaps participating in developing a common vision – is perhaps one of the prerequisites or precursors of the desire to act' (p. 423). Learners did not seem to be concerned about or uncomfortable with approaching their teachers after I had explained to their school head or deputy and teachers what the research entailed and solicited their support.

12.8 Vision Building

Having identified relevant issues of concern in the environmental management activities, learners outlined their visions and suggested solutions to realise these visions. The aim was to enable them to find solutions to the problems they had identified in a democratic way which is the central feature of action competence and which is also consistent with the ESD initiative:

> In doing this, one key role for ESD in an action competence approach becomes that of developing the students' ability, motivation and desire to play an active role in finding democratic solutions to problems and issues connected to sustainable development. (Mogensen & Schnack, 2010:68)

Through questions, learners were encouraged to develop their ideas and perceptions about their envisioned future (Jensen, 1997), participation, their roles and how they could do things in alternative ways to the normalised approaches that had been characteristic of the school culture over the years. One of the learners wrote the group's ideas on a flip chart, while another noted them on paper. I then offered to compile and type the list of theme areas of concern and suggested solutions. Having compiled the list, I suggested that learners approach their teachers and either the school head or deputy to discuss the list further, which they did. I also followed up with the teachers and the head of school to follow up on their discussion.

12.9   Learner-Initiated Activities

This section outlines activities and actions that took place as a result of the direct learners' actions or indirect influence as an initial attempt to fulfil their visions on a social level. In response to their identified problems and in an attempt to fulfil their visions, the activities that learners undertook assumed different forms (Jensen, 2002; Jensen & Schnack, 2006). They started by modelling solutions (see Table 12.2).

12.9.1   Modelling Solutions

Having identified problems, the learners suggested various solutions. As the researcher, I guided and supported them in the focus group discussions, but they were free to adjust and change these solutions. Some examples of this process are captured in the extracts below:

Extract 5 – Learners' suggested solutions:

R: Can you see that we still can't come up with a specific solution for the toilets, but toilets seem to be the biggest environmental management problem you seem to have in the school. How will you ensure that they are clean and fixed? What is the first step that you will take?

L3: We will inform Mrs K or the head-teacher.

R: Good! You can go and tell them about your concerns, the things that you are worried about. And then with the children, what will you do?

L2: We can show them how to use the toilets.

R: You show them how to use the toilets and …?

L1: The toilets can be locked and every class can be given a toilet…

L3: A toilet?

L1: …and the key for that toilet, every class having a toilet for itself so that they should make sure that the toilet is taken care of and lock it. They do that in [another school]

R: That's good. You are coming up with good ideas. So that when the toilets are fixed, each class will have its own toilet. I think what she's suggesting is that every class should be assigned a toilet. It should be given a toilet that they will be using and they keep the key to their toilet. So that they can monitor it. That what you mean isn't it?

L3: Yes, that's a good point.

R: I agree, that's a good point (FL2).

After this, the learners drew up a list of their concerns in consultation with other learners in the school and suggested solutions which they gave to their teachers and heads. They called regular meetings with their peers, and I followed this up and requested that teachers provide necessary support to learners in their activities.

12.9.2 Opening Dialogue with Teachers

Having furnished either their teachers or the school head with the list of their issues and concerns as well as the suggested solutions, learners arranged for meetings and actually met with the teachers who they felt were approachable. The children also met with the school head. These meetings were either preceded or immediately followed by the maintenance of infrastructure, especially toilets. The latter was an indirect influence from learners' submission of the list of their concerns, albeit without further consultation with the learners. But they were given an opportunity to communicate their concerns, and this can be seen as an initial step to open up an interactional dialogue between them and their teachers, something that had never previously occurred in the history of the school (Silo, 2011). This is illustrated in the next extract from the learners' focus group discussions:

Extract 6 – Learners on meeting with their teachers:

R: So after you went to the office, the toilets were fixed?

L1: Yes, last week after we met with our teachers. After we gave Mrs S that paper she arranged a meeting with us and Mrs K and Mrs N [the school head]. They told us that there was money, that the school was given last year by government and it is supposed to be returned in April if it's not used (AFLI).

...

R: So today what can you tell me about what you have been doing?

L3: New doors!

R: What about new doors?

L3: They put new doors and locks in our toilets.

L5: After we met and talked to them about those problems we gave them and they told us that there was money from government they said they will use the money to fix doors and toilets (AFL).

12.10 Evaluation

The evaluation of the activities took the form of continuous reflections by learners on their actions in the new activities. They highlighted both constraints and enabling factors. According to Mogensen and Schnack (2010), from an evaluation perspective, the action competence approach calls for particular attention to self-evaluation. This provides learners with an opportunity in the participation process to assess their successes and barriers in their actions and their own strengths and weaknesses. This evaluation process also involved teachers and other stakeholders in the school community. According to Jensen (2004),

> it is important that a particular action not be viewed as an end-product of an environmental education project. Students must have the opportunity to evaluate, reflect, and restructure their actions – within their project and with their teachers – in order to develop their action-competence. (p. 414)

What was crucial in this phase was for children to demonstrate the power to act (agency) which is a fundamental characteristic of action competence by assessing and reflecting on their actions and activities. They need to consider barriers and constraints as well as enabling factors, and the objective was to allow them not merely to react to but, importantly, to highlight how these activities could have changed their material and social worlds.

The evaluation phase of the children's activities, which involved participant learners, teachers, school head, cleaners and other school children, addressed the various changes that resulted from and were facilitated by the children's activities:

> It is crucial that students have the opportunity to evaluate, reflect on, and restructure their actions – within a certain environmental education project and together with their teachers – in order to develop their action competence. (Carlsson & Jensen, 2006, p. 242)

The evaluation process further assessed whether some of the identified problems were attended to and whether learner visions were realised or not. It did this by placing the barriers and constraints as well as enabling factors into perspective, all of which are contributory factors to action competence development and children's empowerment (Jensen, 2004).

Furthermore, development of social skills and communication between and amongst children and their teachers and other members in the community were evaluated as another crucial aspect of action competence development. The evaluation findings were very mixed but mainly positive about the children's experiences. While generally both learners and teachers spoke warmly of the authentic and action-oriented aspects of the project (Jensen, 2004), they did however highlight a number of barriers that confronted them in its development.

The teachers felt that there was an improvement in the way that children understood environmental management issues. They could now rationalise and make sense of them by being directly involved and by being given an opportunity to talk about them. They now conceptualised waste in a broader and more meaningful sense and understood it in context:

Extract 7 – Teacher evaluation of learners:

> T: *Your project has helped a lot in that you can see that these children are now able to rationalise and understand that when we talk about environmental management what we are actually talking about, not only papers, but food, toilets the neighbourhood etc. (TI)*

The children also observed that because their teachers could see the positive outcomes from the children's initiatives, this had instilled some trust in them:

Extract 8 – Learners' evaluation:

> *L: I think they now enjoy doing what they used to do more because they feel that teachers are not forcing them because they are now being told by their peers whom they are free to talk to and freely voice their concerns, unlike when they are forced to work because they are afraid of teachers whom they cannot question (FLI).*

Furthermore, development of social skills and communication between and amongst children and their teachers and other members in the community were evaluated as another crucial aspect of action competence development which also served to evaluate changes in how learner participation was mediated. While generally both learners and teachers spoke warmly of the authentic and action-oriented aspects of the project (Jensen, 2004), they did however highlight a number of barriers that confronted them in its development.

12.11   Discussion and Findings

This research demonstrated that participation of learners can be rhetorical without dialogue in that teachers tended to view participation of learners in environmental management processes in a very limited and narrow way. There was little connection made between how environmental activities that had all along been undertaken in the school solve the environmental problems at this school, or why they were being undertaken in the first place. In some cases, children, consequently, saw these tasks as hard labour and not as a learning activity. This use of children in this way has its foundations in a culture where children must be submissive to their elders and where it is deemed unnecessary to explain one's motives to them (Maundeni, 2002; Silo, 2011). This is part of a broader problem, too, where environmental education processes are reduced to involving children in technical activities such as clean-ups and litter pick-ups without the necessary learning support to understand *why* such activities are worth undertaking. In this way, children may have been used to promote an agenda which they did not fully understand or support. Even though the curriculum has promoted the importance and relevance of participation of learners in environmental education learning processes to respond to the country's environmental challenges, teachers have come to see it as part of the educational provision of theoretical skills which do not necessarily have any relation to the reality around them (Stevenson, 2007).

If learners are not meaningfully engaging in these activities, critical, reflective participation will fail to develop through which learners could become adults that will cope with current socioecological issues, and future environmental problems could result (Jensen, 2004; Mogensen & Schnack, 2010; Simovska, 2004, 2008). This approach must be seen in connection with whether it is developing learners' will and ability to be involved in environmental management issues in a democratic way, by forming their own criteria for decision-making and action choices. For example, in this study, engaging learners in a dialogue enabled fuller and broader participation by learners in environmental management activities which brought to the surface issues that concerned them in school discourses and environmental management practices.

Another related problem that surfaced was that the teachers' concept of participation lacked the understanding of dialogue as necessary for supporting the engagement and scaffolding of participation of learners. This in turn gave rise to a failure to realise the importance of distributed learning (Russell, 2002) which lies in the learners' ability to analyse the dynamic interpretation of environmental management as both an environmental and a health issue, a gap that teachers should have mediated by understanding, constructing and extending the learners' construction zones of learning (Rogoff, 1995). Learning understood this way considers starting with the learners' understanding of their reality of the environmental issues as they encounter these issues and stretching learners to a higher level of discourse. This stretching should incorporate the scaffolding of the learners' deeper meaning making and levels of participation in ways that broaden their understanding of environmental issues. This takes place mainly through a dialogue, which was a missing aspect in the environmental education teaching processes in the case of the school in this case study. A good strategy would have been to place instruction and learner participation in a meaningful and broader context through which knowledge is continually (re)negotiated (Russell, 2002) between teachers and learners to bring to the fore this interrelationship. Learning within this form of scaffolded participation then is not a neat transfer of information which limits environmental education, to environmental issues only but also includes health issues, development of social skills such as critical thinking and leadership.

The action competence framework aided in developing conceptual tools to understand the need for dialogue and the interrelated nature of these concepts in environmental management and participation of learners in these activities in the school. This research revealed the gap created by a lack of dialogue in the mediation process. For example, children's concern and linkage of health-related issues to environmental issues through their emphasis on unhealthy toilets and insistence on the use of gloves for picking up litter and waste disposal is a link that the action competence approach emphasises (Jensen, 2004; Mogensen & Schnack, 2010; Simovska, 2004, 2008). This raises the need to realise and treat environmental issues and health issues as not only interrelated but also fundamentally connected to social, cultural and political aspects of environmental education as propounded by the SADC-REEP ESD initiative and the RNPE imperative. This is in full harmony with the action competence approach, and aligns well with its broader insistence of understanding environmental problems as societal issues constituted by conflicting interests (Mogensen & Schnack, 2010). A participatory and action-oriented approach highlighted this relationship between health and environmental education by providing basic knowledge and insight around this relationship in the ESD learning processes. This obviously creates important demands and challenges for teachers. They should be in a position to fulfil the consultant and supportive role through dialogue and be able to perceive today's health and environmental conditions from an intersubjective and action-oriented point of view (Jensen, 2004). This in turn, requires giving attention to such issues in teacher education.

Finally, neglect of dialogue and lack of understanding of participation compromise the development of action competence. It emerged from this study that teachers have historically used the self-governing normalising strategies in policy implementation documents (Ketlhoilwe, 2007) to create new technologies of power in the school and in their response to the policy imperative to mediate learner participation. Furthermore, it emerged that teachers, using their authority, emanating from the Tswana cultural influence (Maundeni, 2002; Silo, 2011) attempted to meet the policy imperative through prescription of rules and ascribing roles to children in environmental management activities without any form of dialogue with learners. This gave rise to an elusive objective of children's participation, as the purpose for their participation in these activities was not clear. In fact, it failed to achieve the very object the policy imperative sought to achieve, i.e. participation of children as competent stakeholders in these activities (Barratt Hacking, Barratt, & Scott, 2007). This, coupled with their narrow concept of participation of learners, constrained action competence development in learners. Upon opening up dialogue between learners and teachers and other learners, one achievement was better relationships within the school community. And with improved communication came better ideas to solve environmental management issues that the community still faced on a daily basis. Newly devised solutions were practical and had a broader impact than the litter pick-ups and clean-ups that teachers had always focussed on.

Learners seemed to be developing not only a better understanding of the environment but also the ability to resolve conflict amongst themselves and with their teachers. They felt more confident and more equipped to consider changes in their environment. Because learners were given the chance to share their ideas with their teachers and other learners and saw that they were taken seriously, they also learned that others had a right to be heard and taken seriously, and that they must also be respected (Lansdown, 2001, p. 7). This resulted in enhanced social cohesion which strengthened their action competence development. By engaging in dialogue with teachers and other children, learners became co-catalysts for change in the school community.

12.12   Conclusion

Expanding the ESD learning process by researching and supporting children's participation through opening up a dialogue and action competence development made available a broader range of possibilities and ideas around participation. One significant achievement of this open dialogue was a better relationship within the school community. And with improved communication came better ideas to solve environmental management issues that the school community still faced on a daily basis, such as too much litter, for instance. Newly devised solutions were practical and had a broader impact than the initial ones that teachers had always focussed on. They included mobilising the maintenance of toilets and even re-organising the litter management that had always caused such tension between learners and teachers. Now children seemed to be developing not only a better understanding of the environment but also the ability to resolve conflict amongst themselves and with their elders.

By engaging in a dialogue with children, they become co-catalysts for change in the school community. The research showed that if children's participation is taken seriously, and if opportunities for dialogue exist between teachers and children, positive changes for a healthier environment can be created in the school. Children also appeared to be feeling more confident and more equipped to consider changes in their environment. In reflective conversations with teachers, the research found that this requires critical reflections on teaching practices and support for teachers to identify ways of engaging learners' views on issues in the school in open, dialogical ways. The research also showed that participation in environmental education is more than cognitive changes as proposed in earlier constructivist literature; it involves in-depth engagement with sociocultural dynamics and histories in the school context, such as the cultural histories of teachers, schooling and authority structures in the cultural community of the school.

The environmental management activities in schools presented learners and teachers with opportunities to devise strategies geared towards joint collaborative activity. Teachers could support the learners' participation and capacity development (Jensen, 2004; Rogoff, 1995), using available contextual environmental management issues to advance the level of the learners' actual participation through dialogue. This means that major demands are placed on teachers' abilities to provide the necessary support framework to structure this interaction. This requires careful, close guidance and collaboration with their learners in a way that will develop the required competence and democratic skills (Carlsson & Jensen, 2006) to initiate joint action and dialogue in addressing waste issues in the schools.

References

Barratt Hacking, E., Barratt, R., & Scott, W. (2007). Engaging children: Research issues around participation and environmental learning. *Environmental Education Research, 13*(4), 529–544.

Botswana Government. (1994). *The revised national policy on education*. Gaborone, Botswana: Government Printer.

Botswana Government. (2002). *Environmental education guidelines for primary, secondary and tertiary*. Gaborone, Botswana: Government Printer.

Botswana Government. (2007). *National Environmental Education Strategy and Action Plan*. Department of Environmental Affairs, Ministry of Environment, Wildlife & Tourism.

Carlsson, M., & Jensen, B. B. (2006). Encouraging environmental citizenship: The roles and challenges for schools. In A. Dobson & D. Bell (Eds.), *Environmental citizenship* (pp. 237–261). Cambridge, UK: MIT Press.

Cheyne, J. A., & Tarulli, D. (1999). Dialogue, difference and voice in the zone of proximal development. *Theory Psychology, 9*(1), 5–28.

Greene, S., & Hogan, D. (Eds.). (2006). *Researching children's experience: Methods and approaches*. London: Sage.

Hart, R. A. (1997). *Children's participation: The theory and practice of involving young citizens in community development and environmental care*. London: Earthscan.

Hennessy, E., & Heary, C. (2006). Exploring children's views through focus groups. In S. Green & D. Hogan (Eds.), *Researching children's experience: Methods and approaches* (pp. 236–252). London: Sage.

Jensen, B. B. (1997). A case of two paradigms within health education. *Health Education Research, Theory & Practice, 12*(4), 419–428.

Jensen, B. B. (2002). Knowledge, action and pro-environmental behaviour. *Environmental Education Research, 8*(3), 325–334.

Jensen, B. B. (2004). Environmental and health education viewed from an action-oriented perspective: A case from Denmark. *Journal of Curriculum Studies, 36*(4), 405–425.

Jensen, B. B., & Nielsen, K. (2003). Action-oriented environmental education: Clarifying the concept of action. *Journal of Environmental Education Research, 1*(1), 173–194.

Jensen, B. B., & Schnack, K. (2006). The action competence approach in environmental education. *Environmental Education Research, 12*(3&4), 471–486.

Karlsson, O. (2001). Critical dialogue: Its value and meaning. *Evaluation, 7*(2), 211–227.

Ketlhoilwe, M. J. (2007). Environmental education policy interpretation challenges in Botswana schools. *Southern African Journal of Environmental Education, 24*, 171–184.

Lansdown, G. (2001). *Promoting children's participation in democratic decision-making*. Florence, Italy: UNICEF, Innocenti Research Centre.

Lotz-Sisitka, H. B. (2006). Participating in the UN decade of education for sustainability: Voices in a Southern African consultation process. *Southern African Journal of Environmental Education, 23*, 10–33.

Lupele, J. & Sisitka-Lotz, H. (2012). *Learning today for tomorrow. Sustainable development learning processes in Sub-Saharan Africa*. Howick, New Zealand: SADC-REEP.

Masson, J. (2000). Researching children's perspectives: Legal issues. In A. Lewis & G. Lindsay (Eds.), *Researching children's perspectives*. Buckingham, UK: Open University Press.

Maundeni, T. (2002). Seen but not heard? Focusing on the needs of children of divorced parents in Gaborone and surrounding areas, Botswana. *Childhood, 9*, 277–302.

Mogensen, F., & Schnack, K. (2010). The action competence approach and the 'new' discourses of education for sustainable development, competence and quality criteria. *Environmental Education Research, 16*(1), 59–74.

Palinscar, A. S. (1986). The role of dialogue in providing scaffolded instruction. *Educational Psychologist, 21*(1&2), 73–98.

Rogoff, B. (1995). Observing sociocultural activity on three planes: Participatory appropriation, guided participation, and apprenticeship. In J. V. Wertsch, P. Del Rio, & A. Alvarez (Eds.), *Sociocultural studies of mind* (pp. 139–164). New York: Cambridge University Press.

Russell, D. (2002). Looking beyond the interface: Activity theory and distributed learning. In M. Lea & K. Nicoll (Eds.), *Distributed learning*. London: Routledge.

Schnack, K. (2008). Participation, education and democracy: Implications for environmental education, health education and education for sustainable development. In A. Reid, B. B. Jensen, J. Nikel, & V. Simovska (Eds.), *Participation and learning: Perspectives on education and the environment, health and sustainability* (pp. 181–196). Copenhagen, Denmark: Springer.

Silo, N. (2011). *Exploring opportunities for action competence development through learners' participation in waste management activities in selected primary schools in Botswana*. Unpublished PhD thesis, Rhodes University, Grahamstown.

Simovska, V. (2004). Student participation: A democratic education perspective experience from the health-promoting schools in Macedonia. *Health Education Research – Theory & Practice, 19*(2), 198–207.

Simovska, V. (2008). Learning *in* and *as* participation: A case study from health-promoting schools. In A. Reid, B. B. Jensen, J. Nikel, & V. Simovska (Eds.), *Participation and learning: Perspectives on education and the environment, health and sustainability* (pp. 61–80). Copenhagen, Denmark: Springer.

Stevenson, R. B. (2007). Schooling and environmental education: Contradictions in purpose and practice. *Environmental Education Research, 13*(2), 139–153.

Tabulawa, R. (1997). Pedagogical classroom practice and the social context: The case of Botswana. *International Journal of Educational Development, 17*(2), 189–204.

Tilbury, D. (2011). Are we learning to change? Mapping global progress in education for sustainable development in the lead up to 'Rio Plus 20'. *Global Environmental Research, 14*(2), 101–107.

Tudge, J., & Hogan, D. (2006). An ecological approach to observations of children's everyday lives. In S. Green & D. Hogan (Eds.), *Researching children's experience: Methods and approaches* (pp. 103–122). London: Sage.

Yin, R. K. (2009). *Case study research: Design and methods* (4th ed.). London: Sage.

Chapter 13
Improving the Quality of Education Through Partnerships, Participation and Whole-School Development: A Case of the WASH Project in Zambia

Justin Lupele, Bridget Kakuwa, and Romakala Banda

This chapter shows how the provision and integration of water, sanitation and hygiene (WASH) education in schools, through institutional partnerships, participatory approaches and whole-school development (WSD), can bring about significant gains in improving learning environments and learner achievement. The chapter focuses on the US Agency for International Development (USAID)-funded WASH in school programme – Schools Promoting Learning Achievement Through Sanitation and Hygiene (SPLASH) in Zambia. SPLASH is a 4-year project involving over 616 schools and reaching 250,000 learners. The implementation of SPLASH in Zambia demonstrates that adequate and improved WASH facilities improve teacher-pupil contact time as well as enrolment and attendance, especially for adolescent girls as a result of improved menstrual hygiene management. However, this can only be possible through institutional partnerships, participatory approaches and whole-school development.

13.1 Background to the School Water and Sanitation Situation in Zambia

Water and sanitation challenges facing education systems of developing nations are common. Studies show that water, sanitation and hygiene (WASH) facilities in schools improve class attendance, teacher-pupil contact time, teacher deployment

The authors worked for the SPLASH project. They write in their personal capacity to reflect on the project in the context of improving quality of education through partnerships, participation and whole-school development.

J. Lupele (✉) • B. Kakuwa • R. Banda
Schools Promoting Learning Achievement through Sanitation and Hygiene (SPLASH), Lusaka, Zambia
e-mail: lupelejustin@yahoo.com; kakuwabridget@yahoo.com; romakala2@gmail.com

© Springer International Publishing Switzerland 2017
H. Lotz-Sisitka et al. (eds.), *Schooling for Sustainable Development in Africa*,
Schooling for Sustainable Development 8, DOI 10.1007/978-3-319-45989-9_13

and retention and also create an enabling learning environment for children (Duran-Narucki, 2008; SPLASH, 2015). WASH also creates a healthier environment for learners by reducing the incidence of diarrhoeal diseases, helminth infection, schistosomiasis and other diseases. Duran-Narucki observed that schools with improved physical conditions are conducive to learning and will stimulate academic performance in foundational subjects such as reading and mathematics. Healthier pupils and teachers will spend more time in class and on task (Taalat et al., 2011). Adolescent girls are more likely to go to and stay in schools with adequate sanitation facilities (SPLASH). All the above should lead to learning achievement in the target academic subjects. In their systematic literature review, Jasper, Lee and Bartram (2012) argued that inadequate water and sanitation facilities in schools may be a major hindrance to the achievement of the Millennium Development Goals (MDGs). They suggest that schools in developing and developed countries that lack adequate water and sanitation services are associated with potential detrimental effects on health and, as a result, school attendance. This systematic review, which included 47 studies, came to the following major conclusions:

- Inadequate drinking water contributes to inadequate hydration, which may be associated with decreased physical activity and cognitive capacity, weight gain and urinary tract infections.
- Post-pubescent school girls and female teachers face challenges to school attendance during menses, due to unavailability and affordability of sanitary materials and the lack of adequate school facilities (e.g. lack of gender-segregated toilets, no running water, broken toilet doors).

The prevailing situation for the provision of WASH facilities in Zambian schools is below the Ministry of General Education standards of a pupil/toilet ratio of 50 pupils to 1 toilet for both girls and boys. Further, the (Zambian) National Development Plan (adopted in 2011 and revised in 2014) identified the need for safe drinking water, good governance as well as water resource management as necessary for sustainable development (Ministry of Finance [MOF], 2011). This is along with recognition of the need for an integrated approach towards water resource management and WASH more generally. The integrated approach calls for more partnerships and coordinated effort. For example, the Zambia National Food and Nutrition Commission (NFNC) recognises the need to strengthen the partnership between the Commission and the water sector. According to NFNC (2015), children who lack access to clean water do not compare positively with those that have this access (NFNC). The national policy on education, *Educating Our Future* (Ministry of Education [MOE], 1996), noted that half the rural schools in Zambia do not have safe drinking water and adequate sanitation facilities. Schools in urban areas are overpopulated. The school enrolment outstretches the available resources. It is for this reason that the Ministry of Education encourages participation and partnerships in the education provision in Zambia (MOE). The project works within the Ministry of Education (MOE) and partners with other line ministries such as the Ministry of Local Government and Housing (MLGH), the Ministry of Community Development Mother and Child Health and the Ministry of Health (MOH).

13.2   Project Description

The US Agency for International Development (USAID)-funded Schools Promoting Learning Achievement Through Sanitation and Hygiene (SPLASH) project in Zambia was designed and implemented in line with the USAID education policy and the goals of the water and sanitation sector in Zambia, as a response to the water and sanitation situation in schools in four districts (Mambwe, Chipata, Lundazi and Chadiza) of the Eastern Province of Zambia. SPLASH operated within the School Health and Nutrition (SHN) programme framework of the Ministry of General Education, which articulates a relationship between health, nutrition and education. SPLASH supported the WASH component, which is a critical but under-represented aspect of SHN. It helped to overcome limitations of inadequate human and financial resources and offered a model school WASH programme by implementing these five strategies:

1. Instal and rehabilitate improved drinking water, sanitation and hygiene infrastructure in schools.
2. Improve the hygiene behaviours and health of learners, teachers and subsequently their communities.
3. Strengthen local governance and coordination of WASH in schools.
4. Engage those who set policies at the national, provincial and district levels to support WASH in schools.
5. Strengthen the capacity of small-scale service providers and the private sector to deliver WASH goods and services to both schools and communities on a sustainable basis.

The implementation of SPLASH shows that investing in WASH infrastructure in schools improves learners' opportunities to learn and, subsequently, education quality. By its nature, school WASH fits well within sustainability discourse, raising questions of participation, behaviour change and value commitments of democracy, equity and inclusion.

In line with education for sustainable development (ESD) principles (Lupele, 2002), SPLASH worked with strong partnerships and participatory approaches in schools in the target districts. The project involved its partners in a range of activities including participatory planning, teacher education, school community linkages, whole-school development, curriculum integration and behaviour change outreaches. These practices aimed to improve sustainably access to safe water, adequate sanitation, hygiene information and health practices, thereby improving learning environments and educational performance in primary schools.

This chapter focuses on the following features of SPLASH:

- Participatory approaches
- Whole-school approaches
- Partnerships
- Levels of intervention

13.3 Participatory Approaches

In most education and development work, lip service is all that is paid to participation. Stakeholders are said to contribute to the programmes/project through participation at very superficial levels. Barrow and Murphree (2001) outlined six 'levels' of participation in the development programme context as shown in Table 13.1.

Based on this typology, Barrow and Murphree (2001) noted that the common forms of participation are consultative and functional. However, from the trends in development work and in dealing with government institutions, there is always a temptation to engage in the passive, information giving and consultative (Barrow & Murphree). Globally, development work has been centred on experts providing information on how things run or work and perhaps what tools and avenues are available for project implementation. The local beneficiaries (and to some extent project staff) are rarely involved in decision-making processes (Lupele, 2002). It is assumed that the experts, 'often' foreigners to the land, culture and context of the people they wish to serve, know the problem and have the solution.

However, within SPLASH, stakeholders were fully engaged at various levels of project implementation. The vast range of SPLASH partners adopted different roles and responsibilities in the participatory processes.

SPLASH engaged its partners in the last three 'levels' of Table 13.1, namely, functional, interactive and self-mobilising participation. While it can be problematic to 'neatly categorise' participatory processes in this way, we use Barrow and Murphree's (2001) table as a way of reflecting on the nature of participatory processes in the SPLASH project. The mixture of functional, interactive and empowerment orientations to participation is reflected at different levels of the education system in Zambia, i.e. at central Ministry of Education, provincial, district, zone and school levels. The project engaged community members, traditional leaders, private companies, NGOs and other government workers, especially in the mobilisation of resources such as labour and building materials. For example, institutional structures

Table 13.1 Typology of participation (Adapted from Barrow & Murphree, 2001)

Participation typology	Roles assigned to local people
Passive	Told what is going to happen or already happened
Information giving	Answer questions from extractive researchers. People not able to influence analysis or use
Consultative	Consulted. External agents listen to views. Usually externally defined problems and solutions. People not involved in decision-making
Functional	Form groups to meet predetermined objectives. Usually done after major project decisions made therefore initially dependent on outsiders, but may become self-dependent and enabling
Interactive	Joint analysis and actions. Use of local institutions. People have stake in maintaining or changing structures or practices
Self-mobilisation or empowerment	Take decisions independent of external institutions. May challenge existing arrangements and structures

and their functionality and commonly held beliefs of what works were subjected to ongoing critical reflection. These participatory processes yielded results in the construction of water and sanitation facilities in schools in the project areas. Through these and other processes, and by means of joint interactions and objective analysis, SPLASH and its partners therefore made significant learning gains.

Freire's (1970) teaching principle of working from people's own concerns and knowledge as a participatory approach to development is one that was demonstrated within the SPLASH project. This approach calls for project implementers that have knowledge of the sociohistorical and cultural contexts within which participating schools are situated. Where this knowledge is lacking, buy-in of expertise is always key to the success of participatory processes.

By working with policy makers at various levels (national, provincial, district and traditional), SPLASH learnt the need to engage more interactively beyond superficial propaganda or development-related social marketing. This was achieved through stakeholder participation in joint planning of tasks (during quarterly planning meetings), actual construction, planning of the pedagogical processes of learning and teaching and especially the Ministry of Education's in-service continuous professional development system.

As part of the organisation's learning culture (Senge, 1990), SPLASH learnt that propaganda[1] as an approach to participation may not be helpful to sustainable development and WASH behavioural change (Parker & Adler, 2013). While propaganda seeks to use events to win people's uncritical acceptance of key WASH messages, the project and its key partners have realised the need to work with ESD principles as a key dimension of the multifaceted concepts of quality education. Thus SPLASH's 'WASH messaging' of events take more educational approaches to attitude and habit formation, supported by evidence of changing hygiene behaviours. SPLASH approaches take into account the cultural concerns of, for example, males and females sharing a latrine, or the elderly sharing a latrine or defecation bushes with younger children, or concerns over myths and taboos associated with menstrual hygiene management. SPLASH approaches aimed to encourage critical thinking, correction of misconceptions, input and dialogue. This is in contrast to the development of uncritical skills and behaviour change, instruction, dictation and specialisation that propaganda-based approaches may encourage (Cooke & Kothari, 2001). McKernan (1991) noted that participation on 'the other's terms' can be disempowering, and Cooke and Kothari (ibid.) argued that the 'environment' in which participation is employed must enable participants to critically engage with the assumptions of those who want them to participate.

---

[1] Propaganda is a term for communication that is aimed at achieving people's uncritical agreement with certain messages.

13.4 Whole-School Development

Whole-school development (WSD) can be seen as a mechanism to improve the academic, infrastructural, social and safe learning environments in schools. It aims to ensure that schools have the necessary management and community leadership to support an environment conducive to excellence in teaching and learning (Shallcross, Loubser, Le Roux, O' Donoghue, & Lupele, 2006). In most development work, children are seen as beneficiaries who are at the receiving end of donor support and community initiatives. While this may be true, it is generally accepted by most development agencies that there is more ought to be done to involve children beneficiaries in problem-solving and project activities that concern them or work for them. Whole-school approaches imply that the concern shown for environmental problems such as WASH in the formal curriculum is, whenever possible, reflected in day-to-day practice in a school's non-formal curriculum. This means that values and attitudes (such as handwashing) taught in the classroom help to form habits in the daily actions of teachers, pupils and support staff. WSD places emphasis on the fact that schools should practise the values and habits they teach (e.g. Shallcross et al., 2006).

SPLASH worked with the concept of WSD by exploring the potential of children as change agents in the promotion and supporting of a comprehensive WASH programme in the intervention schools (Bresee, Lupele, & Freeman, 2014). Pupils organised themselves into WASH clubs under the support and supervision of a teacher patron/matron. The club activities included dialogue and critical reflection on issues such as how to store and distribute WASH consumables like soap, toilet paper and sanitary pads in schools and other topics of interest. An interesting dynamic of the club activities was how pupils used games, songs, dance and poetry to deliver WASH messages to their peers and parents on special occasions such as *Global Handwashing Day*. The WASH clubs proved to be a good platform for grounding WASH-related topics taught during official class time and for assessing and encouraging behaviour change around what was taught in class.

Teachers viewed the WASH activities in the SPLASH schools as their own; a great sense of ownership had arisen from the way they were involved in project implementation processes. Evidence of this could be seen through teachers' development of their own learning and teaching aids that integrated WASH themes. They also included WASH themes/topics in the teacher professional development programme – the School Programme of In-service for the Term (SPRINT). Initial training workshops and classroom teaching of WASH in school adopted participatory practice-centred approaches.

Based on the principle of ESD from the UN Earth Summit in Rio de Janeiro – 'We are all learners and educators' (UNEP, 1992) – the learning and teaching materials, developed collaboratively with participants, were further developed into booklets and teacher WASH guides. This enabled the project to draw from teachers' rich resource of knowledge and experience, as well as localised WASH problem-solving approaches such as action research (Kemmis & McTaggart, 1988). A key common method that teachers used to deliver and discuss WASH was a form of an action research approach called 'lesson study approach'. Lesson study is a participatory problem-solving process of teacher professional development (MOE, 2010). As part of this approach, teachers supported integration of WASH topics in classroom teaching, aided by the MOE syllabi. Small groups of teachers of fewer than ten met at least once a month to plan, implement, evaluate and revise lessons collaboratively. While this approach traditionally only involves teachers, SPLASH encouraged teachers and community members to work together to deliver challenging and sensitive topics such as menstrual hygiene management. Here, parents were mobilised to engage the children to practise what they were taught (e.g. the importance of using clean and hygienic latrines against the practice of open defecation).

13.5 Partnerships

The SPLASH overarching strategy was to work within and strengthen the existing system at a scale using a systems approach that created or strengthened relationships among partners and stakeholders to increase impact (SPLASH, 2012). The diagram below illustrates the various partners and project team that worked with the project.

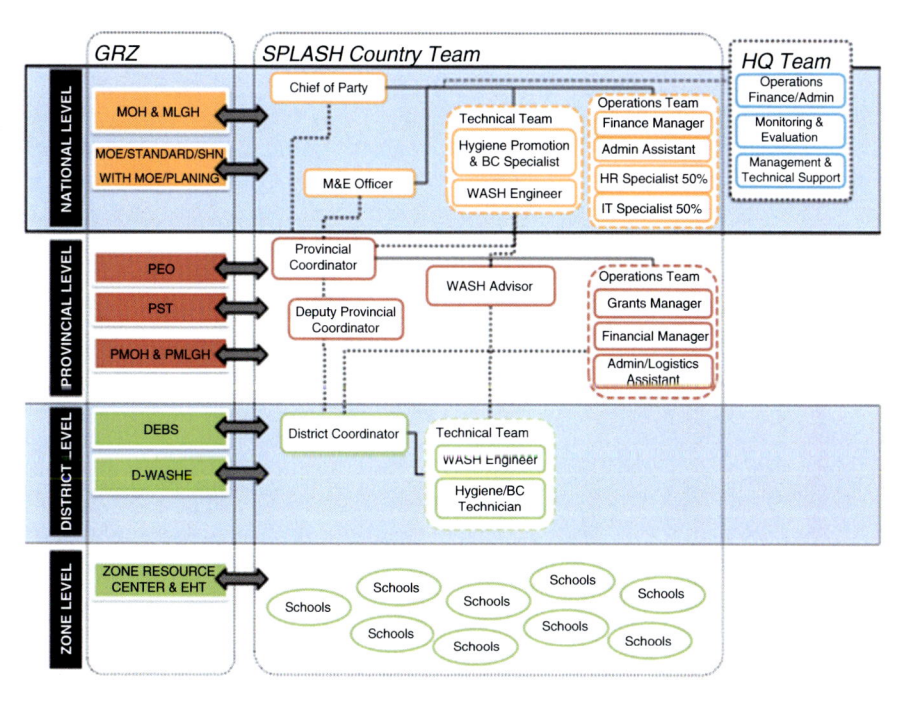

To achieve the national goal of 80 % WASH coverage for schools by 2015 at provincial levels (MLGH, 2006), SPLASH collaborated and worked in synergy with both traditional and more unusual partners. The project also aimed to strengthen and increase government leadership in school WASH programmes while responding to demands and needs identified by schools and the MOE using creative problem-solving and an infusion of expertise. The beneficiaries and those concerned with school WASH programmes determined *why*, *what* and *how* SPLASH activities were developed (USAID Water and Development Strategy, 2013-2018). Similarly, experience and research within the programme show that successful reform requires heightened, informed and concerted engagement with stakeholders. The project's strategy for maximising partnership was one of working with/within existing government and partner structures and systems at national, provincial, district and school levels, and partnerships were forged with a range of public, private and non-governmental organisations. The project worked with partners at national, district, sub-district (zonal) and school levels. However, there was a thin line between the boundaries of each level.

13.6 Government Partners

Government departments and SPLASH staff work collaboratively together on WASH activities. For example, the project worked with the Ministry of Education officials and teachers to integrate WASH messages in the national curriculum. Thus, the new curriculum contains water, sanitation and hygiene practice content and teaching methods. SPLASH also worked collaboratively with the Ministry of Education and the Ministry of Local Government and Housing to commemorate global WASH-related days such as *World Water Day*, *World Toilet Day*, *Global Handwashing Day* and *Menstrual Hygiene Day*, among others. The partnership was strong because each depended on the other for input, for the success of the programme. The Ministry of Community Development Mother and Child Health and the Ministry of Health collaborated with SPLASH in water quality monitoring, hygiene education and campaigns through the environmental health technologists.

SPLASH and its partners carried out hygiene education and reinforced the curriculum that teachers and children were applying in schools through the district resource centres. The existing continuous professional development programmes such as School Programme of In-Service for the Term (SPRINT) provided a structure which brought school-based educators together as professional learning communities to develop their practice and learning support materials on WASH (Lave & Wenger, 1991). SPLASH supported the teacher education CPD programmes through the resource centres at the province and district levels to build a system that incorporated regular meetings and workshops for teachers, head teachers and resource centre coordinators for the purpose of WASH in-service and material development. This process was in line with the MOE policy of localised curricula to address specific needs of a given region or area (MOE, 2006).

SPLASH promoted partnership by working with primary schools, surrounding communities, small-scale service providers, masons and line government departments. These partners were involved in all the project phases such as planning, implementation of activities, monitoring and evaluation. Parents in the school catchment area were mainly engaged through structured community involvement via gender-balanced committees and an unstructured involvement through whole-community meetings such as School-Led Total Sanitation (SLTS) and hygiene awareness-raising meetings. Participation took place through ongoing dialogue and reflection underpinned by respect for culture, community dynamics such as those of power relations, gender and indigenous knowledge of the local people. The outcomes and lessons drawn from these discussions informed the training material development, information and feedbacks into programme implementation. Opportunity for collectively strengthening some of the practices, power play around gender and a focus on girl children formed some of the most meaningful parts of ongoing deliberations.

Nonschool participants such as community leaders and local municipalities were engaged through the District Water, Sanitation and Hygiene Education (D-WASHE) Committees. These committees act as overall supportive inter-sectoral WASH technical resource to the district councils, and all district government authorities are members. The district councils are mandated to provide services for water supply and sanitation within their jurisdiction (MLGH, 2006).

In addition, the project worked with local artisans for most of the construction work at school level. These were trained by means of onsite trainings/orientation by the SPLASH and the DEBS teams. A qualified site supervisor was attached to every construction site to guide the local artisans in all aspects of construction. Further, parents played a major role in raising upfront building materials such as bricks, sand and crushed stones for construction of latrines. This accounted for 35 % of the total cost of the construction.

13.7 Public-Private Partnerships

In order to forge successful public-private partnerships, SPLASH learned the need for the project to properly package products for the private sector. There was some marketing involved in the art of setting up WASH in school partnerships. Ultimately, partners 'buy something' if well packaged. We learned that it was essential to prepare tailored packages for each partner, to perfectly match their philanthropic corporate social responsibility (CSR) or core business agenda. These packages ideally blended the project activities, not only to best match partner CSR and core business strategies but also to manifest a state-of-the-art approach to development and reflect a greater understanding of the challenges and deep concern for and greater proximity to beneficiaries.

Over the years, SPLASH effectively partnered with Yash Pharmaceuticals to support WASH in schools. Yash supported menstrual hygiene management in schools through the sewing of eco-friendly re-usable pads on a large scale for sale in schools throughout the country. This literally meant that girls would not miss classes due to lack of pads during their menses. Other partners from the corporate world included Colgate Palmolive and Country Choice Chickens who supported the project during global WASH-related events such as the *Global Handwashing* and the *World Toilet* days. UNICEF also collaborated with the project to provide water and sanitation facilities in Chadiza District.

13.8   Conclusion

This chapter has demonstrated how SPLASH worked with the partnerships and participatory and whole-school development approaches to support comprehensive WASH programmes in Eastern Province of Zambia, to provide water and sanitation in schools. The impact of this work improved attendance, teacher retention and teacher-pupil contact time, among other benefits. This conclusion is corroborated by the results of the project's School WASH Outcome Study which sought to estab-lish the effect of the provision of water, sanitation and hygiene education on learner/pupil attendance and teacher-pupil contact time as proxy indicators for learners' improved academic performance (SPLASH, 2015). The study was a quasi-experimental cluster design which was conducted in two similar geographical areas (intervention matched with a control group) over a period of ten months. The two were 64 schools in the Eastern Province that received a full WASH programme and 64 control schools without a full WASH package in Lusaka Province. Partnerships and participatory and whole-school development approaches (as key principles of ESD) helped SPLASH to achieve its targets and shared the cost of implementing a comprehensive WASH in school programme. This has subsequently contributed to the quality of education in the schools where the project intervened.

References

Barrow, E., & Murphree, M. (2001). Community conservation: From concept to practice. In D. Hulme & M. Murphree (Eds.), *African wildlife and livelihoods: The promise and perfor-mance of community conservation*. Oxford: James Currey.
Bresee, S., Lupele, J., & Freeman, M. (2014). *Children as agents of change for WASH in rural Zambia*. Atlanta, GA: Emory University.
Cooke, B., & Kothari, U. (2001). *Participation: The new tyranny?* London: Zed.
Duran-Narucki, V. (2008). School building conditions, school attendance and academic achieve-ment in New York City Schools: A Meditational Model. *Journal of Environmental Psychology, 28*, 278–286.
Freire, P. (1970). *Pedagogy of the oppressed.* New York: Seabury Press.
Jasper, T., Lee, T., & Bartram, J. (2012). Water and sanitation in schools: A systematic review of the health and educational outcomes. *International Journal of Environmental Research and Public Health, 9*(8), 2772–2787.

Kemmis, S., & McTaggart, R. (1988). *The action research planner* (3rd ed.). Geelong, Australia: Deakin University Press.

Lave, J., & Wenger, E. (1991). *Situated learning: Legitimate peripheral participation*. Cambridge, UK: Cambridge University Press.

Lupele, J. (2002). Ambivalent globalising influences in a local context: The case of an environmental education practitioner's experience in Zambia. In E. Janse van Rensburg, J. Hattingh, H. Lotz- Sisitka, & R. O'Donoghue (Eds.), *Environmental education, ethics and action in Southern Africa: EEASA Monograph*. Pretoria, South Africa: Human Resources Research Council.

McKernan, J. (1991). *Curriculum action research: A handbook of methods and resources for the reflective practitioner*. London: Kogan.

Ministry of Education (MOE). (1996). *Educating our future*. Lusaka, Zambia: Ministry of Education.

Ministry of Education (MOE). (2010). *School-based continuing professional development through lesson study*. Lusaka, Zambia: Ministry of Education/Japan International Cooperation Agency.

Ministry of Finance (MOF). (2011). *Sixth National Development Plan*. Republic of Zambia.

MLGH. (2006). *Natural water supply and sanitation programme*. Lusaka, Zambia: Ministry of Local Government and Housing.

National Food and Nutrition Commission. (2015). Retrieved from http://www.nwasco.org.zm/index.php/news.

Parker, J., & Adler, S. (2013). Module 2 study guide: Values and participation in EfS – from local to global: Distance learning centre. London: South Bank University.

Senge, P. M. (1990). *The fifth discipline: The art and practice of the learning organisation*. New York: Doubleday.

Shallcross, T., Loubser, C., Le Roux, C., O Donoghue, R., Lupele, J. (2006). Promoting sustainable development through whole school approaches: An international, intercultural teacher education research and development. *Journal of Education for Teaching, 32*(3), 301.

SPLASH. (2012). *Programme implementation plan*. Washington, DC: WASHplus.

SPLASH. (2015). *SPLASH school outcome study: The effect of WASH in schools on educational outcomes: Absenteeism and teacher-pupil contact time*. Washington, DC: WASHplus.

Taalat, M., et al. (2011). Effects of hand hygiene campaigns on incidence of laboratory-confirmed influenza and absenteeism in schoolchildren, Cairo, Egypt. *Emerging Infectious Diseases, 4*, 619–625.

UNEP. (1992). *Agenda 21*. Rio de Janeiro, Brazil: UNCED.

USAID Water and Development Strategy 2013–2018. Retrieved from http://www.usaid.gov/sites/default/files/documents/1865/USAID_Water_Strategy_3.pdf.

Part IV
Teachers and Teacher Education

Chapter 14
Learning as Connection: Pedagogical Innovations to Support ESD Learning Processes in Science Teacher Education Settings

Overson Shumba and Royda Kampamba

This chapter argues for 'learning as connection' in teacher education in order to influence the pedagogical knowledge and practices of science teachers to become education for sustainable development (ESD) oriented. It articulates pedagogical innovations needed to effectively support ESD learning processes and to strengthen 'learning as connection' among science student teachers. The argument for this is developed by interrogating western metaphors on pedagogy and a theory of pedagogical discourse. This chapter draws examples from ongoing attempts at the Copperbelt University to engage science students in ESD projects and to integrate ESD principles and approaches into research and teaching.

One thing the authors of this chapter have learned over the international Decade of Education for Sustainable Development (DESD) is that learning for sustainable development cannot be achieved without the fundamental reorientation of educational policies and practices. Our times of accelerating change are characterised by uncertainty, risk, and ambiguity and require new competencies to be developed among educators, teachers and learners for ESD learning today, tomorrow and forever. We need learning-based responses to the challenge of (un)sustainability; we need to 're-orient, re-connect and re-imagine' (Wals & Corcoran, 2012). Wals and Corcoran have pointed out that 'while most of the conversation about how to mitigate human impacts on the planet focuses on technology, economics and choice, the more fundamental change that must occur is the realm of learning' (p. 15). This chapter argues for institutionalising the metaphor 'learning as connection' in science teacher education. This will influence the ESD pedagogical knowledge and practices of science teachers so that they may bring about this fundamental change in learning. The arguments are illustrated by reflecting on a research project that tried to mainstream ESD and ESD change projects in a university teacher education setting (Shumba & Kampamba, 2012/2013).

O. Shumba (✉) • R. Kampamba
School of Mathematics and Natural Sciences, Copperbelt University, Kitwe, Zambia
e-mail: oshumba@yahoo.co.uk; r.kampamba@cbu.ac.zm

© Springer International Publishing Switzerland 2017 189
H. Lotz-Sisitka et al. (eds.), *Schooling for Sustainable Development in Africa*,
Schooling for Sustainable Development 8, DOI 10.1007/978-3-319-45989-9_14

14.1   Rationale Behind the Innovation

The research project sought to mainstream ESD concepts and principles into a teacher education programme offered at the Copperbelt University, the second largest public university in Zambia, with an estimated 9,000 students. The teacher education programme discussed is offered by the Department of Mathematics and Science Education (DMSE) housed in the School of Mathematics and Natural Sciences (SMNS). The programme prepares its students for the bachelor and master of science degrees with specialisation options in Biology Education, Chemistry Education, Mathematics Education, and Physics Education.

The researchers made certain observations and assumptions in planning the ESD mainstreaming research project. In many developing countries around the world, science education fails to impact much on the personal lives of students and on the quality of life in their communities. In a World Bank Paper, it was pointed out that the 'curriculum hardly relevant to the daily lives of students' was one of the many challenges for science, mathematics and technology education in sub-Sahara Africa (Ottevanger, van den Akker, & de Feiter, 2007). The students emerged from educational institutions largely unprepared for the responsibilities they are expected to play as citizens of their local communities, their nations and the global community.

Turning to science education and science teacher education, it is often the case that learning predominantly entails symbolic and technical abstraction more than it does learners' experiences based on the real-world life in the community and in society or the treatment of socio-scientific issues. Socio-scientific issues represent complex social dilemmas based on applications of scientific principles and practice (Sadler, 2004), such as pollution, global warming and the use of natural resources, poverty reduction and democracy. They go beyond such environmental issues to include questions about quality and relevance of science education and thus often are concerned with the dynamic interactions of science and society. Tackling socio-scientific issues can be one way to articulate the science as well as the social, political, economic and moral challenges people in communities face every day. It may be assumed therefore that an ESD framework and mainstreaming it into research projects would increase quality and relevance of courses and the professionalism of both researchers and practitioners.

In designing the present project, the team heeded perennial criticisms of quality and relevance of education systems particularly in sub-Sahara Africa. For example, Lotz-Sistika (2010) criticised education systems of Southern Africa for their outdated syllabus content and forms of pedagogy that were 'decontextualised and disembodied from local history, experience, culture and aspiration' (p. 24). This is often reflected in a pedagogy transfixed on technical aspects of the science and success in examinations and thus, relatively speaking, a pedagogy that either pays lip service to addressing or fails to address personal and societal needs. The team conjectured that orienting science education towards ESD would effectively contribute towards meeting personal and societal needs and transform learners into responsible citizens. This is possible because ESD can transfer knowledge and change attitudes, values and behaviour while at the same time developing learners' capabilities and opportunities to engage with sustainability issues and empower them with alternatives. It was worthwhile therefore to reflect on the *Bonn Declaration* (UNESCO, 2009a) that called for concrete actions for the development of effective pedagogical approaches, teacher education, teaching practice, curricula and learning materials.

Further, the team examined the implications of the pedagogical content knowledge models (Shulman, 1986) with respect to science education and models of science education that overly emphasise learning demands inherent in the subject matter (Bucat, 2004). Geddis (1993) (cited in Bucat) asserted that 'many of the pedagogical skills of the outstanding teacher are content-specific' which reinforces pedagogical content knowledge and the potential exclusion of societal issues and responsible citizenship goals in science education. The authors concur with Yager's (2002) point that '*context* – rather than *content and process skills* – is essential for learning science'. 'Science in context' for us implied 'learning as connection,' and this helped forge our perspective on the need to mainstream ESD. It is pertinent for educators and their students to tackle real-life issues, often reflecting consequences and risks for society, e.g., global warming, climate change, biodiversity loss, and disaster risk.

14.2   Highlighting the Research Experience

As researchers tried to mainstream ESD into science teacher education in the Department of Mathematics and Science Education, they needed to interrogate and reflect on the quality and the rationality of the curriculum and its dominant pedagogy. They evaluated the curriculum and the pedagogy informed by educational and developmental thinking in which ESD principles and practices are framed. The research team evaluated national education policies in Zambia, analysed a localised science (Chemistry 5070) syllabus, assessed a university teaching method course and surveyed 54 third-year mathematics and science education students on their views of mainstreaming ESD into their courses. The results are reported in the *Southern African Journal of Environmental Education* (Shumba & Kampamba, 2012/2013). In summary, the findings showed the following:

1. Education policies including *Educating our Future* (Ministry of Education, 1996) and *Educate the Nation* (Ministry of Education, 2005) promulgated the rationale for education as tackling poverty and creating responsible citizens who would contribute to sustainable development.
2. The secondary science curriculum, exemplified by the results of analysing the Chemistry 5070 syllabus, stipulated that environmental and sustainability issues be tackled as 'issues of national concern' (see Box 14.1), and its Unit 13 was explicitly titled *Chemistry, Society and the Environment* (Curriculum

Development Centre, 2000). In various other topic areas, the syllabus stipulated that teachers must 'refer' to local issues, examples and applications. The syllabus also stipulated that learners need to appreciate the relation between scientific thought, action and technology and the sustenance of quality of life, tolerance and valuing other people's liberties, rights and views.

> **Box 14.1: Issues of National Concern in Zambian in Chemistry 5070**
> 'The syllabus also addresses issues of national concern such as Environmental Education, Gender and Equity, Health Education and HIV/AIDS, Family Life Education, Human Rights, Democracy, Reproductive Health, Population Education, Entrepreneurship and Vocational Skills, Life and Values Education' (Curriculum Development Centre, 2000, p. v).

3. The university teaching and learning methods course (TM 310) prepared students to teach Chemistry 5070 and other science syllabuses. The analysis of TM 310 suggested that the course did not cater for ESD and did not prepare students for integrating ESD and other issues in Box 14.1 in the teaching of Chemistry 5070. The analysis established that from the 2006/2007 academic year to 2011, TM 310 had undergone numerous changes to make it more focussed on the delivery of science content. None of the changes took cognisance of ESD and its emergence as a framework for quality education. However, the ESD- and action-oriented approaches introduced in 2009 as part of the research process proved valuable leading to 'action research' as a major topic in the research method course PE 330. The overall analysis of TM 310 was that it was not adequate to provide sound pedagogical content knowledge for Chemistry 5070 and to enable teachers of this high school course to integrate ESD in their teaching of it.
4. The survey of third-year students in 2011 showed positive perceptions for integrating ESD into their courses; more than three quarters of the students agreed to the inclusion of sustainable development in the educational sciences, i.e., teaching methods, principles of education and educational research methods, and two thirds to the inclusion of sustainable development in science courses.

Overall, the findings showed that the policy and curriculum frameworks were pointing science education towards inclusion of sustainable development issues, but the science syllabuses exemplified by the syllabus for Chemistry 5070 articulated assessment objectives aligned to the technical content and did not provide adequate pedagogical guidance on how to integrate ESD concepts and issues in teaching or in assessment. The university teaching and learning methods course was not designed to support integrating ESD into the Chemistry 5070, and changes that were being made to it did not take into account ESD as an emerging framework for quality and relevance. Our innovation sought to influence pedagogical practice in our courses to shift fundamentally by mainstreaming ESD perspectives, but the team also recognised the resilience of teaching and learning practices in the department. There was relative inertia in taking up new pedagogical concepts as required in shifting to an

ESD framework. As the department was located within a natural sciences faculty, there was an evident preference to adhere strictly to the nomenclature of 'pure' science built on the belief in the delivery of purely technical sciences unadulterated by social issues and concerns. The research team also noted lack of vertical articulation between the school Chemistry 5070 and chemistry courses delivered in our teacher education programme. Topics in the high school curriculum are not deliberately taught as the core of the science teacher education programme.

Overall, what was evident in the findings is the contradiction between innovation in policy and school science and the persistence of traditional pedagogy in the teacher education setting. How could these contradictions and resilience of traditional pedagogy be explained? In order to gain insights into this question, the next section provides a critical interrogation of the teacher education setting basing on Bowers' metaphors of pedagogy (2008, 2009) and Bernstein's theory of pedagogical discourse (Sadovnik, 2001; Singh, 1997).

14.3 Metaphors of Pedagogy and Pedagogical Discourse Theory

Bowers criticised pedagogy because it persisted in being informed by outdated western metaphors of progress and modernisation which were deficient as models to change consumption and production behaviours that have contributed to the derogation of planet Earth (Bowers, 2008, 2009). These metaphors lack or ignore awareness of environmental limits and other cultural ways of knowing; they advance a mechanical way of thinking about the world and promote a view of development as linear and progressive and a view of relationships with nature in an anthropocentric way. The root metaphors also promote the view of science as the most powerful, universal and legitimate knowledge source to the exclusion or denigration of other ways of knowing and other forms of knowledge (Shumba, 1999, 2011). These metaphors put severe constraints on innovations and pedagogy to address ecojustice issues and ESD learning. They contribute to the marginalisation of knowledge systems in Africa resulting in their exclusion in mainstream education, especially formal education. Yet, as acknowledged by Shumba, African indigenous knowledge and culture provide useful models for addressing sustainability challenges. Western root metaphors lacked inclusivity necessitating our efforts for an inclusive concept, 'learning as connection'.

Basil Bernstein's concepts of classification, framing and contextualisation in the theory of pedagogical discourse (Sadovnik, 1991, 2001) provide a basis for reflecting on the resilience of pedagogical practices in our teacher education setting. They provide a view of curriculum and pedagogy and evaluation as they constitute the structure and processes of school knowledge, transmission and practice (Sadovnik). As Bernstein (1973) (cited in Sadovnik, p. 4) noted: 'Curriculum defines what counts as valid knowledge, pedagogy defines what counts as valid transmission of knowledge, and evaluation defines what counts as a valid realization of the knowledge on the part of the taught'. Classification relates to the organisation of the curriculum by defining the degree of boundary maintenance between contents. Sadovnik further explained that classification is concerned with the insulation or boundaries between the different curricula or areas of knowledge and subjects. In this vein, strong classification refers to a curriculum that is highly differentiated and separated into traditional subjects, while weak classification refers to a curriculum that is integrated (Sadovnik, 1991, 2001).

A strongly classified curriculum will be reflected by strong representation as separate subjects or disciplines, and a weakly classified curriculum will show as integrated subjects. This may relate to the teacher education setting where colleagues showed commitment to 'pure' sciences (strong classification), while in the project we sought mainstreaming societal issues, an interdisciplinary and integrating effort (weak classification). As noted by Sadovnik (2001), Bernstein suggested that a shift from collected curriculum (strong classification) to an integrated curriculum (weak classification) 'represents the evolution from mechanical to organic solidarity (or from traditional to modern society), with curricular change marking the movement from the sacred to the profane' (p. 3). This imagery is important to deal with fundamentals needed for mainstreaming ESD learning in a context of resilience of pedagogic traditions and lack of innovation in our science teacher education setting.

The lack of innovation may be related to Bernstein's concept of framing that is related to the transmission of knowledge by regulating the form and substance in the message communicated and learned. Like its relation 'classification', framing can be strong or weak reflecting the degree of control teacher and pupil possess over the selection, organisation, pacing and timing of the knowledge transmitted and received in the pedagogical transaction (Sadovnik, 2001). As such 'strong framing' refers to a limited degree of options between teacher and students; 'weak framing' implies more freedom. Singh (1997) stated:

> The principles of classification and framing may be either strongly or weakly regulated depending on the negotiating power of teachers and students. Weak regulation constitutes the possibilities for transformation rather than reproduction or resistance of power relations. (p. 6)

The science teacher education programme that is the subject of this chapter is located in a university, the highest institution of learning, where freedoms in academic and curriculum decisions are supposed to be exercised leading to innovations. There is relative freedom on what to bring into the curriculum and pedagogy implying a context of weak framing that should enable flexibility, innovation and transformation. Bernstein's principle of recontextualisation is therefore important for reflecting on this science teacher education setting.

Recontextualisation allows researchers to analyse how practices of pedagogic communication directly or indirectly relay dominant power and control relations and thus regulate cultural reproduction and change (Singh, 1997). It is a pedagogic discourse, an ensemble of rules or procedures for the production and circulation of knowledge within pedagogic interactions. Recontextualisation can be viewed as a field with rules and procedures to mediate between the fields of knowledge production (usually the field of universities and research institutions) and the fields of reproduction (typically the field of schooling and teacher education institutions that circulate this knowledge). As such, recontextualisation governs how educational knowledge is moved from the production field to the field of reproduction; thus as noted and explained by Singh:

> This process of movement of curricular knowledge opens a space for changes in power and control relations and thereby a change in ideological meaning. Ideologies are not merely carried as surface features of the knowledge, but are structured into the selection, organization, transmission and acquisition of curricula. (p. 7)

The Department of Mathematics and Science Education at the university was considered to be a pedagogic recontextualising field responsible for actualising the official curriculum as issued by the statutory organisation constituting the official recontextualising field, in this case Curriculum Development Centre and the Examinations Council of Zambia. As a department in the university, it enjoys relative freedom in the manner that it designs the curriculum and its pedagogical practices. It was expected to be at the forefront of designing innovations including mainstreaming ESD learning. With this relative freedom, more possibilities of curriculum and pedagogical transformation than possibilities of stagnancy and reproduction were expected. The mainstreaming of ESD requires interdisciplinary thinking and thus weakening symbolic boundaries between subjects. This does not necessarily need to involve diluting of subject matter. The notion of 'pure' science held in the teacher education setting was a real barrier to ESD learning as it represents strong framing.

14.4 Innovation and Framework for Mainstreaming ESD

As researchers privileged to be operating within a pedagogic recontextualising field, the Department of Mathematics and Science Education, it was pertinent to explore opportunities to mainstream and embed ESD into the teacher education setting. In order to do this, the team worked with a group of student teachers in their third year in 2011. Together, researchers and students carefully examined the content, objectives and content sections of the syllabus, Chemistry 5070, for the case study. This resulted in many ideas for mainstreaming ESD issues into 12 out of 13 units in Chemistry 5070. Unit 13 entitled *Chemistry, Society and the Environment* is the most directly related to ESD. The researchers with the students explored how to structure and input ESD content into the rest of the units in the syllabus Chemistry 5070. Figure 14.1 provided the guiding framework. This framework identifies a topic, sustainable development issues relevant to the topic, and proposes projects and value clarification activities for inclusion on the topic. It was envisaged that specification of these would provide more informative guidance to chemistry

Fig. 14.1 Framework for
mainstreaming ESD into
Chemistry 5070

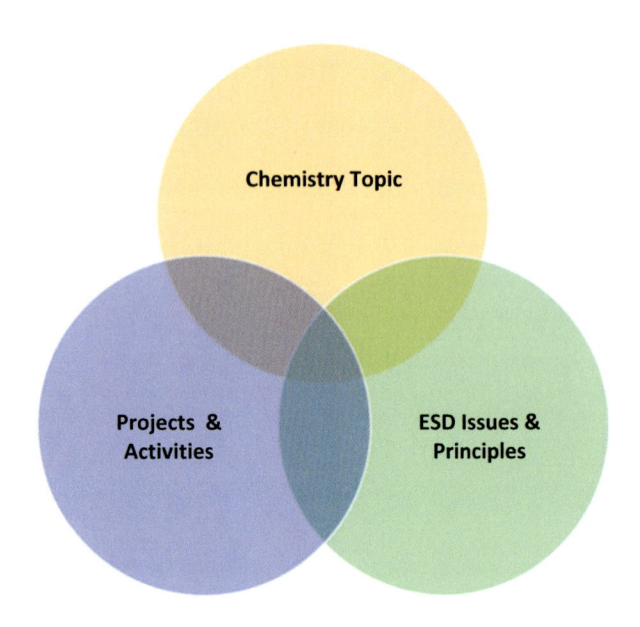

teachers seeking to mainstream ESD. The area of overlap in the circles reflects what
the authors envisaged to be ESD pedagogical content knowledge relevant to chemistry instruction.

Overall, the framework shown in Fig. 14.1 was expected to guide school chemistry teachers to see appropriate points to incorporate sustainable development
issues and at the same time show them possible projects to engage their students. In
the university science teacher education setting, it was anticipated that it could influence student teachers in the same way and also the projects that they undertook and
their appreciation of socio-scientific issues. The results of applying this framework
are shown in Tables 14.1 and 14.2 with examples of units 1 and 8 from Chemistry
5070. The first column shows the units and content drawn from the syllabus. The
second column shows environmental, sociocultural and economic knowledge areas
formulated by researchers that reflect the ESD focus content. Examples of action
research projects and value clarification activities for ESD learning are shown in the
third column.

The frameworks in Fig. 14.1 and Tables 14.1 and 14.2 have the potential to
sharpen the ESD focus in various units of the school Chemistry 5070 syllabus. This
framework permits addressing the subject matter content for the specific topics
while at the same time orienting the curriculum and pedagogy towards sustainable
development, action and value clarification. It enables addressing the chemical concepts without dilution while at the same time dealing with the personal and societal
sustainability issues and concerns. How has this made a difference in the science
teacher education setting?

Table 14.1 Proposals for mainstreaming EE/ESD issues in unit 1 of Chemistry 5070

Unit	Proposed ESD issues for inclusion	
	ESD issues	Projects and value clarification
1.0 Introduction to chemistry	Environmental	Action projects
	Environmental education	Survey local industries, e.g. paint, sugar, copper processing
Branches of chemistry	Air, soil and water pollution	
	Environmental disasters/chemical spillages; disposal of chemicals	Develop a newsletter on chemistry and chemicals in the home and the community, e.g. chemistry of baking, brewing, etc.
Importance of chemistry	Sociocultural	
	History of chemistry; African chemists; ethno-chemistry, e.g. brewing; beauty enhancement products	Assess safe and unsafe use and disposal of chemicals; plastic bags and bottles
	Family life education, reproductive health and population education	Transportation of industrial chemicals, e.g. acid, caustic soda, gas industry
	Beauty enhancement products; contraception; abortion	
		The use of carbon calculators to estimate local community's greenhouse gas emissions (GHGs)
	Economic	
	Branches of chemistry and careers	Value clarification
		Gender representation in careers in chemistry
	Agricultural, industrial and technological applications of chemistry	The use of chemicals in agriculture
		Alcohol, drugs and narcotics; chemistry in war
		Personal, family and institutional lifestyles and contribution to GHG emissions

14.5   Influence of the Innovation in the Science Teacher Education Setting

The findings of the research project described in this chapter provide further support for thinking that the science teacher education setting can be an active site of ESD learning processes and can proactively enable 'learning as connection'. Learning as connection is best in real-world contexts. ESD is a good framework for learning for a real-world context because it provides for simultaneously tackling complex sustainability issues. In the teacher education setting at the centre of reflection in this chapter, these two important observations have found expression in projects undertaken by students in partial fulfilment of the requirements of their BSc and MSc degrees in science education. Students conduct these projects after they have completed the research methods course PE 330 for undergraduates and PE 520 for

Table 14.2 Proposals for mainstreaming ESD issues in unit 8 of Chemistry 5070

| Unit | Proposed ESD issues for inclusion | |
	ESD issues	Projects and value clarification
8.0 chemical reactions and chemical changes	Environmental	Action project
	Environmental education: burning fuels, nuclear energy, fuel cells and biofuels; Renewable energy sources; clean energies; deforestation, climate change and biodiversity	Energy and or water conservation projects
Energy sources		Develop an energy yield chart for fossil fuels, nuclear energy, fuel cells and biofuels
Products of energy changes in the environment		
	Sociocultural	Create photo albums and portfolios on local and international energy issues
Photography	*Human rights*: poverty, food and biofuel crops; land-rights and gender issues	
		Value clarification
	Economic	Relevance and impact of photography and information and communication technologies (ICTs) on local lifestyles
	Transportation and energy	
	Alternative energy sources and biofuels; availability of technologies	Assessing alternative energy sources for the community
	Charcoal industry and natural products	Health, transport and pollution

postgraduates. Some of these projects are profiled below to illustrate how the metaphor of 'learning as connection' is activated in teaching, learning and researching for sustainability in the real-world context.

In these projects the students experiment with some innovative teaching and learning methods to teach chemistry in a real-world context. For example, some students have conducted field trips, excursions and outdoor lessons and inquiry projects to explore pollution in the community and at the same assessed learners' achievement in the target science concepts and learners' attitudes towards the local environment and other sustainable development issues in chemistry. Students have also explored the use of teaching and learning approaches drawn from experiences in other disciplines and experimented with them as they taught science in a real-world context. Examples for chemistry education are given in Box 14.2 and those for biology education in Box 14.3.

The examples in Box 14.2 show efforts to experiment with active and participatory methods in exploring a real issue in the community and thus connecting chemistry learning to everyday life. This action-oriented approach introduces learners directly to real-life situations and socio-scientific issues which in turn act as a motivation by reinforcing learning and creating lasting impressions which could never occur in a normal learning set-up. For example, project C4 led the student teacher

Box 14.2: Sample Classroom Action Research Projects in Chemistry Education

BSc level projects

C1 – Field trip on land pollution: improving learners' performance and attitude towards sustainable development in chemistry (2012)

C2 – Impact of project method on pupils' achievement and attitude in learning the chemistry of acids and their impact on the environment (2013)

C3 – Improving learners' achievement on water pollution using the local environment as a teaching aid (2013)

C4 – Investigating the impact of using field trip in teaching the organic chemistry: a case of alcohol functional group in a traditional beer (2014)

MSc level project

C5 – Impact of Biological Science Curriculum Study 5e instructional model on grade 11 students' comprehension and attitudes towards Chemistry 5070 acid-base concepts taught in real-world contexts

Box 14.3: Sample Classroom Action Research Projects in Biology Education

BSc level projects

B1 – Integrating concepts of climate change into unit 13 (Ecology) of the Zambian syllabus (2012)

MSc level projects

B2 – Documenting Names of Insects in Zambian Local Languages for Entomological Literature and for the Use in School Curricula and Instruction (2014)

B3 – Engaging biotechnology students in culturing and identification of cyanobacteria by the 16 s ribosomal RNA gene and testing the cyanobacteria's effectiveness as bio-fertiliser of select crop plants (2014)

and her learners to explore fermentation and functional groups (target chemistry concepts) and at the same time the impact of alcohol abuse (social issue) in the community. In the MSc project C5, the teacher-researcher provided a real-world context for his students to conduct experiments and projects with materials, foodstuffs and soils in the community. The projects in Box 14.3 for biological education students are collaboratively supervised by biologists and the lead author of this chapter (Shumba).

Project B1 integrates a global issue, climate change, into the curriculum. Project B2 documented the 55 names of insects in Tonga (a local Zambian language) with participation of local community people in order to be able to include these in curriculum and instruction. The study in Project B3 engaged the biology educator and her students in the biotechnology course in exploring the potential of a biotechnological application to solve a societal problem of food security and poverty while at the same time evaluating the extent biotechnical and ESD skills and knowledge were acquired in the context of working in real-time biological environments. In addition to learning the biotechnological knowledge and applications, the biology educator was expected to be learning and developing pedagogical competencies for effective teaching in a real-world context.

14.6   Synthesis and Discussion

The types of research projects conducted permitted student teachers and teacher-researchers to become more reflective about relevance in the curriculum. The students became more proactive in experimenting with teaching and learning approaches that brought sustainability issues into the science curriculum. Students appeared to become more engaged in learning how science concepts at the centre of instruction were connected to the wider personal and social contexts. This is particularly valuable experience for educators and students especially when engaged in collecting and examining data they have collected in the real-world setting of local communities. It is evident in the above cases that sustainability issues specific to local communities and immediate environments, culture and resources are important to serve as entry points for scientific investigations and for ESD learning. This contributes to increased motivation and effort on student teachers (BSc) and teacher-researchers at MSc level to experiment with innovative ESD-oriented teaching and learning approaches as they investigate issues relevant to their science courses in the local community and environment. It is contended that ESD learning will depend on the extent of infusion of local content and examples and being innovative in doing so. It will depend too on the extent of analysing the official curriculum and finding in it opportunities for ESD learning, as illustrated with Chemistry 5070 in this chapter. It then becomes possible to demonstrate the importance of balancing the needs of learners (learner-centeredness), the needs of society (focus on societal issues) and the learning of subject matter (focus on acquisition of knowledge and skills in a particular subject). Students are learning by exploring topics with sustainability content, investigating them scientifically and exploring implications for education and for sustainable development. In the final analysis, they are learning to 'make the connection' between, for example, chemistry and life at the individual, familial, community and societal levels. This represents relevant and significant learning, and this finds a natural rationale in our metaphor of 'learning as connection'.

In the first global monitoring and evaluation report on the UNDESD titled *Review of Contexts and Structures for Education for Sustainable Development* (UNESCO, 2009b), ESD is described as:

> ESD is about – through education and training – engaging people in sustainable development (SD) issues, developing *their capacities* to give meaning to SD and to contribute to its development and *utilizing the diversity* represented by all people –including those who have been or feel marginalized – in generating innovative solutions to SD problems and crises. (p. 7)

Our innovation shows the potential for teacher education to play a critical role in guiding teaching and learning institutions and communities towards sustainability. Based on the metaphor 'learning as connection', the aim is to reflexively transform science teacher education for inclusivity and for sensitivity towards mainstreaming ESD using local community contexts as entry points. There are a variety of ways of doing this as the experiences in the innovation in this chapter has shown. As noted by Lupele and Lotz-Sisitka (2012), ESD enables 'learning as connection' which expresses the relationship between meaning making, context and concept, making education relevant and situated in local cultures and local environments. Critically, ESD learning processes will build learner capabilities, critical thinking and agency relevant to engage with sustainability issues in both local and global contexts.

However, learning as connection cannot be superficially adopted in science teacher education where there is a conflation of many knowledge forms and disciplines. Teacher education draws from the educational sciences, the natural sciences and other disciplines; it is multi-disciplinary. As such, it requires a deeper reflection on the competencies that teachers must possess and how they must be developed, especially if teacher educators are going to engage their students in ESD and transformative learning. As shown here, they will need to connect science to humanistic, social and personal aspects and societal issues. It is therefore suggested that in addition to the view of learning as connection, their pedagogical content knowledge must include ESD knowledge and competencies (Shumba, 2012). Teachers need knowledge of specific sustainable development issues and the pedagogical approaches to integrate ESD into their teaching; they need ESD issue-specific pedagogical content knowledge (ESD PCK).

14.7   Conclusion

ESD PCK implies a science teacher needs to connect subject matter and concepts to personal and societal issues and controversies. This chapter has shown that science teacher education settings are places with plentiful possibilities and responsibility for educating for sustainability thinking and action. As Shumba and Kampamba (2012/2013) have asserted, for this to happen, social and humanistic issues in the sustainable development discourse must not remain at the periphery of science learning. For 'learning as connection', ESD issues and suitable pedagogical approaches must become part of the science pedagogical content knowledge. The authors of the chapter continue to believe that it is imperative, in thinking about ESD PCK and learning as connection (Shumba, 2012), to reflect upon how learning in natural sciences must instigate and catalyse a sense of responsibility and agency for sustainable development among educators and learners alike.

References

Bowers, C. A. (2008). *Toward a post-industrial consciousness: Understanding the linguistic basis of ecologically sustainable educational reforms*. Retrieved January 20, 2010 from http://cabowers.net/CAPress.php.

Bowers, C. A. (2009). *Educating for ecological intelligence: Practices and challenges*. Retrieved January 20, 2010 from http://cabowers.net/CAPress.php.

Bucat, R. (2004). Pedagogical content knowledge as a way forward: Applied research in chemistry education. *Chemistry Education Research and Practice, 5*(3), 215–228.

Curriculum Development Centre. (2000). *Chemistry high school syllabus 5070 grades 10–12*. Lusaka, Zambia: Curriculum Development Centre.

Lotz-Sisitka, H. (2010). *Conceptions of quality and 'learning as connection': Teaching for relevance*. Paper presented at the UNESCO EFA High Level meeting, Amman, Jordan.

Lupele, J., & Lotz-Sisitka, H. B. (2012). *Learning today for tomorrow: Sustainable development learning processes in Sub-Saharan Africa*. Howick, New Zealand: SADC REEP. Retrieved January 21, 2016 from http://sadc-reep.org.za/uploads/books/91103%20Learning%20Today%20For%20Tomorrow%20(2).pdf.

Ottevanger, W., van den Akker, J., & de Feiter, L. (2007). *Developing science, mathematics, and ICT Education in Sub-Saharan Africa patterns and promising practices*. Washington, DC: World Bank. Retrieved January 21, 2016 from http://siteresources.worldbank.org/INTAFRREGTOPSEIA/Resources/No.7SMICT.pdf.

Ministry of Education. (1996). *Educating our future*. Lusaka, Zambia: Ministry of Education.

Ministry of Education. (2005). *Educating the nation: Strategic framework for implementation of education for all*. Lusaka, Zambia: Ministry of Education.

Sadler, T. D. (2004). Informal reasoning regarding socioscientific issues: A critical review of research. *Journal of Research in Science Teaching, 41*(5), 513–536.

Sadovnik, A. R. (1991). Basil Bernstein's theory of pedagogic practice: A structuralist approach. *Sociology of Education, 64*(1), 48–63.

Sadovnik, A. R. (2001). Basil Bernstein (1924–2000). *Prospects: the Quarterly Review of Comparative Education, XXXI*(4), 687–703. Retrieved May 14, 2016 from http://www.infoamerica.org/documentos_pdf/bernstein03.pdf.

Shulman, L. S. (1986). Those who understand: Knowledge growth in teaching. *Educational Researcher, 15*(2), 4–14.

Shumba, O. (1999). Critically interrogating the rationality of Western science vis-à-vis scientific literacy in non-Western developing countries. *Zambezia, XXV*(i), 55–75.

Shumba, O. (2011). Commons thinking, ecological intelligence and the ethical and moral framework of Ubuntu: An imperative for sustainable development. *Journal of Media and Communication Studies, 3*(3), 80–83.

Shumba, O. (2012). Learning for sustainability in science education in Africa: 'learning as connection' an imperative for transformation. In A. E. J. Wals & P. B. Corcoran (Eds.), *Learning for sustainability in times of accelerating change (Chapter 27)*. Wageningen, Netherlands: Wageningen Academic Press.

Shumba, O. & Kampamba, R. (2012/2013). Mainstreaming ESD into science teacher education courses: A case for ESD pedagogical content knowledge and learning as connection. *Southern Africa Journal of Environmental Education, 29*, 151–166.

Singh, P. (1997). *Review essay: Basil Bernstein (1996). Pedagogy, symbolic control and identity.* London: Taylor & Francis. Retrieved May 14, 2013 from eprints.qut.edu.au/2864/1/2864.pdf.

UNESCO. (2009a). *Bonn declaration: A call for action.* Paris: UNESCO.

UNESCO. (2009b). *Review of contexts and structures for educational for sustainable development.* Paris: UNESCO.

Wals, A. E. J., & Corcoran, P. B. (Eds.). (2012). *Learning for sustainability in times of accelerating change.* Wageningen, Netherlands: Wageningen Academic Press.

Yager, R.E. (2002). *Achieving the visions of the national science education standards.* Retrieved January 26, 2009 from www.education.uiowa.edu/iae/iae-z-op-yager-1-4.pdf.

Chapter 15
Strengthening Teachers' Knowledge and Practices Through a Biodiversity Education Professional Development Programme

Zintle Songqwaru and Soul Shava

What constitutes adequate teacher professional development support that enables teachers to engage meaningfully with ESD learning processes? In an attempt to answer this question, this chapter focuses on how continuing teacher professional development programmes can support teachers of Life Sciences to teach biodiversity as a grounding concept to strengthen educational quality and relevance of Life Sciences education. It reflects on how continuing teacher professional development programmes may be designed and implemented to support South African teachers to work creatively with a content and assessment-referenced national school curriculum. The chapter focuses on what content knowledge, teaching and assessment approaches to include as well as teachers' reflections on the impacts of such a programme.

Environmental issues are emerging as a major global concern, and their prominence in education curricula is becoming more evident. The current South African school curriculum for Grades R–12 (students aged 6–18 years) reveals ample evidence of this trend across all subject disciplines.

South Africa has seen many curriculum revisions in the schooling system since the move to democracy in 1994. The first revision came in 1997, when an outcomes-based curriculum, Curriculum 2005 (C2005), was developed in South Africa with environmental education integrated as a phase organiser in all the subjects. This meant that all the teachers in the education system were required to have an environmental focus in their teaching. In 2001 a revised national curriculum was developed, the Revised National Curriculum Statement (RNCS). This revision brought changes in how environmental education was represented in the curriculum. One of the

Z. Songqwaru (✉)
Environmental Learning Research Centre, Rhodes University, Grahamstown, South Africa
e-mail: z.songqwaru@ru.ac.za

S. Shava
Department of Science and Technology Education, University of South Africa,
Pretoria, South Africa
e-mail: soul.shava@gmail.com

H. Lotz-Sisitka et al. (eds.), *Schooling for Sustainable Development in Africa*,
Schooling for Sustainable Development 8, DOI 10.1007/978-3-319-45989-9_15

principles that underpinned the RNCS was the relationship between social justice, a healthy environment and inclusivity, showing an emphasis on environmental issues.

In 2001, with the revision of C2005 to RNCS, the National Environmental Education Project for General Education and Training (NEEP-GET) for environmental education in the formal curriculum was established to support teachers in developing lesson plans for environmental learning (Irwin & Lotz-Sisitka, 2004). In 2012 the Curriculum and Assessment Policy Statement (CAPS), a content and assessment-referenced curriculum, was implemented in the country for the first time. Throughout this curriculum reform from 1997 to 2012, there has been a shift from a curriculum that puts emphasis on learning outcomes to a content knowledge-driven curriculum. Environmental education still remains an underpinning principle of the current curriculum due to the large amount of environmental content in some subjects.

However, having environmental topics in the curriculum does not equate to these topics being adequately addressed in teaching and learning processes. There are emerging environmental issues that are new and unfamiliar to teachers who do, however, need to grapple with the progressive and the controversial or contested nature of knowledge.

Research suggests that teachers have a poor understanding of sustainable development and thus little capacity for integrating environmental issues and sustainable development into teaching and learning (Lotz-Sisitka, 2011). Hence, there is a need for capacity building to enable teachers to engage with these concepts in their teaching.

The development of the Fundisa for Change (FfC) teacher education programme was a response to the failure of numerous efforts to have a systematic impact on strengthening environment and sustainability in teacher education (ibid.). The Fundisa for Change programme is a national collaborative programme that was established in 2011 to strengthen environmental learning in the national school curriculum, the Curriculum and Assessment Policy Statement (CAPS).

Recently, Songqwaru (2012) identified some of the emerging challenges in teaching environmental subject content knowledge in the South African curriculum. In teachers, the following were observed:

- Lack of adequate content knowledge
- Lack of familiarity with teaching methods that can be used to teach environmental and sustainability content knowledge
- Inability to link teaching with assessment
- Lack of adequate teaching and learning resources
- The need for in-service training

These challenges reflect 'teaching gaps' that need to be addressed. In responding to the above challenges, the aim of the Fundisa for Change (FfC) environmental teacher development programme is to provide teacher support to enable a deepening of their environmental content knowledge and how they teach it. In line with the above, the FfC programme focuses on providing good quality environmental teaching and learning resources as well as in-service training to support teachers. The teacher professional development courses and resource materials cover three main aspects:

1. Know your subject
2. Improve your practice
3. Improve your assessment practice (Fundisa for Change, 2013)

The thematic teacher modules or educative curriculum materials (Davies & Krajcik, 2005) are resources intended to develop teacher environmental knowledge through adequate coverage of the subject areas. This is meant to enable the teachers to have the necessary content knowledge as well as teaching and assessment approaches to improve their teaching practice. They are also intended to give educators an in-depth grasp of the environmental topics beyond meeting basic curriculum requirements. In the next section, a biodiversity module developed for Life Sciences Grades 10–12 under the FfC is presented and analysed.

15.1 Developing a Biodiversity Education Module for Life Sciences Grades 10–12

It is noteworthy that in South African national examinations from the Life Sciences curriculum for Grades 10–12, environmental content comprises more than 50 % of what is assessed (Songqwaru, 2012). Biodiversity-related knowledge aspects make up a considerable amount of this environmental content. A biodiversity module (Shava & Schudel, 2013) is one among many subject-specific and thematic modules developed under the FfC environmental teacher professional development programme in response to the knowledge gap on environmental issues among teachers (see www.fundisaforchange.co.za). The aims of the biodiversity module are to:

1. Develop and expand content mastery, enabling teachers to gain deep knowledge of the subject, thereby creating confidence in teachers in their ability to teach biodiversity (key concepts, definitions, principles, issues).
2. Promote teacher familiarity with the South African Curriculum Assessment Policy Statement (CAPS) for Grades 10–12 and its requirements with regard to biodiversity content knowledge and assessment.
3. Provide access to other relevant environmental subject content knowledges on biodiversity.
4. Expose teachers to possible teaching, learning and assessment approaches associated with biodiversity in order to enable their application in teaching and learning processes.
5. Explore biodiversity issues in relation to the local and global context.
6. Promote the use of local biodiversity contexts through exploring local biodiversity and engaging the biodiversity knowledge of learners and the local community.
7. Raise awareness on environmental issues related to biodiversity.
8. Enable them to develop and re-examine environmental attitudes, values, ethics, practices and educational ends and purposes in relation to biodiversity issues.

The above aims are aligned with the Fundisa for Change programme objectives, which are discussed above (Fundisa for Change, 2013).

In developing the biodiversity module, several aspects guided the development process. These included the following:

1. A focus on knowledge on the broader concept of biodiversity and key biodiversity issues in relation to local and global contexts (i.e. the bigger picture)
2. Incorporating aspects and activities that enabled the educators to engage with, think and reflect on biodiversity issues from the micro to the macro level (local and global contexts)
3. Providing a diverse range of teaching approaches and appropriate assessment activities that are applicable in the local context of the educator (with particular emphasis on the utilisation of the local environment in biodiversity education processes)
4. A continual reference to (revisiting of) the key concepts and issues on biodiversity in order to develop mastery of (understanding of) core knowledge on biodiversity concepts and issues
5. Providing interdisciplinary connections on biodiversity issues and concepts and links to additional resources
6. Organising the knowledge/information and activities in order to enable a progressive sequence of learning and knowledge development (scaffolding)
7. A critical engagement with controversial issues and current/changing trends in biodiversity in order to reveal the uncertain and progressive (rapid advancement) nature of biodiversity knowledge
8. Drawing on knowledge and practices at the margins (indigenous knowledges)
9. Linking biodiversity to the broader environmental issues of human well-being, equity and social justice in the everyday society context

The approach to developing the biodiversity module was to provide first some grounding content knowledge on biodiversity based on such features as key concepts, definitions, principles and issues (i.e. variety of life and life systems, levels of biodiversity (genetic, species and ecological), species, population, genus, community, habitat, ecological niche, ecosystem, trophic levels and biomes). This content knowledge was then aligned to the curriculum topics related to biodiversity at the selected level in order to create teacher familiarity with the required curriculum content to be covered. The knowledge was structured in a logically sequenced format, using a conceptual framework to link systematically the key concepts, definitions, terms and principles. The conceptual framework was to introduce first the key concepts pertaining to biodiversity, including the terminology, the levels of biodiversity, the South African biomes and the important role of taxonomy in defining the extent of biodiversity in a particular context. The second aspect of the framework was to provide an overview of different roles of biodiversity and the threats to biodiversity, with contextual examples. The final aspect was to provide emerging responses to biodiversity loss, with examples in the local context. This approach was intended to achieve content coherence and progression. The conceptual framework and its link to CAPS Life Sciences Grades 10–12 is given in Table 15.1.

Table 15.1 Extract from Fundisa for Change biodiversity module, pp. 6–7

Sub-theme	CAPS Grades 10–12 content		Grade	Term
Definitions of key concepts	Biodiversity – variety of life and life systems		10	3, 4
	Species			
	Population		11	3
	Genus			
	Community			
	Habitat			
	Ecological niche			
	Ecosystem			
	Trophic levels			
	Biomes			
Levels of biodiversity	Genetic (variation within species)		10	4
	Species (species diversity, indigenous and endemic species)			
	Ecological (diversity of ecosystems)			
Ecosystems	Aquatic (marine and freshwater), terrestrial (forests, coastal, savanna, grasslands, desert, fynbos, etc.)		10	3
		Species composition of ecological communities: fauna and flora		
		Trophic energy levels		
		South African biomes		
Taxonomy	Basis of classification – homologous (shared) features		10	4
	Hierarchical (graded order) system of increasing complexity:			
	(1) Across levels of biological organisation		11	1
		Prokaryotes (acellular)> Eukaryotes		
		Unicellular>Multicellular		
		Cell>tissue>organ systems		
	(2) Diversity of species			
	Kingdoms of living organisms:			
		Viruses, bacteria, protists, fungi, plants, animals		
	Binomial classification system:			
		The species concept		
		Nomenclature (Latin names) – Carl Von Linne		
	Plant taxonomy/classification – the plant kingdom:			
	Hierarchical system of increasing complexity			
		Levels of organisation (cellular>tissue>organ systems)· algae>mosses/liverworts>ferns>gymnosperms>angiosperms		
		From aquatic > terrestrial habitats		
		From H_2O-dependent reproduction>wind pollination>insect and animal pollination		

(continued)

Table 15.1 (continued)

Sub-theme	CAPS Grades 10–12 content	Grade	Term
Ecosystem services and human well-being	Life support systems and processes	10	1,3
	Biodiversity resources		
	Livelihood sustenance	1	
Impacts of human activities on biodiversity	Biodiversity loss	11	4
	Overexploitation		
	Extinction		
	Invasive aliens		
	Ecosystem change		
	Desertification		
Biodiversity conservation	Environmental education and education for sustainable development		
	Protected areas, CBNRM, in situ and ex situ conservation		
	Invasive alien control		
	Policy and legislation		
	Multilateral environmental agreements/international conventions (CBD, Ramsar, CITES, World Heritage Convention)		
	Low carbon and green economy		

In the first unit of the module, biodiversity is described and defined from species and genetic level up to biomes. The use of taxonomy to identify species is explained. There is a particular focus on South African biomes (marine, freshwater and terrestrial). In the second unit the environmental interrelationships between biodiversity and humankind in real-life contexts are explored, specifically the roles of biodiversity in ecosystem services and human well-being (economic, social and cultural roles). The third unit focuses on the human impacts on biodiversity and human responses (various forms of human agency) to address biodiversity loss. Included in the content are controversial issues relating to biodiversity. Some controversial issues with regard to biodiversity are also addressed under human impacts on biodiversity and responses to biodiversity loss in the module. These include the impact of colonial conservation approaches that removed indigenous communities from their land and fenced-off conservation areas, the implications of international conservation policies such as the Convention on International Trade in Endangered Species (CITES) and decisions on elephant overpopulation in southern Africa, the effects of intellectual property rights (IPRs) on traditional medicinal plant patenting and conservation and the introduction and impacts of alien species.

The module also provides the educators with a diverse range of teaching approaches (see Rosenberg, O'Donoghue & Olvitt, 2008) for use in both indoor and outdoor contexts. The aim of the emphasis of the outdoor activities is to work with biodiversity in the local environment for education processes, while also linking this to the bigger picture (regional, national and international biodiversity). An example of an activity from the module is given in Table 15.2.

Table 15.2 Extract from the Fundisa for Change biodiversity module, p. 28

Activity 5
Biodiversity is everywhere
Outdoor practical investigation – learning by doing
Learning focus:
Identifying, grouping/classifying and developing classification tools
Resources:
School grounds and/or garden
(Grade 10, term 4)
Task the learners to explore the grounds and/or garden within your school in groups. Each group should attempt to identify (using common or local indigenous names) and make a list of all living organisms that that exist within the garden. They should then attempt to classify these organisms based on their observed similarities. They should represent this in a table and present it before the entire class
(Grade 11, term 1)
For the higher grades, each group should then attempt to classify the living organisms further by trying to differentiate them more according to their similarities and differences (using distinguishing features such as shape, size or height, mobility, colour, habitat use). They should also represent this as a table. Each group must try to develop a simple key that can be used to identify these different organisms. Ask them to present their findings to the class

In addition to teaching approaches, various examples of assessment activities, mainly targeting high-order questions, are provided as a starting point for teachers to develop their own assessment activities on the biodiversity theme. The aim of the assessment examples was not summative but rather to support and enhance the learning processes (Shepard, 2000). The subject content knowledge, teaching and assessment approaches presented in the teacher module together comprise an integrated teaching system or programme.

15.2 Orienting Workshops for the Biodiversity Education Teacher Development Programme

Since 2012, several Fundisa for Change biodiversity workshops have been organised in six provinces: Eastern Cape, Mpumalanga, KwaZulu-Natal, Gauteng, Limpopo and Western Cape. These workshops were first targeted at teacher educators and subject advisors[1] under the Training of Trainers course. Subsequent workshops were held for the subject teachers, but some subject advisors also participated in the certified workshops and received certificates. A total of 93 Life Sciences teachers and 41 Life Sciences subject advisors have participated in Fundisa for

---

[1] Subject advisors are also called curriculum advisors or subject education specialists. They are officials in the Department of Basic Education who support subject teachers in schools in their districts.

Change biodiversity training to date. The other provinces have opted to be trained on other collaboratively authored Fundisa for Change modules.

Training sessions are undertaken by Fundisa for Change stakeholders[2] (partners from institutions of higher education, the environmental sector (including Department of Environmental Affairs (DEA)) and the Department of Basic Education (DBE) officials). The duration of the workshops was 5 days: 3 days in the first session and 2 days of follow-up to cover all aspects of the training. Between the first session and the follow-up session, participants implement what was covered in the first session and reflect back to the group during the follow-up session and to finalise the portfolio assessment tasks. The date for submission of the portfolio was negotiated with participants.

The model of teacher education being used is a multiplier model, and each of the training teams involved DBE officials, environmental sector partners and teacher education partners (who accredit the training). In this way, capacity is built for integrating the Fundisa for Change training into ongoing professional learning communities that are supported by the environmental sector, teacher education and DBE partner organisations.

Besides the workshop activities that involved engaging with environmental and sustainability content knowledge, pedagogical content knowledge and assessment, participants handed in a portfolio of evidence at the conclusion of the training, demonstrating their engagement with the materials in their teaching practice. Successful completion of the portfolio resulted in the presentation of a certificate of competence.[3]

---

[2] Department of Environmental Affairs, Department of Basic Education (DBE), Department of Water Affairs, Wildlife and Environment Society of South Africa (WESSA), South African National Biodiversity Institute (SANBI), South African National Parks (SANParks), Delta Environmental Centre, Environment Learning and Teaching, GreenMatter, South African Council for Educators (SACE), German corporation (GIZ), Murray & Roberts, The Lewis Foundation, Rhodes University, University of Stellenbosch, University of South Africa (UNISA), University of KwaZulu-Natal, University of Fort Hare, University of Cape Town, University of Zululand, North-West University and University of Witwatersrand (Wits). Other partners are joining as the programme grows; more universities have shown interest in the programme and are coming on board as the programme unfolds.

[3] **Outcomes to be achieved:**

- *Subject knowledge and learning (theory)*: show improved understanding of environment and sustainability subject knowledge in the context of a specific subject and phase.
- *Pedagogical knowledge and learning (practice)*: show ability to use appropriate teaching methods and assessment approaches for a specified subject knowledge, skills and values.
- *Situated knowledge and learning (context)*: relate subject-specific environment and sustainability content knowledge with broader issues and contexts.
- *Practical knowledge and experience (work experience)*: design, develop, implement and reflectively review a lesson plan/sequence of lesson plans relevant to an environmental and sustainability content focus in a particular subject and phase/grade.

15.3   KwaZulu-Natal Educators' Experiences and Reflections on the Programme

15.3.1   Fundisa for Change Training

The programme seeks to support the professional development of knowledgeable and skilled teachers.[4] It does this through *supporting and extending* the environment and sustainability content, teaching methods and assessment practices that are outlined as a 'minimum' in the CAPS (Fundisa for Change, 2013).

The purpose of each theme-based module, e.g. biodiversity, is to develop *deep subject knowledge*, with support for *how to teach* and *assess the subject knowledge* (ibid.).

Trainees bring their curriculum documents and the textbooks they use. One of the sessions involves examining these to see how environmental and sustainability concepts have been represented. This enables teachers to see possible content gaps between what is specified in the curriculum document and the textbooks they use and the foundational knowledge they need to teach the content specified. It also enables them to see the relevance of the course and how it aims to strengthen their understanding of what they are expected to teach.

The content specified in the curriculum and the extended concepts in the biodiversity module are engaged with, and teaching methods teachers may use are also discussed. Some of the teaching methods are modelled to the teachers, especially fieldwork. Teachers then plan a lesson with colleagues which they present to the whole group for critique.

Research (Songqwaru, 2012) has shown that South African students perform poorly in literacy and numeracy and so the programme prioritises teaching methods that involve listening with intent, reading and writing to learn, using numbers and mathematical concepts, active learning, developing critical and creative thinking skills.

The 2013 South African diagnostic reports on Grade 12 results showed that learners struggle to respond to higher-order questions, which has a negative impact on their overall performance. During Fundisa for Change training sessions, teachers work with final examination past papers to identify higher-order questions. One of the activities also requires them to design higher-order questions and to discuss how learners can be assisted to decode higher-order questions.

In a focus group interview, teachers stated that the Fundisa for Change training was 'more advanced' compared to any CAPS training they had previously attended. It was more interactive and gave them confidence in implementing the CAPS curriculum. They also mentioned that a teacher curriculum workshop should be more than 1 day and it should be subject focused. One of the teachers said the training

---

[4] Raw data in Songqwaru (2012), unpublished master's thesis available from http://www.ru.ac.za/media/rhodesuniversity/content/elrc/documents/historical/SONGQWARU-MEd-TR13-261.pdf

enabled her to work in-depth with and to critically analyse the CAPS curriculum document (Songqwaru, 2012).

Teachers reflected that they learned a variety of teaching methods they would use in their teaching of biodiversity, which would make their lessons more interesting for their learners. They also stated that they were encouraged to do more fieldwork with their learners and to take learners' prior and indigenous knowledge into consideration when they teach biodiversity concepts. Teachers claimed the sessions on assessment gave them ideas on various methods to use when they assessed their learners. They added that this would benefit their learners since they have different learning styles.

Most of the teachers in the Eastern Cape reflected that they still struggled with working with dichotomous keys and teaching genetic diversity. The teachers' struggle with taxonomy aspects led to the development of another related module, focusing specifically on taxonomic issues.

The teachers also valued the space that the training provided for them to share their practice with colleagues and to learn from each other. They felt that sharing what they had implemented in their classrooms and getting feedback from colleagues assisted their learning. This points to the possibility of developing teacher communities of practice for specific subject disciplines in the different sites where pilot training has been undertaken.

Subject advisors stated that creative and innovative ways of unlocking the content for learners were demonstrated during the training and that personal views were changed – no longer will Environmental Studies be viewed as a boring part of the syllabus. With enriched knowledge and experience in practical work, they will be able to support and develop Life Sciences teachers in their attempt to unlock the content for the learners and to stimulate their interest in the environment. They argued that every Life Sciences teacher needed to attend the Fundisa for Change programme, based on their experience from the training workshops and the resource materials.

15.3.2 Fundisa for Change Biodiversity Module

During all the biodiversity training sessions, content knowledge and key concepts on biodiversity specified in the national school curriculum are covered. Use is made of the Fundisa for Change biodiversity module which expands the content knowledge specified in the national school curriculum. Table 15.3 is an extract from the module.

Teachers found the biodiversity module they received during the training valuable as it covered most of the key concepts (such as biodiversity, species, genus, population, community, ecosystem, biome, habitat and taxonomy) they had to teach in Life Sciences. They also mentioned that they would use the module to enrich what is in the textbooks which are sometimes limited in terms of the content. One

Table 15.3 Extract from the Fundisa for Change biodiversity module, pages 4–5

How do these units support teaching and learning about biodiversity?
These units attempt to draw together biodiversity aspects in the curriculum in order to develop a progressive understanding of the topic
The three units focus on the following topics:
What is biodiversity?
What roles does biodiversity play?
What causes biodiversity loss and how do humans respond?
These three biodiversity units have been developed to expand teachers' knowledge and expertise in ways that also support teaching the CAPS Life Sciences curriculum for Grades 10–12 (details provided in the tables below). The sections do not follow the sequence of the CAPS; instead the progression is from exploring the concept of biodiversity, the roles it plays, the causes of biodiversity loss, and finally human responses to this loss

of the teachers stated that the module provided content which was new to them but which they had to teach.

One of the subject advisors who had not attended the training but had looked at the biodiversity module advised teachers in her district to use the module as a standard when teachers have to choose books they will order for themselves and for their learners to teach Life Sciences. Teachers agreed that they struggled to choose which textbooks to use and sometimes chose textbooks for their learners which they later discovered were not good enough.

The biodiversity module also provides teachers with additional references, enabling teachers to follow links to get more content knowledge on the topic they want to teach. This was a 'lifeline' for most teachers as they had stated that they found it difficult to find suitable resources to enrich their content knowledge. One of the teachers described having developed a new appreciation of biodiversity which would be passed on to learners.

The subject advisors for Life Sciences felt that the biodiversity module made them aware of the lack of knowledge they had of the existing policies to save the planet and unique biodiversity in South Africa. They were astonished by the great volume of information that was available but also realised that the current curriculum (CAPS) is not keeping up with the content change with regard to emerging environmental issues.

15.3.3 Portfolio of Evidence

In order for teachers to receive certificates for the training, they had to complete and submit a classroom practice-based assignment task. The portfolio task required them to focus on four aspects of their daily practice: know your learners, know your subject, improve your teaching and improve your assessment. The approach to working with the assignment involved participants working together during the training and a 'working away' task in the classroom. The assignment was in four parts which had to be completed over the duration of the course.

Having worked through the assignment, teachers felt that many of the new concepts they had to deal with in the Life Sciences had been explained. They reflected that they had learned about conceptual progression and connections within and across the grades and the value of research and finding resources for teaching instead of relying only on one textbook.

They stated that the assignment task encouraged them to work with the CAPS document and helped them to develop new activities for their learners. They also acknowledged that the work on the assignment was valuable in that they applied what they had planned.

In completing one of the assessment tasks, teachers were able to identify content and skill gaps their learners brought to class from previous grades, which then enabled them to plan appropriately on how and what to cover in the current grade. One of the biggest challenges with completing the tasks for the portfolio was setting higher-order questions, although some teachers acknowledged that they learned more about assessment during the training than they had previously. Some teachers also felt anxious due to the time it took to complete the tasks with all the other pressures on their time.

15.4   Discussion

Teacher professional development plays a critical role in educational reform. Garet, Porter, Desimone, Birman and Yoon (2001) stated that core features of professional development include a focus on content knowledge, opportunities for active learning and coherence with other learning activities. The FfC environmental teacher professional programme promotes transformative environmental learning processes through developing teachers' environmental subject content knowledge, teaching and assessment approaches. Some core aspects of the FfC programme are discussed below.

15.4.1   Subject Content Knowledge

A key aspect of the FfC programme is improving teacher subject content knowledge (such as knowledge of biodiversity concepts, role of biodiversity, threats to biodiversity and related responses covered in the module). The teachers' subject content knowledge significantly impacts on their teaching practice and learning outcomes delivery (Supovitz & Turner, 2000). This reveals the need for teachers to understand extensively the subjects that they teach, including acquisition of emerging and unfamiliar content, and the need to situate content-based knowledge in the broader landscape of teacher professional knowledge.

15.4.2 Curriculum-Based Educative Materials

Since the curriculum generally guides the teacher on what content to teach, the development of the FfC teacher modules has been guided by environmental content within the South African curriculum (CAPS). Curriculum-based educative (teacher resource) materials can serve as an effective tool for professional development (Callopy, 2003).

15.4.3 Teacher Training Workshops

The teacher training workshops were intended to reinforce the application of the subject content knowledge, teaching and assessment practices covered in the teacher modules. The importance of teachers practising by practically testing out and implementing the learned content knowledge, teaching and assessment practices was a very significant aspect to the programme as it demonstrated evidence of the effectiveness of their professional development. This finding was also substantiated by Penuel, Fishman, Yamaguchi and Gallagher (2007). However, the impact of the professional development process on students' learning outcomes is yet to be determined.

Teacher training workshops also provide the opportunity for the formulation of professional communities of practice for sharing knowledge and practice experiences, as was evidenced in some provinces (such as KwaZulu-Natal). Such professional communities can contribute to improving teaching practice and, subsequently, student achievement (Darling-Hammond & Richardson, 2009). It is hoped that the groups trained in the different provinces of South Africa will continue to work together as communities of practice for their subject disciplines.

15.5 Conclusion

The Fundisa for Change environmental teacher professional development programme originated from the realisation that although there is a high proportion of environmental content across subject disciplines in the South African context, teachers are not adequately prepared to teach this content. The programme focuses on three key aspects: subject content knowledge, pedagogical knowledge and assessment practice.

The FfC programme has undertaken to assist teachers through the provision of topic-specific, curriculum-based teacher resource materials and in-service training workshops. Teachers have found the resource modules informative and relevant, addressing their content knowledge gap, and the participatory learning interactions during the workshop beneficial. The requirement for teachers to develop portfolios of evidence provided an opportunity for them to test out practically and apply the knowledge gained from the workshop and the resource modules.

References

Callopy, R. (2003). Curriculum materials as professional development tool: How mathematics textbooks affected two teachers' learning. *The Elementary School Journal, 103*(3), 287–311.

Darling-Hammond, L., & Richardson, N. (2009). Research review/teacher learning: What matters? *Educational Leadership, 66*(5), 46–53.

Davies, E. A., & Krajcik, J. S. (2005). Designing educative curriculum materials to promote teacher learning. *Educational Researcher, 34*(3), 3–14.

Fundisa for Change Programme. (2013). *Introductory core text.* Grahamstown, South Africa: Environmental Learning Research Centre, Rhodes University.

Garet, M. S., Porter, A. C., Desimone, L., Birman, B. F., & Yoon, K. S. (2001). What makes professional development effective? Results from a national sample of teachers. *American Education Research Journal, 38*(4), 915–945.

Irwin, P., & Lotz-Sisitka, H. (2004). A history of environmental education in South Africa. In C. Loubser (Ed.), *Perspectives on environmental education in southern Africa* (pp. 34–56). Pretoria, South Africa: Van Schaik.

Lotz-Sisitka, H. (2011). *National case study. Teacher development with an education for sustainable development focus in South Africa: Development of a network, curriculum framework and resources for teacher education.* Paper prepared for the Association for the Development of Education in Africa Conference, February 2012.

Penuel, W. R., Fishman, B. J., Yamaguchi, R., & Gallagher, L. P. (2007). What makes professional development effective? Strategies that foster curriculum implementation. *American Educational Research Journal, 44*(4), 921–958.

Rosenberg, E., O'Donoghue, R., & Olvitt, L. (2008). *Methods and processes to support change-oriented learning.* Grahamstown, South Africa: C.A.P.E. CEP, Rhodes University. Distributed through Share-Net, Howick.

Shava, S., & Schudel, I. (2013). *Teaching biodiversity* (Fundisa for Change Programme). Grahamstown, South Africa: Environmental Learning Research Centre, Rhodes University.

Shepard, L. A. (2000). The role of assessment in a learning culture. *Educational Researcher, 29*(7), 4–14.

Songqwaru, N. Z. (2012). *Supporting environment and sustainability knowledge in the Grade 10 Life Sciences Curriculum and Assessment Policy Context: A case study of the Fundisa for Change Teacher Education and Development Programme Pilot Project.* Unpublished master's thesis. Rhodes University, Grahamstown.

Supovitz, J. A., & Turner, H. M. (2000). The effects of professional development on science teaching practices and classroom culture. *Journal of Research in Science Teaching, 32*(9), 963–980.

Chapter 16
Education Quality and the Introduction of New Teaching Methods into Teacher Education

Evaristo Kalumba

A large number of teachers in schools are not aware that the use of a wider range of teaching methods can improve the quality of teaching and learning. This chapter shows how teachers who at first were using a limited number of teaching methods and who were later introduced to the use of a wider range of preferred new teaching methods became more reflective and effective practitioners. The chapter further shows how the use of an additional new sociocultural and structural educational quality framework has potential to improve teachers' reflective practice and enhance the quality of environmental learning, making it also more relevant to the African teaching and learning context.

16.1 Quality Education and ESD

Quality education is a concern in Zambian education (Carmody, 2004; SACMEQ, 2000, 2006, 2011). The study reported on in this chapter (Kalumba, 2012) was conceptualised as a contribution to the current debate on questions of educational quality and sought to contribute to addressing some of the causes of the declining quality of education in schools. In particular, the study was interested in finding out how perspectives from ESD and improved use of teaching methods could enrich debates about quality and relevance in Zambian schools.

It is evident that the performance levels of learners in Zambia and other SADC countries in mathematics, reading, English language and other subjects at grade 7 are below average as regularly reported on in the Southern and Eastern Africa Consortium for Monitoring Educational Quality (SACMEQ) studies I, II and III (Carmody, 2004; SACMEQ 2000, 2006, 2011). Other quality-related concerns are

E. Kalumba (✉)
Natural Sciences Department, Mufulira College of Education, Mufulira, Zambia
e-mail: ekalumba695@gmail.com

© Springer International Publishing Switzerland 2017
H. Lotz-Sisitka et al. (eds.), *Schooling for Sustainable Development in Africa*,
Schooling for Sustainable Development 8, DOI 10.1007/978-3-319-45989-9_16

inadequate provision of learning materials, large class sizes, inadequate attention to critical thinking and inadequately trained teachers for science and mathematics (ibid.). This ongoing concern about educational quality was echoed by the Zambian Education Minister in his November 2014 inauguration speech at the newly established Chalimbana University Council established to investigate the reasons for the decline in the quality of teaching and learning in Zambian government schools.

This chapter discusses a small-scale research project that was conducted in Zambia in 2012 to investigate how the use of a wider range of preferred teaching methods could potentially improve the teaching and learning of environmental learning in schools. Environmental learning, reflecting ESD principles, when integrated into a number of diverse subjects, has the potential to foster improved engagement with contextual dynamics and to situate learning and improve meaning making (Lupele & Lotz-Sisitka, 2014).

Through a review of debates on different approaches to educational quality which took place in the study, but also in the SADC Regional Environmental Education Programme ESD research programme (Lupele & Lotz-Sisitka, 2014), the study identified and considered three major paradigms or discourses on educational quality, namely, the human capital, human rights and the social justice and capability approaches to educational quality. A human capital approach to educational quality emphasises economic growth, performance and efficiency. It favours standardised assessments to assess quality (e.g. SACMEQ tests). A human rights approach to educational quality emphasises the role of the state in guaranteeing basic rights to education. The focus and interest of this paradigm is on rights to education, rights in education and rights for education. A social justice/capability/ sociocultural approach emphasises redistribution, recognition, participation, meaning making and capabilities (people's own valued beings and doings) and is influenced by works of social justice theorists such as Fraser (2008) and Sen (1999) and sociocultural theorists such as Vygotsky (1978), Rogoff (1990) and Daniels (2008). This framework promotes curricula and pedagogies that recognise and value histories, lifestyles and pedagogic texts; it also supports the concept of localised curricula, but does not narrow all learning to the local only. It proposes that social arrangements such as education should permit all to participate as peers in social life. Each of these three paradigms of education is important in quality education (Lotz-Sisitka, 2013; Tikly & Barrett, 2011, 2013), and given my interest in ESD, I was particularly interested in the latter conceptualisation of educational quality.

The UNESCO Global Monitoring Report on Educational Quality (UNESCO, 2005, p. 9) stated that 'the teaching and learning process is a key arena for human development and change'. It further claimed that 'it is within the teaching and learning *process dynamics of quality* where the impact of curricula is felt' (ibid., my emphasis). Quality is also felt within the processes when a teacher's method works well, amongst other factors (UNESCO, 2005). Style of teaching, referred to as 'preferred teaching methods' in this study, affects educational quality (UNESCO, 2005). Although indirect enabling inputs such as learning time, teaching materials, assessment, feedback, incentives, etc. are closely related to the quality dimension, the actual teaching and learning processes in the classroom are the most important in the transmission of knowledge, skills and values needed to improve quality education (Barret et al., 2007; UNESCO, 2005). It is with this in mind that UNESCO (2005) has recommended that more attention be given to teaching and the teaching process in educational programmes focusing on quality. It is also this discussion related to educational quality that motivated me to focus on the introduction of new preferred teaching methods and then to review the outcomes of this process from an educational quality perspective.

Given this interest, the second core concept and focus of this study is teaching methods. Teaching methods are perceived in different ways by different teachers locally, regionally and internationally. In some cases, teachers use the terms 'teaching methods', 'teaching strategy' and 'teaching approach' to mean the same concept. In other cases, the three terms mean three different things. In this paper the term 'teaching method' means the method of instruction. This is in line with the definition by WWF Tanzania Environmental Education Programme (Tanzania, Ministry of Education [MoE], 2001) where the teaching method has been defined as a way in which the teacher decides what the learners will learn and how the learning will be structured, supported and mediated.

16.2 Methodology and Approach

Using action research and an interpretative methodological framework, a series of research activities were undertaken to generate research data because the focus of the study was on investigating teachers' practice with a view to probe change and to analyse the findings. The series of activities that teachers undertook included a sequence of (1) attending a workshop in which they reflected on and described existing methods that they were using in the classroom, (2) engaging with a methodology book which described a wider range of teaching methods that could be used for ESD teaching and learning (Rosenberg, O'Donoghue, & Olvitt, 2007), (3) selecting two new methods to try out in their lessons and (4) teaching the lessons in their classrooms. The process was documented using questionnaires, observations and teacher interviews.

The study was based out of Mufulira College of Education, a college in Zambia in the Copperbelt Province where secondary school teachers in Mathematics, Science, Home Economics, Music, Art and Physical Education are trained. The Primary Teachers' Diploma by Distance Learning (PTDDL) is one of the courses offered by distance education. The teachers who participated in the research were graduates of the PTDDL because they had acquired the knowledge, skills and teaching methods of environmental education module 5 of the PTDDL course. The study expected them to use the teaching methods with a view to integrating environmental issues in Mufulira district into the core teaching subjects.

Common environmental issues faced by students in the Copperbelt Province include air pollution from the mining industry, water pollution, deforestation and land pollution. The mining activities largely contribute to the air and water pollution in the area because of the discharge from copper smelting and refineries. Waste management is another common environmental issue. Deforestation is the result of poverty-stricken residents who rely on charcoal because they cannot afford electrical power. Unemployment also forces people to cut trees for furniture as a form of self-employment. Land pollution is mainly due to mine tailing dumps and uncollected garbage by the municipality councils. This creates a rich context for integration of environmental concerns into mainstream subjects, and teachers participating in the study included concerns such as pollution, waste and deforestation in their lessons.

To analyse the data, the Nikel and Lowe (2010) fabric of dimensions of educational quality was adapted and used to explore whether teachers included dimensions of quality in the teaching process. Nikel and Lowe suggested the following features of educational quality be reviewed in relation to each other, making up a 'fabric' of educational quality: effectiveness, efficiency, equity, responsiveness, reflexivity, relevance and sustainability. In adapting this framework and based on the work of Cornbleth (1990) and the emphasis on the capability approach to educational quality, I added two additional features of quality to this 'fabric', namely, sociocultural and structural dimensions, as shown in Table 16.1. I used this as an analytical tool to review the work that teachers did with a set of preferred new methods.

Like the review by Southern African researchers on educational quality (Lupele & Lotz-Sisitka, 2014), the study was able to identify that little work has been done on the relationship between the use of teaching methods and the sociocultural and structural dimensions of educational quality (Johannessen, 2006; Kelly, 1999; Zambia [MOE], 2000). The study therefore offered insights into the significance of taking account of sociocultural and structural dimensions of educational quality in improving quality dynamics of teaching and learning.

16.3   Findings

The initial phase of the study found that the teachers who participated in the study had a limited view of educational quality. They saw educational quality as being linked to teaching by using those methods that were most common (e.g. lecture, question and answer, demonstration, discussion and experiments). They also thought that quality teaching was achieved when the prescribed daily routine of teaching as laid down in teachers' textbook was followed and when the objectives of the prescribed lesson plans were achieved. They believed that quality teaching and learning took place in schools equipped with good textbooks, computers and staffing, thus reflecting mainly views of efficiency and effectiveness in the Nikel and Lowe model of educational quality. This also resulted in them using a limited number of teaching methods, i.e. those with which we are most familiar. The problem of teachers' use of a narrow range of teaching methods can also be traced to the

Table 16.1 Summary of dimensions of quality (example here is from lesson 2)

Nikel and Lowe (2010) dimensions of quality							Southern African dimensions	
Effectiveness	Efficiency	Equity	Responsiveness	Reflexivity	Relevance	Sustainability	Sociocultural	Structural
✓	✓	✓	✓	✓	✓	✓	✓	✓
All objectives achieved; mathematical concepts successfully learned	All steps followed and lesson completed successfully in time allocated	Gender sensitive, reduced marginalisation of local knowledge and recognised the value of other forms of knowledge (not only school knowledge but also community knowledge)	Had relevance to own life conditions, and mathematical knowledge could be applied in home context. Also referred to how learners can get rid of waste at their homes	Planned lesson with arrangements with community members. Teacher was able to reflect on how the method added value and new relevance to mathematics teaching (compared to previous methods used)	Use of community as guest teachers and using mathematical knowledge in ways that can be applied in other/home contexts	Discussed with each house owner the concept with real examples	Discussed with house owners cultural aspect of garden, crops and animals domesticated; use of local language; negotiation of cultural beliefs (e.g. sharing of household knowledge with learners)	Administrative permission to do fieldwork interviews and arrangements with community group work to give more learners a chance to participate

Table 16.2 Teachers in their groups and preferred choice of new teaching methods

Pair	Category of teaching method preferred	Selected and preferred complementary teaching method option 1 to be used by teacher A in each pair	Selected and preferred complementary teaching method option 2 to be used by teacher B in each pair
1	Experiential	T1: music, poetry, artwork (drawing)	T2: role play and value clarification (will choose one of the two); artwork
2	Investigative	T3: fieldwork (household interviews)	T4: field trip (household interviews); field trip/investigative
3	Learning by doing	T5: practical action (tree planting)	T6: action taking (tree planting)
			Exploring indigenous ways of knowing
4	Deliberative	T7: social learning method (human resource, presentations, storyline)	T7: drama and theatre
			T8: story methods

T refers to teacher; teachers also worked in pairs from the same school to ensure that they could support each other

colleges where they were trained, and it seems that colleges are promoting only a limited range of methods as was found by Moose (2009) in the Zambian context.

The initial phase of the study also found that prior to the methods workshop where teachers were exposed to a wider range of teaching methods, they did not use any quality education framework in their choice of teaching methods because they were not aware that any existed. They were also unaware of the wide range of methods that *could* be used for ESD-related teaching. The booklet that they received in the workshop included a wide range of different methods clustered into groups of information transfer, experiential, investigative, learning-by-doing and deliberative methods. Teachers explained that they usually used commonly known teaching methods because they were easy to prepare and work with and they consumed little time. Kelly (1999, p.148) explained that this led to a situation where Zambian education is dominated by rote learning and traditional teacher-centred teaching methods in primary schools in what he described as a context of 'obsolescence and irrelevance' in relation to Zambia's educational content and teaching methods. His primary point is that teaching methods as a mode of knowledge transmission affects the quality of education.

Following the workshop, teachers were exposed to a range of new methods and were asked to select two new preferred methods that they would like to try out in their ESD-related lessons. The new teaching methods used by teachers in the study are outlined in Table 16.2 above.

Following observations and interviews in which I both observed the teachers using the new methods in their ESD-related lessons and interviewed teachers on their reflections, the study found that when teachers used a quality education framework (see Table 16.2) together with preferred new teaching methods, they became more reflective of their practice. Here, it is important to emphasise *preferred new teaching methods*, as by choosing and selecting a preferred method, a teacher needs to consciously and reflexively consider his or her choice and why it is being made.

In the workshop and in planning their new lessons, teachers had to consider certain factors and reasons for preferring or choosing one teaching method and style from another such as age, gender, cultural reasons, ability of learners, availability of teaching materials and context and environment in which the lesson is taught and the season during which the topic is taught. This they could not do when they were using only one traditional teaching method. This process supported teachers to become more deliberate regarding their teaching. In their reflections, teachers said that being deliberate made them consider why they do something in a particular way. Thinking about the reasons one has for doing something allows one to act in a more deliberate manner. The study found that teachers who act in a deliberate manner no longer do something merely because that is what the textbook says or that is what they have done in the past. Rather, teachers who act in a deliberate manner teach the lesson in a particular way for a specific reason, and as a result all the lessons were more strongly and carefully situated and aligned with local needs and learners' realities and qualities.

One of the more interesting findings related to how the process of becoming more reflective appeared to free teachers from routine behaviour. Teachers explained that at first they thought teaching was an easy routine practice of simply following steps in a teachers' textbook. At first they taught lessons in the same way they taught them in the past with little effort and without any reflection on the lessons. They even confessed that their lessons were ineffective since they did little to modify them for a specific class. After working through the process of using new methods and reflecting on them, teachers described how being a reflective teacher frees one from following a routine rigidly and can result in more creative and effective lessons.

Becoming more reflective enabled teachers to create more original ideas for their teaching; for example, one teacher said she had to find a real plant which learners practically planted to make the lesson effective instead of using a chart with a plant diagram. She said the new teaching methods opened her mind to create seven teachers' and seven learners' activities, something she could not do if she had used only one teaching method. Importantly too, the expansion of the use of the range of teaching methods offered learners a variety of new ways of discovering knowledge, knowing, doing things, thinking, expressing themselves differently and wider learning opportunities to apply critical thinking (Daniels, 2008; Sen, 1999). This is in line with research findings by Rusbult (2007) who stated that there are several advantages of using a wider range of teaching methods. The first advantage is that the use of a wider range of teaching methods would cater for multiple learning styles and help students retain information and strengthen understanding (Janse van Rensburg & Lotz - Sisitka, 2000). The second advantage is that it ensures that all students have equal opportunities to learn. The third advantage is that it caters for both slow and fast learners. The fourth advantage is that it develops the ability to engage in thought-provoking learning processes and supports the learners not only to become critical thinkers but also to become responsible for their learning (as shown in the four lessons). The fifth advantage is that it increases interest in learners, which was also shown in the four lessons. All these advantages improve the quality of education (Rusbult, 2007) and the quality of environmental learning. This is supported by Barret et al. (2008) who claimed that quality learning is the result of the teacher meeting diverse learners' needs through the provision of learner-centred pedagogies.

16.4 Towards an Indicator Framework for Further Research into ESD and Education Quality and Relevance in African Countries

While the study has shown that the use of a quality education framework together with new teaching methods firstly helped teachers to become more reflective of their practice and, secondly, helped them to improve the quality of their teaching and learning in schools, it also offered possible indicator framework for guiding further research into ESD and educational quality and relevance in African countries which foregrounds the sociocultural and structural dynamics of teaching and learning (see Table 16.1).

Including sociocultural and structural dimensions of education quality in the reflection process with teachers (the adapted Lowe and Nikel framework discussed above; see Table 16.1) appeared to be an important focus of more fully understanding quality-related aspects of ESD-related teaching and learning as shown by this explanation of one of the reflections that the teachers made in a lesson on deforestation and planting of trees:

> Learners constantly noted that trees are important for medicine and that they were related to poverty (since people used the fruits and products from trees for livelihood purposes). The teacher reflected that including social-cultural aspects into her lessons was new to her. She reflected on the lifestyle of many Zambians which involved heavy reliance on charcoal for domestic energy. She said she did not cover that part in the lesson plan. Thus, an important contextual dimension of the deforestation issue was simply excluded from the lesson, as the teacher chose to work with more 'traditional' school content knowledge, which is often not contextualised. She also did not discuss the name of the specific tree that was being planted, or its social or cultural significance, thus losing another potentially enriching teaching and learning opportunity. After participating in the workshop and reflecting on this issue, she noted that she would slowly adapt to integrating the sociocultural and structural issues in teaching and learning. (Kalumba, 2012, p. 81)

As can be seen from the above, sociocultural and structural dimensions help to bring out regional aspects of life such as the customs, lifestyles and values that characterise the society in which education is taking place and thus add meaning to education and facilitate 'learning as connection' (Janse van Rensburg & Lotz-Sisitka, 2000; Lotz-Sisitka, 2011; Namafe, 2008; Wood, 1998). Sociocultural factors include anything within the context of Southern African society that has potential to affect teaching and learning (UNESCO, 2014). In this study, these included giving attention to cultural norms, language, societal practices and important knowledge, which would otherwise have been marginalised in the teaching and learning setting. Structural dimensions are equally significant to consider in attaining educational quality, and, as found in this study, these include teacher-pupil ratios, class size, teachers' qualifications, the physical environment, nutritional level and socio-economic status (amongst others). As shown in this study, structural factors such as large class sizes or inadequate teaching and learning materials can affect the quality of the teaching and learning.

This study has also shown that there is value in including these two dimensions in observing quality teaching and learning in the Southern African context. Inclusion of these features into the 'fabric' quality model resonates with what Tikly and Barrett (2009) argued for, when they explained that efficiency and human rights-based approaches to quality can construct schools which are set apart from the local context. Through these approaches to quality, schools are ascribed an insulating role for providing effective, safe and gender-sensitive learning environments (ibid.). They, together with Southern African researchers (Hogan, 2008; Ketlhoilwe, 2008; Lotz-Sisitka, 2008; Namafe, 2008), argued that although children in schools should take for granted support and freedom to facilitate learning in schools, it should also be recognised that schools exist in specific sociocultural contexts and quality education must be responsive to the lived realities of learners and educators in those contexts. Hogan's (2008) and Namafe's (2008) research show that this is an important epistemological question, and Lotz-Sisitka (2011, 2013), working with colleagues in the SADC REEP Educational Quality research programme, argued that this is also a pedagogical issue which involves meaning making and 'learning through connection'. Tikly and Barrett (2009) have explained that human rights approaches are an entitlement related to cultural rights and therefore require that quality education meets learners' needs, including those stemming from their various cultural identities; but experience in Southern Africa is that such cultural identity perspectives are most often marginalised in modern education discourses (Hogan, 2008; Namafe, 2008). This adds to the argument put forward here to include these dimensions more explicitly in quality education research.

The study has also shown that although several teaching methods are used in teaching and learning, quality is not realised if the combination of the different frameworks of quality dimensions is not included in the planning, in the implementation and in teachers' reflections on the lessons taught. In this study, teachers reflected on the importance of including sociocultural and structural dimensions into quality discussions and indicated that this was not a common practice, but that it should be included in teacher education. This has implications for the way in which the intersection of the three broad orientations to education quality being reframed and developed in the SADC REEP (referred to above; see also Lotz-Sisitka & Lupele in Chap. 1 of this book) can come together at a practical level. The argument being put forward in the SADC REEP is that all the education quality orientations should come together to improve educational quality, but there was little guidance on how this could be done at a practical level. This study has provided some tools and approaches for doing this.

Another important observation from the study was that it supported the current emerging international understanding that all three orientations in Fig. 1.1 in Chap. 1 of this book (the preventative, responsive and proactive modes) have some validity within educational quality, and it is the intersections of all three that are significant (Nikel & Lowe, 2010; Tikly & Barrett, 2011, 2013). The study has opened up a new ground for more research in the understanding of education quality by asking further research questions such as: Can there be an educational quality framework designed suitably for Southern Africa that foregrounds sociocultural dynamics and structural factors more explicitly which can lead to a new way of understanding ESD relevant to African countries? As shown in this study, the reflective use of such quality frameworks in teacher education is also important for enhancing actual improved educational quality practices. These questions arise because it seems that whatever education framework we have used as a lens to understand the dynamics of teaching and learning was not originally designed for African schools, but was rather a copy of the Western research which does not always take account of local dynamics of pedagogy influenced by different socioecological, sociocultural and structural dimensions of educational quality.

16.5   Conclusion

The use of a quality education framework, together with a wider range of new teaching methods, firstly helped teachers to become more reflective in their practice. Secondly, it improved the quality of teaching and learning. Thirdly, the Zambian research offered an expanded framework for developing indicators for reflecting on education quality that includes sociocultural and structural dimensions of quality. It can be deduced from this study that it is not all or any one type of educational quality framework and model that can be applied to understand patterns or dynamics of teaching and learning to improve the quality of teaching and learning, but that it is the intersection of all that is important. The study also showed that focusing on teaching methods in quality enhancement is a fruitful way to move forward with aspects of enhancing quality education and that it is important in this process to support teachers to become more deliberate and reflective of their practice in the process.

References

Barret, A.M., Ali, S., Clegg, J., Hinostroza, E., Lowe, L., Nikel, J. (2007). *Initiatives to improve the quality of teaching and learning: A review of recent literature 2008*. EdQual research programme on implementing education quality in low income countries. EdQual working paper no. 11. Retrieved November, 10, 2010 from www.edqual.com.

Barret, A., Ali, S., Clegg, J., Hinostroza, E. J., & Lowe, J. (2008). *Initiatives to improve the quality of teaching and learning: A review of recent literature* (Background Paper for the Global Monitoring Report, EdQual Working Paper No. 11). University of Bristol, UK. . Accessed on Dec 2010.

Carmody, B. (2004). *The evolution of education in Zambia*. Lusaka, Zambia: Bookworld Publishers.

Cornbleth, C. (1990). *Curriculum in context*. London: Falmer.

Daniels, H. (2008). *Vygotsky and research*. London: Routledge.

Fraser, N. (2008). Abnormal justice. *Critical Inquiry, 34*(3), 393–422.

Hogan, R. (2008). Contextualization of formal education for improved relevance. A case of the Rufiji wetlands. Tanzania. *Southern African Journal of Environmental Education, 25*, 44–58.

Janse van Rensburg, E., & Lotz-Sisitka, H. (2000). *Learning for sustainability: An environmental education professional development case study informing education policy and practice*. Johannesburg, South Africa: Learning for Sustainability Project.

Johannessen, E. M. (2006). *Basic education – also a question of quality*. Flyktninger, Norway: Save the Children Norway Research Fund. Educare.

Kalumba, E. (2012). *Improving the quality and relevance of environmental learning through the use of a wider range of preferred teaching methods. A case of primary schools in Mufulira District in the Copperbelt province in Zambia*. Unpublished masters (Education) thesis, Grahamstown, South Africa: Rhodes University.

Kelly, M. J. (1999). *The origins and development of education in Zambia*. Lusaka, Zambia: Image Publishers.

Ketlhoilwe, M. J. (2008). *Supporting environmental education and education for sustainable development in higher education institutions in Southern Africa*. Howick, South Africa: SADC Regional Environmental Education Programme.

Lotz-Sisitka, H. (2008). Editorial: Environmental education and educational quality and relevance. Opening the debate. *Southern African Journal of Environmental Education, 25*, 1–13.

Lotz-Sisitka, H. (2011). *Education for sustainable development and educational quality and relevance*. Unpublished research programme report, Grahamstown, South Africa: Rhodes University.

Lotz-Sisitka, H. (2013). Conceptions of quality and 'learning as connection': Teaching for relevance. *Southern African Journal of Environmental Education, 29*, 25–38.

Lotz-Sisitka, H., & Janse van Rensburg, E. (2000). *Learning for sustainability contextual profile*. Johannesburg, South Africa: Learning for Sustainability Project.

Lupele, J., & Lotz-Sisitka. (2014). *Learning today for tomorrow: Sustainable development learning processes in Sub-Saharan Africa*. Howick, South Africa: SADC REEP.

Moose, J. (2009). *Recontexualizing issues in the "NISTCOL" environmental education curriculum module for primary diploma by distance learning in Zambia*. Unpublished master's (Education) thesis, Grahamstown, South Africa: Rhodes University.

Namafe, C. (2008). What selected schools of western province in Zambia are best at environmental and sustainability education. *Southern African Journal of Environmental Education, 25*, 59–80.

Nikel, J., & Lowe, J. (2010). Talking of fabric: A multi-dimensional model of quality in education. *Compare a Journal of Comparative and International Education, 40*(5), 589–605.

Rogoff, B. (1990). *Apprenticeship in thinking: Cognitive development in social context*. New York. Oxford University Press.

Rosenberg, E., O'Donoghue, R., & Olvitt, L. (2007). *Educational methods and processes for responding to environmental and sustainability concerns*. Howick, South Africa: Share-Net.

Rusbult, C. (2007). *Teaching methods*. Retrieved December 20, 2010 from http://www.asa3.org/ASA/education/tech/active.htm.

SAQMEC. (2000). *SAQMEC Country report for Zambia*. Retrieved October 10, 2010 from http://www.sacmeq.org/reports.htm.

SAQMECII. (2006). *SAQMEC II report*. Retrieved October 10, 2010 from http://www.sacmeq.org/reports.htm.

SAQMECIII. (2011). *SAQMECIII report*. Retrieved October 20, 2015 from http://www.sacmeq.org/reports.htm.

Sen, A. (1999). *Development as freedom*. Oxford, UK: Oxford University Press.

Tanzania, M. O. E. (2001). *Environmental education for teacher educators*. Dar-es-Salaam, Tanzania: E & D Limited.

Tikly, L., & Barrett, A. M. (2009). *Social justice, capabilities and the quality of education in low income countries* (EdQual research programme consortium on implementing education quality in low income countries. EdQual working paper no. 18a). Bristol, UK: University of Bristol. Retrieved November 3, 2010 from www.edqual.com.

Tikly, L., & Barrett, A. M. (2011). Social justice, capabilities and the quality of education in low-income countries. *International Journal of Education Development, 31*(1), 3–14.

Tikly, L., & Barrett, A. (Eds.). (2013). *Education quality and social justice in the global south: Challenges for policy, practice and research*. London: Routledge.

UNESCO. (2005). *United Nations decade for education for sustainable development: framework for a draft international implementation scheme*. Draft document. October, 2004 (see also www.unesco.org/desd).

UNESCO. (2014). *Global monitoring and evaluation report, shaping the future we want. UN decade of education for sustainable development (2005–2014)*. Paris: UNESCO.

Vygotsky, L. S. (1978). *Mind in society*. Cambridge, MA: Harvard University Press.

Wood, D. (1998). *How children think and learn* (2nd ed.). Oxford, UK: Blackwell.

Zambia, M. O. E. (2000). *Zambia primary school syllabus*. Lusaka, Zambia: Curriculum Development Centre.

Chapter 17
Reflecting on Innovative ESD Pedagogies in the Context of Teacher Education in Lesotho

Tšepo Mokuku and Mantoetse Jobo

This chapter presents selected innovative pedagogic strategies in environmental learning in teacher education programmes in Lesotho. Two main pedagogic strategies are discussed: conceptualisation of a holistic environment and outdoor learning. They constitute an important critique of the teaching and learning landscape in Lesotho that is predominantly classroom based, teacher centred and disconnected with the local context. The chapter illustrates how the concept of 'holistic environment' is conceptualised by engaging student teachers with their local school environment and perspectives. Students conceptualise 'holism' through an ecosystem experiment. The actual and potential benefits of outdoor learning in teacher education are also explored, based on selected local case studies.

Lesotho's intention to educate for sustainable development is stated in a number of national documents including the Constitution of Lesotho (Kingdom of Lesotho, 1993), the *Environmental Education Strategy Towards 2014: A Strategic Plan for Education for Sustainable Development in Lesotho* (Lesotho Government, 2009) as well as *Curriculum and Assessment Policy* (Ministry of Education and Training, 2009). These documents provide a framework for teacher education in the country to ensure reorientation of teaching to address the real and escalating socio-economic and environmental problems that result from overdependence on natural resources. Teacher education is playing a major role in addressing environmental challenges in a country as well as teaching citizens how to live sustainably.

T. Mokuku (✉)
Faculty of Education, Science Education Department, National University of Lesotho, Roma, Lesotho
e-mail: tmmokuku@yahoo.com

M. Jobo (✉)
Faculty of Science, Applied Science Department, Lesotho College of Education,
P.O.Box 1393, Maseru. 100, Roma, Lesotho
e-mail: mantoetse_ejobo@yahoo.co.uk

© Springer International Publishing Switzerland 2017
H. Lotz-Sisitka et al. (eds.), *Schooling for Sustainable Development in Africa*,
Schooling for Sustainable Development 8, DOI 10.1007/978-3-319-45989-9_17

There are two teacher training institutions in Lesotho. The diploma certificates at pre-primary-, primary- and secondary-level teachers through a pre-service and distance learning mode are offered at the Lesotho College of Education (LCE). The National University of Lesotho (NUL) offers graduate and postgraduate programmes through a pre-service mode for secondary and an in-service mode for primary-level teachers.

This chapter presents selected innovative pedagogic strategies in environmental learning in the teaching of certain courses in teacher education programmes at NUL and LCE. Two main pedagogic strategies are discussed in terms of their potential to actively support students conceptualise a holistic view of environment and to engage with the local biophysical environment and indigenous knowledge. The specific strategies discussed are conceptualisation of a holistic environment and outdoor learning. In this context, they constitute an important critique of a teaching and learning landscape that is predominantly classroom based, teacher centred and disconnected with the local context. Firstly, the chapter illustrates how, in one course offered to postgraduate diploma students, the concept of 'holistic environment' is conceptualised by engaging student teachers with their local school environment and perspectives. The teaching of the same concept is further demonstrated in the college context, where students conceptualise 'holism' through an ecosystem experiment. The outdoor learning strategy is explored in the literature to articulate its meaning and significance in the Lesotho context; secondly, we demonstrate actual and potential benefits in teacher education based on selected local case studies.

17.1 Conceptualising a Holistic Environment

In this section we consider how a holistic view of environment is conceptualised in the context of two teaching strategies. In one strategy, at NUL, students engaged with the literature definitions and perspectives of environment and their local school environment to broaden their understanding of environment; at LCE students were exposed to a classroom-based experiment, which they interrogated to conceptualise a holistic functioning of the environment – the planet Earth.

In the introductory part of the Postgraduate Diploma in Education (PGDE) course entitled *Environmental Education* (SCE506), offered at the National University of Lesotho, students are exposed to different environmental world views and how they shape and determine our interactions with the environment. The course is offered part-time as an elective to bachelor's degree holders, who mostly teach at secondary school level, without a teaching qualification. The students are introduced to a 'holistic' perspective of environment as a dynamic interaction of the biophysical, social, economic and political systems drawing on the works of Barrow (1995), le Roux (2000), Sandell, Öhman and Östman (2003) and Loubser (2011, 2014), amongst others. To relate the literature to the local context, in conceptualising the environment, they are first engaged to conceptualise 'holism' in their school environment, through the following assignment:

> *Describe your institution environment in terms of a 'holistic' perspective according to O'Donoghue and others. Give specific examples in your explanation to illustrate your points. Your response should clearly illustrate the <u>dynamic interactions of the systems</u>/dimensions of the environment, and how <u>they shape</u>/impact on the biophysical system.*

This assignment tended to broaden the students' thinking in terms of their conception of environment. Their responses often outlined the various examples of instances, activities and structures that fit within the four dimensions of environment in their school environment. There were usually two striking dimensions in the students' responses. First, there was their inability to articulate the dynamic interactions between the pillars of environment and to demonstrate how the interactions precipitate into an impact or an environmental issue on the biophysical system. Drawing students' attention to these processes, prior to their reflection on their local school environment, was a key element in the development of a holistic conceptualisation of the environment. A second important issue was to engage students with indigenous environmental perspectives from the Lesotho context. These are shared with the learners for discussion and critique. The non-scientific and spiritual local perspectives often came through as comparatively distinct and generated great excitement and debate amongst the learners.

The materials shared with the students drew on the local research undertaken in the Lesotho highland communities and reflected evidence of world views that are an amalgam of scientific and non-scientific knowledge forms in biodiversity conservation (Mokuku & Mokuku, 2004), people's conception of environment (Maloti-Drakensberg Transfrontier Project [MDTP], 2006) and wetlands (Mokuku & Taylor, 2015). In these communities it had been established, for instance, that organisms are respected for their apparent ecological functions in the food web, some for their supernatural and lethal powers often expressed when destroyed and others for their ability to communicate messages and to convey blessings to humans when seen and encountered and when performing certain movements (Mokuku & Mokuku). These were organisms in the taxa of amphibians, reptiles, birds, insects and plants (ibid.). The affirmative presentation of the knowledge tended to create a climate wherein students, especially those with rural background, comfortably shared knowledge that would normally be denigrated and denied space in the classroom.

Another indigenous perspective of environment that students engaged with, which is perhaps downplayed in the four pillars of environment, is the recognition of *Molimo* or God as the power that brought environment itself into existence. In a study that explored how the highland communities would describe 'tikoloho' or environment, it was defined as the creation of God, a place 'where animals, plants and humans that are created by God are found'/*Ke moo liphoofolo, limela le batho ba hloliloeng ke Molimo ba fumanehang* (MDTP, 2006). In one community, wetlands were viewed as animate and responsive to external pressure (Mokuku & Taylor, 2015).

Table 17.1 Characteristics of the environmental world view emerging from research in Lesotho context

Humans are connected in intricate and spiritual ways with other others, other life forms and ancestors
Plants and animals are respected and revered for the powers they possess and abilities to convey blessings and communicate important messages to humans
Revealed and spiritualised knowledge constitutes an important body of viable knowledge, e.g. knowledge revealed through dreams and spontaneity
Natural systems can be viewed as animate and responsive, e.g. wetlands
Scientific knowledge constitutes an important body of knowledge, but it does not dominate

When stressed by anthropogenic pressures, such as the construction of a well, wetlands were said to 'move or migrate' to another place (ibid.). Drawing on these indigenous world views, Mokuku (2012) has advanced a theory of *lehae-la-rona*, which means 'our home'. In terms of this theory, our world is sensed as *lehae* or a home – a place of plurality of knowledge, cultures and life forms that are physically and spiritually interconnected. In this amalgam of knowledge, viable non-scientific knowledge can be revealed to people, through *lipono*,[1] and there is consciousness of oneness with 'distant' people, cultures, life forms and places. The key ideas in relation to the indigenous world view emerging from the above-mentioned local studies carried out in the Lesotho context, which the students were exposed to for interrogation, are summarised in Table 17.1.

The exposure of students to a holistic view of environment and the indigenous conceptions in the form of readings and class discussions were intended to broaden their conception of their environment beyond scientifically rational understanding. It also aimed to create an epistemologically and ontologically pluralistic learning space by recognising and affirming viable world views on environment from the learners' own culture in the classroom.

In order to facilitate students' conception of the Earth as a holistic system of natural cycles that make the planet function, students at LCE worked with a functional experimental model. This began with two colleagues who taught agriculture working on a 'spaceship' metaphor using a 5-l jar with a lid, moist soil, *plantain* plant and sunlight (see Fig. 17.1). The jar was placed where it gained full access to any sunlight on a daily basis. They worked with student teachers and a *plantain* of about 5 cm was transplanted into a jar on 9 August 2013. The jar was closed with a lid that has not been opened to date. The *plantain* grew over 5 months and reached the lid of the jar, bore flowers and ultimately died. Before the *plantain* died, another plant, *Clover* species, whose seeds came with soil, germinated and began to grow and continued when the *plantain* died. The plants that have been growing in the jar have been recycling moisture through transpiration. Droplets of moisture on the

---

[1] *Lipono* is a Sesotho language word for 'revealed knowledge', which could occur through dreams, intrusive insights and other means.

Fig. 17.1 Ecosystem jar – plants have been growing for over a year and a half, and the lid of the jar has not been opened since the day of planting

wall of the jar from condensation processes were seen every morning and would run back into the soil. Nutrients and air were also recycled in the jar.

This jar has been exchanged between lecturers in different courses, including biology, to engage their learners with sustainability concepts. The recycling processes of moisture, air and nutrients that were seen in this jar have illustrated to the students the 'spaceship' metaphor in which the planet Earth is likened to a jar that has all systems that support life as it is in the jar. Thus, students and teachers all learn about how our world functions holistically as one unit and consider their responsibilities to sustain the systems, on which all life forms depend.

The new teaching strategy of using an experimental set-up to illustrate a concept that has traditionally been taught in an abstract way came as a response to the long journey of traditional disciplinary teaching and learning followed in teacher education in Lesotho (Jobo, 2013). It presents itself as a new 'paradigm' for teacher trainers at LCE.

17.2 Outdoor Learning

In Lesotho teachers hardly ever use the school grounds for teaching, and exploration of local neighbourhoods and excursions is rare. Outdoor activities usually involve occasional whole-school long-distance school trips. These trips require teachers to apply for permission from the Ministry of Education and Training (MOET) through a bureaucratic procedure of application to the Ministry. These school trips usually involve a whole school or several combined grades taking a tour locally or to the neighbouring countries to view the sea, sites of historical interest, animals at a zoo or nature reserve. These trips usually have little or no educational planning linking them to the curricula. The activities could thus be described as more reflective of *outdoor education* than *outdoor learning* (Beames, Higgins, & Nicol, 2012). The concept of *outdoor education* has its roots in Western countries and usually refers to activities that are undertaken with the learners outside the classroom, away from school, but which do not necessarily facilitate learning of the subject content:

> Outdoor education in many English-speaking 'Western' cultures has, in the last 50 years or so, become increasingly focused on adventurous activities conducted in highly controlled environments (e.g. ropes courses). These often take place far from the school, have few connections to the curriculum, and are provided by instructors trained to facilitate these activities using specialised equipment. (Beames et al., 2012, p. 4)

The school trips in Lesotho take a form of 'tour adventures' and, although not identical to the outdoor education described above, are comparable in that they engage learners in outdoor activities with little or no outdoor learning. Outdoor learning, on the other hand, concerns activities that take place outside the classroom, with the intention to meet the curriculum objectives and achieve associated learning outcomes. Outdoor learning is a concept used to refer to all kinds of learning that might take place outside the classroom (Beames et al., 2012) and may occur in the following zones: school grounds; local neighbourhood – can be explored by foot or by using public transport; day excursions (field trips) – usually require group transport; and residential outdoor centres which involve expeditions away from home, overnight. Nielson argued that classroom-based teaching can be oppressive and that 'classroom walls are strong in keeping students comfortable with the idea that nature is outdoors and humans are separate from environment' (2009, p. 140). Beames et al. (2012, p. 7) claimed that outdoor learning should be a regular feature of lessons:

> Rather than being regarded as an infrequent, recreational disruption to learning, taking classes outdoor should be seen as an extension of, or indeed integral part of classroom activities and used to meet the curricula and other needs of students… Outdoor learning is a key way of integrating curricula content that, depending on age and stage, is often traditionally taught in separate subject areas (e.g. geography, literature, ecology, and history).

In Nielson's (2009) view, taking lessons outdoors exposes the class to numerous factors that shape their environment and the wider society and allows the teacher and the learners to be collaborative learners. In her view, going outdoors bridges the 'separation' between the teacher and the learner, as the teacher collaboratively explores new knowledge with the learners. In the context of this joint outdoor work, learners become aware of the impact of their own actions on the environment, as well as reconnect with the natural environment (Beames et al., 2012). Below we analyse three cases of outdoor learning at NUL and LCE.

17.2.1 Case Study 1: Outdoor Learning in School Grounds

This first case is of a study[2] that was undertaken in an urban-based primary school in Lesotho to generate knowledge for inclusion in teacher education programmes in respect of *outdoor learning* and *peer tutoring* (Mokuku, Ramakhula, & Jobo, 2012/2013). A team of NUL, LCE and Durham University (UK) researchers collaboratively conducted the study under the aegis of Development Partnerships in Higher Education (DelPHE) programme. The study was guided by the following research questions:

1. What are the learners' experiences of peer tutoring in the context of their participation in outdoor learning activities?
2. To what extent is peer tutoring appropriate for teaching and learning in a primary school setting in Lesotho?

In this context, the introduction of outdoor learning was an exploration of a rare practice of engaging learners with their local environment in schools that are characterised by predominantly classroom-based teaching. Thus, the research team expected the findings of the study to make a significant contribution to ESD pedagogy in teacher education programmes at the participating local institutions. The investigation of peer tutoring was seen as a response to an unintended consequence of a free primary education policy that the government introduced in 2000. In addressing a sustainable development issue of many children who could not attend school due to their families' inability to afford school fees, the government of Lesotho introduced the policy and thus inadvertently burdened schools and teachers with huge numbers of learners (World, 2005). Large class sizes are one of the several classroom challenges for which teachers are ill-prepared to work with in their pre-service training.

The outdoor activities involved learners telling stories based on litter they collected from the school environment, colour observation and identification in the local school environment to develop concepts such as camouflage in animals, as well as litter reuse and recycling activities. Peer tutoring is a strategy of learning in which learners in the same class or in another class in the same school provide one-on-one support or one learner assists a small group of learners (Fitzgibbon, 1992). The study involved a peer-tutoring trial for 2 months in a primary school located in Maseru that had a population of about 800 learners (classes 1–7). The tutoring was arranged so that 104 year 6 learners tutored 86 year 2 learners. The older learners were expected to help younger ones and to benefit equally from the tutoring process. After 2 months, focus group discussions were conducted with 20 tutors and 20 tutees, with five members per group. Both the tutors and the tutees responded

---

[2] In 1997–2010, the National University of Lesotho led a Development Partnership in Higher Education (DelPHE) study in partnership with the Lesotho College of Education and Durham University in the UK: a British Government sponsored project aimed to investigate effective and relevant teacher education in Lesotho, in relation to numeracy, English literacy and Education for Sustainable Development (ESD).

Table 17.2 Lessons on joint use of outdoor learning and peer tutoring

ESD knowledge gains
Learners (tutors and tutees) can feel knowledgeable enough to advise (e.g. on others on pollution)
Willingness to act in response to environmental issues (e.g. pollution)
Learners (tutees) can develop awareness of the implications of their actions on the environment (e.g. littering)
Learners can begin to own the school environment as a 'learning environment', which eventually results in students' improved performance
Play-like outdoor activities register particularly positively with young tutees
Learners (tutors and tutees) can place little value on the more reflective and structured indoor part of the outdoor activity
The deeper meaning, embedded in outdoor learning activities, can easily be lost in play-like activities; teachers should guard against this
Outdoor pedagogy and peer tutoring
Tutors can recognise and appreciate outdoor learning as a significant shift from the traditional textbook learning to learning by doing things outdoors
Tutors can feel nervous about their new supportive role of facilitating outdoor activities and tutorship, and teachers should guide them accordingly
The act of tutoring outdoors can create a nexus of interactions between and amongst the tutors and tutees, thus enabling meaningful learning for all participating learners
Tutors, with assumptions the nature of lessons as teacher centred, may feel disrespected or inadequate and stifle dynamic learning, when engaging with the tutees in the power relations even outdoor set-up
The tutors and tutees can have positive social relations, providing tutees with attention and freedom rarely received in large classes, thus facilitating learning
The use of mother tongue, Sesotho, can become a spontaneous expression of thought amongst learners appreciated by learners and a freedom from the use of English language which the teachers use in structured classroom-based lesson
Tutees may seek to be more controlled by the tutors, as in the traditional setting, negating the freedom that outdoor peer-tutoring learning environment provides
Development of social skills
Through peer tutoring tutors can learn a variety of social skills including the following: take care of the tutees and know them better and increase confidence to talk and participate in class

positively to the study indicating a number of pedagogic benefits of both outdoor learning and peer tutoring (see Table 17.2). The two strategies have been introduced in an environmental education course (SCE506) for PGDE students, at the National University of Lesotho, and inform teaching and learning at LCE.

17.2.2 Case Study 2: Outdoor Learning in the Teacher Training College Context

The Lesotho College of Education has had opportunities in collaboration with SADC-REEP to develop the capacity of some of its teacher educators in ESD, in line with the policy framework of the country. In one case, a teacher trainer who participated in the SADC programme affirmed changing her pedagogical approaches from old to new styles of teaching. In her case she experienced such a mind-set shift that she described herself as a 'new teacher'. This agriculture teacher educator participated in a 'change project' that required her to develop a specific plan on how she would teach a chosen course differently, followed by a training workshop. As a result of her reflexive implementation of the planned project of teaching the agriculture course differently to the primary teacher trainees, she encountered 'levers and mechanisms that enable ESD learning processes to evolve and flourish' (Lupele & Lotz-Sisitka, 2012, p. 9) in her courses. She worked with student teachers using outdoor learning methodologies in which she involved student teachers working in teams through a project-based learning approach to identify soil erosion problems around the college campus. After learning about theories of soil conservation, student teachers undertook projects to conserve soil in the college campus using recycled materials such as waste plastic to barricade areas under protection. She avoided the traditional teaching approaches whereby knowledge is delivered to students and expounded by the teacher. This approach lends itself to a very narrow knowledge base since it originates from the teacher as a single source of knowledge.

This teacher educator's students worked in teams, seeking out possible solutions and implementing them to reclaim the eroded land using the built knowledge base that they learned from the theories. This exercise stimulated much learning and excitement both for students and the teacher trainer herself. In this teacher educator's approach, students' teamwork allowed for broader knowledge production by the students when deliberating on real issues, applying solutions and observing results while they had time to analyse and reflect on solutions. According to this teacher educator, students drew on all forms of knowledge, including indigenous knowledge, in their deliberations on possible causes and solutions to problems at hand. As such, they learned experientially how best they could also teach in schools. They gained confidence in teamwork. The teacher educator stated that, at first, teamwork seemed a big challenge. She also indicated that student teachers developed critical thinking skills when they analysed newly formed, experienced knowledge against previous knowledge they had learned earlier in life, which they had not applied. Outdoor exploration of real-life problems under the teacher educator's facilitation enhanced students' confidence. This teacher trainer claims that her students portrayed action competence as they visualised solutions from different actions and based on a variety of options with good reasons for what they valued most. Lotz-Sisitka and Russo (2003, p. 2) argued that actions such as this enable 'environmental education in the region to become more deliberative, interactive and action-centred'.

17.2.3   Case Study 3: Outdoor in the University-Community Context

The National University of Lesotho students who are in their third year of their Bachelor of Science with Education (BSc. Ed.) degree programme and Diploma in Agriculture Education (DipAgric. Ed.) programme take a course entitled

Development of Science and Technology in Society (*STS*), coded SCE 303, in the Science Education Department. The overall aim of the course is to deepen students' understanding of the nature of science and technology and their application in communities to solve real-life problems. At the end of the course, students should demonstrate the following outcomes:

Comprehension of the meaning and nature of science and technology:

(a) Ability to distinguish scientific rationality from other forms of thinking (e.g. superstitious thinking) in their society, as well as appreciate its value
(b) A critical understanding of science and technology developments and their impact on society
(c) Teaching science in ways that enhance pupils' ability to think critically and rationally, take decisions and solve problems that affect them and their society

In this course, distant outdoor learning excursions are often organised. For the first time in 2014, a class of 58 students visited a community with which the Department, within the framework of the Regional Centre of Expertise-Lesotho, is engaged in a collaborative project to conserve biodiversity and improve people's livelihoods. The community is situated north of Lesotho, in the Botha-Bothe district, about 250 km from the University, in the scenic Tlokoeng Valley. Initiated in 2012 with the support of Global Environment Facility's Small Grants Programme, the project aimed to conserve birds, wetlands and related biodiversity in the valley. The Southern Bald Ibis is confined to Lesotho, north-eastern South Africa and western Swaziland with an estimated population of 8000–10,000 individuals, of which approximately half involve breeding birds (Allan, 1997). In 2000, an assessment of their global status estimated a decline of 20 % over the next three generations if habitat loss to afforestation, mining and increasing human population continues (Barnes, 2000). Two decades ago it had been suggested that their numbers may be declining in Lesotho (Bonde, 1993), but no formal assessment has been made to date. The Southern Bald Ibis is currently considered to be vulnerable to extinction owning to its small population which is projected to decline in the future due to habitat loss and degradation (Barnes; IUCN, 2013). The project has turned out to be an exciting experimental site for knowledge generation, indigenous knowledge development, eco-tourism skills development and community-school partnership exploration.

Some achievements of the project to date include development of nature trails, training of community environmental education facilitators and the production of environmental educational materials. In recognition of the value of scientific rationality in conservation, the project team innovatively introduced biodiversity-related scientific concepts in the environmental education materials. The education policy in Lesotho dictates the use of Sesotho as a medium of instruction for the first 3 years of schooling only, based on the widely held assumption that the language is incapable of communicating complex scientific ideas. On the contrary, the research succeeded in developing Sesotho materials that articulated scientifically related concepts in relation to the Southern Bald Ibis and other birds of the valley and the processes of landscape formation. The scientific content includes description of the structure, function and behaviour of the ibis and owl species, the food chains associated with the birds and their adaptation to the environment. In addition, the research team explored and documented knowledge on *botho* philosophy in relation to nature and integrated it into the environmental education activities. The application of the philosophy has proved to be complementary to the scientific knowledge taught to the community by providing an ethical base for conserving other life forms. The students' exposure during the field trip included a community youth-led presentation on *botho* philosophy and trails focused on biodiversity. They experienced and reflected on various aspects of environmental education and on the manifestation of some key concepts associated with SCE303. To guide them in achieving this, they were given a worksheet to submit after the field trip.

The specific concepts that the students were to learn during the outdoor activity concerned the following aptitudes in respect of the application of science in the community: *action taking to solve real community problems, application of science to solve at least one real-life problem, decision-making issues that affect the community, critical thinking, clarification of values and active participation in democratic decision-making processes.* These aptitudes had previously been taught in an abstract way and many students struggled to conceptualise their meaning. Their worksheets reflected a marked improvement in their comprehension of various concepts in the context of Tlokoeng community. Some examples of their responses are reflected in Table 17.3. However, many of the students' answers were rather shallow and vague, suggesting that they still struggled to think at a higher order relating their fieldwork experiences with the mentioned aptitudes and skills.

17.3 Conclusion

In this chapter we have demonstrated innovative teaching strategies in relation to conceptualisation of a holistic view of environment and outdoor learning. We illustrated the significance of engaging with and analysing the local environment and indigenous perspectives to deepen and broaden the conventional 'holistic' definition of environment as a complex of interactions of the economic, social and biophysical systems. Explicitly engaging students with local indigenous perspectives is important in this post-colonial environment, where indigenous knowledge and perspectives have historically been excluded from formal learning contexts. Using a scientific classroom-based experiment to deepen students' understanding of holism in terms of functioning of the Earth tends to develop their appreciation of the interconnectedness of the natural cycles and a holistic function of the planet.

Outdoor learning has been illustrated as relevant for its potential to break new ground, in a predominantly teacher-centred and classroom-based pedagogical landscape in Lesotho. Through this approach learners and students can collaboratively engage with the local environment. In the process the approach clarifies the subject matter concepts and generates new knowledge. Combining outdoor learning with peer tutoring can promote environmental learning, particularly in large classes, and develop learners' social skills and aptitudes.

Table 17.3 Excursion worksheet: development of science and technology in society

Aptitude/skill	Briefly explain evidence from your field trip experiences (examples of students' responses)
(a) Action taking to solve real community problems	'Wild animals are not killed since they are important for future generations'
	'People at Tlokoeng are competent; they are able to identify the existing problem which is unemployment and at the same time be able to solve with the available natural resources. For instance, they conserved birds of which the tourists are paying for the services'
	'The people of Tlokoeng undertook the process of conserving birds that were under threat of extinction by being killed by their community or neighbouring communities'
(b) Application of science by the community to solve at least one real-life problem	'Science is applied in the manner that they noticed that there is a problem of land degradation; thus, they used silt traps to control soil erosion'
	'The use of telescope to look at birds that are far'
	'Tlokoeng community realised that traditional hunting of animals will lead to high rate of loss of biodiversity; therefore, they think scientifically that the wildlife like birds must be conserved in order to make or encourage sustainable food chain or food web in the ecosystem'
(c) Decision-making on important issues that affect the community	'The decisions are made by the association, chief and the entire community collectively. This means when the association has something that is beneficial or that can affect the villagers, they consult the chief. The chief then gather the villagers together and they all negotiate together'
	'They decided to conserve the natural resources so that future generations can still have access to them. This was done by the community basing themselves on the herdboys hunting the birds'
	'The Tlokoeng community people made decision that they had to establish a project, and the name of the project is Thaba-Khubelu Conservation and Tours, which will protect the wildlife especially birds and owls and protect the natural habitats for birds and owls'
(d) Critical thinking by community members	'They believe (cit.) that if they continue killing the animals, one day they will go extinct; therefore, the coming generation won't have an idea of the existence of such animals'
	'Tlokoeng community has applied critical thinking whereby they were able to solve the problem of rats destroying the grains, by using owls to control the population of the rats, because they know that they can control rats without necessarily buying the chemicals'
	'In zoos animals are fenced expensively and people give them food, but in Looking they are conserved by means of educating the community which is cheaper and fairly critical'

(continued)

Table 17.3 (continued)

Aptitude/skill	Briefly explain evidence from your field trip experiences (examples of students' responses)
(e) Clarification of values/morals/ethics that are important for the community	'People should live in peace and harmony with each other, should happily help each other as much as they can. They should also have a feeling of humanity towards the animals by actually taking care of them; children should obey their parents'
	'Tlokoeng community people have commitment to participate in activities focused at addressing environmental problems like loss of biodiversity. They have willingness to work cooperatively with others to solve the problems, respect for life of animals'
	'They understand that people in the community ought to work together, care for each other and their environment. They live by proverbs like "A person is a person by other persons", which means for a person to be a person, they have to think about other people'
(f) Active participation in democratic decision-making processes	'Everyone in the community is involved in the decision-making such that youth and elders like chief's representative are engaged in the process of conserving the environment. The chief does not make decisions alone on solving problems but the community itself decides'
	'The Tlokoeng conservation and tours members assigned each some positions and some roles to play within the community to improve their environment. Each member is playing a certain role to ensure the growth of the association and the community in general, either economically or socially…'
	'The chief (through represented by his delegate) is incorporated in the scheme without actually taking a lead. They also had handicapped member whom they allowed to be within the committee'

References

Allan, D. G. (1997). Bald ibis *geronticus calvus*. In A. Harrison et al. (Eds.), *The atlas of southern African birds* (Vol. 1, pp. 104–105). Johannesburg, South Africa: BirdLife South Africa.

Barnes, K. N. (2000). Bald Ibis *Geronticus calvus*. In K. N. Barnes (Ed.), *The Eskom red data book of birds of South Africa, Lesotho and Swaziland* (pp. 69–71). Johannesburg, South Africa: BirdLife South Africa.

Barrow, C. J. (1995). *Developing the environment: Problems and management*. London: Longman.

Beames, S., Higgins, P., & Nicol, R. (2012). *Learning outside the classroom. Theory and guidelines for practice*. London: Routledge.

Bonde, K. (1993). *Birds of Lesotho. A guide to distribution past and present*. Pietermaritzburg, South Africa: University of Natal Press.

Fitzgibbon, C. (1992). Peer and cross-age tutoring. In M. Elkin et al. (Eds.), *Encyclopedia of educational research* (6th ed.). New York: Macmillan.

IUCN (2013). *Geronticus calvus*. Retrieved July 22, 2014 from http://www.iucnredlist.org/details/22697496/0.

Jobo, M. E. (2013). *Developing a revitalising resource material for enabling Lesotho teacher educators to integrate EE/ESD into curriculum*. Howick, South Africa: Share-Net, SADC REEP.

Kingdom of Lesotho. (1993). *Constitution of Lesotho*. Maseru, South Africa: Government Printers.

le Roux, K. (2000). *Environmental education processes: Active learning in schools*. Pietermaritzburg, South Africa: University of Natal Press.

Lesotho Government. (2009). *Environmental education strategy towards 2014: A strategic plan for education for sustainable development in Lesotho*. Maseru, South Africa : Government Printers.

Lotz-Sisitka, H., & Russo, V. (2003). *Development, adaptation and use of learning support materials in support of environmental education processes. SADC regional environmental education programme. Course developers' toolkit*. Howick, South Africa: Share-Net.

Loubser, C. P. (2011). *Environmental education: Some south African perspectives*. Pretoria, Pretoria: Van Schaik.

Loubser, C. P. (2014). *Environmental education and education for sustainability: Some south African perspectives* (2nd ed.). Pretoria, South Africa: Van Schaik.

Lupele, J., & Lotz-Sisitka, H. (2012). *Learning today for tomorrow: Sustainable development learning processes in sub-Saharan Africa*. Howick, South Africa: SADC-REEP.

Maloti-Drakensberg Transfrontier Project. (2006). *State of environmental education/awareness programmes in Lesotho*. Maseru, South Africa: MDTP.

Ministry of Education and Training. (2009). *Curriculum and assessment policy*. Maseru, South Africa: Ministry of Education and Training.

Mokuku, T. (2012). Lehae la rona: Epistemological interrogations to broaden our conception of environment and sustainability. *Canadian Journal of Environmental Education, 17*, 159–172.

Mokuku, T., & Mokuku, C. (2004). The role of indigenous knowledge in biodiversity conservation in the Lesotho highlands: Exploring indigenous epistemology. *Southern African Journal of Environmental Education, 21*, 37–49.

Mokuku, T., Ramakhula, L., Jobo, M. (2012/2013). Learners' experiences of peer-tutoring in the context of outdoor learning: The case of a primary school. *Southern African Journal of Environmental Education, 29*, 180–192.

Mokuku, T., & Taylor, J. (2015). Tlokoeng valley community's conceptions of wetlands: Prospects for more sustainable water resources management. *Journal of Education for Sustainable Development, 9*(2), 196–212.

Nielson, A. L. (2009). The power of nature and the nature of power. *Canadian Journal of Environmental Education, 14*, 138–148.

Sandell, K., Öhman, J., & Östman, L. (2003). *Education for sustainable development: Nature, school and democracy*. Lund, Sweden: Studentlitteratur.

World Bank (2005). *Building on free primary education, primary and secondary education in Lesotho. A country status report*. Africa region human development Paper Series No. 101.

Chapter 18
Enhancing Agency and Action in Teacher Education in Zimbabwe

Caleb Mandikonza and Cecilia Mukundu

In this chapter we use a case study of teacher education in Zimbabwe to review a professional development model for mainstreaming environment and sustainability education (ESE). The process is driven by a collaborative and practice-based course design where participants work towards a change project in their professional work context. An institutional change project approach was developed within a work-together/work-away process in a Rhodes University teacher education course undertaken in partnership with the Southern African Development Community Regional Environmental Education Programme (SADC REEP).

The chapter is introduced with an examination of the education context covering policy, compliance structures and agency-based processes that informed the inclusion of a participant-led intervention project as part of a professional development course. This scopes the expansion of teacher education in Zimbabwe, delving into the historical context and competency specification. It then looks at the course, specifically at the mediated actions in the on-course sessions and how these translated into work-based actions.

The course structure and the mediating processes provided teacher educators with an opportunity to implement ESE practices that were primarily based on, but not necessarily limited to, a social change process in their work context. The chapter is thus developed as a reflection on change-oriented learning in a teacher education course. In this way it examines the emerging opportunities and challenges that shaped efforts to mainstream ESE using this change project-based model in teacher professional development.

C. Mandikonza (✉) • C. Mukundu
Environmental Learning Research Centre, Rhodes University, Grahamstown, South Africa
e-mail: calebmandikonza@yahoo.co.uk; ctsopotsa@yahoo.co.uk

© Springer International Publishing Switzerland 2017
H. Lotz-Sisitka et al. (eds.), *Schooling for Sustainable Development in Africa*,
Schooling for Sustainable Development 8, DOI 10.1007/978-3-319-45989-9_18

18.1 The Importance of Professional Development in Teacher Education

Modern education is commonly orientated towards progress and development. Education is seen as contributing to a good life by offering solutions to problems of society while preparing people to cope with adversities and risks. It is also expected to open up opportunities for employment, especially in marginalised communities. Good education has high standards and enables meaningful learning with the prospect of social change, and this is the reason why the world continues to strive to achieve high-quality education. Consistent with these suppositions, education and training are seen as very important components in the endeavour to pursue quality education in Africa, and it follows that all efforts to educate imply the development of more meaningful and purposeful learning.

Teacher educators and, subsequently, teachers have an important role to play in education. The teacher is a member of society with the role of supporting learners and contributing to society. Teachers and teacher educators are, in this view, more than mere facilitators of knowledge acquisition. They work towards achieving the wider purpose of learning-led change amongst individual learners and in the community at large. In recognition of this role in science education, McGinnis et al. (2012) called for reforms in teacher academic professional development in order to enhance educators' knowledge and pedagogy continuously. Academic professional development should also recognise teacher educators as individuals interacting with other practitioners in intersubjective spaces of learning and social change (Kemmis et al., 2014). Forms of professional development therefore need to be sensitive to the provision of intersubjective conditions for learning interactions to take place. This chapter suggests that teachers have to be well prepared to implement an expanding educational agenda and that the teaching and learning should include innovative approaches and pedagogic renewal (Dembe'le' & Miaro-II, 2013). The approach to academic teacher professional development examined here claims to work towards enabling teachers to see themselves as members of the community they serve and to use classroom experiences to promote and engage in meaningful social change.

The massification of education in Zimbabwe has come to include its recognition as a societal strategy for alleviating poverty by providing the basis for employment. Here, quality teacher education includes change-orientated attributes of quality education. Teacher education institutions have guidelines of competences as indicators of performance that are expected of their staff. These competences are constantly reviewed in relation to powerful contemporary narratives of change and sustainable development influencing the sector. In southern Africa, quality and relevance in education, environmental education and education for sustainable development are some of the major narratives influencing the provision of academic professional development initiatives for teachers. The section that follows briefly reviews competences derived from global and national policies and guidelines in a Zimbabwean context, noting how these are framing new forms of teacher education practice that include a concern for change practices in an institutional context.

18.2 Institutional and Policy Influences on Teacher Education Competences

The University of Zimbabwe exists in the context of the challenging socio-economic conditions characterising the country. These include under-skilled and unskilled manpower in the majority of the population, housing shortages, differential incomes and opportunities, illiteracy, inadequate health and sanitation provision, underdevelopment of rural areas, HIV and AIDS, gender disparities, environmental degradation, worsening poverty and reduced access to ecosystem services. With the proclamation of education as a human right in 1990, there was compulsory free primary education for all Zimbabwean citizens. Since then, education has been regarded as a vehicle for tackling the aforementioned challenges.

The ballooning of primary school enrolment meant an increased number of schools leading to a shortage of teachers. This development was subsequently followed by expansion of teacher education. To begin with, institutions that were already training teachers were expanded from four government-led and four church-led institutions to 14 teacher training colleges working in association with the University of Zimbabwe. This expansion was accompanied by a rapid expansion in enrolment figures in these institutions and the challenge of giving teacher educators the capacities to function in the teacher training colleges. Although there has been an increase in the number of teacher education institutions, enrolment figures are still very high today as the need for teachers remains. The resulting large class sizes in colleges have implications for the quality of the training, hence the development of a list of expected competences for teacher educators and for the educators coming out of the training. These competences were for university personnel and for teacher education lecturers in the colleges. The link between the university and teacher training colleges was defined by the associate relationship.

Specific competences for teacher education are classified into four categories: knowledge and understanding, educational practice and skills, values and ethics and interpersonal and intrapersonal abilities (University of Zimbabwe, n.d.). The competences were derived in order to achieve the Faculty of Education's vision of being a premier institution in producing quality teachers and teacher educators and doing so through accrediting and supporting affiliated colleges of education. The notion of quality has a steering influence on teacher education practices reflected in the imperative of:

> Enabling our clients and customers to make meaningful contributions to sustainable development in Zimbabwe. To this end we provide high quality education, training and advisory services on a needs oriented basis. We guarantee the above by maintaining excellence in Teaching, Learning, Research and Service to the community. (University of Zimbabwe's Mission Statement, n.d.)

Table 18.1 Excerpt of teacher education-specific competences of the University of Zimbabwe

Knowledge and understanding:
The local and international social, political, economic, cultural and environmental contexts of education
National and institutional policies relating to education
Educational practice and skills:
Select, adapt and use appropriate teaching methods and learning activities
Use a range of assessment skills to set, mark and grade learners' achievement
Manage learners both inside and outside formal classroom contexts
Create conducive learning environments that encourage learning
Use language appropriately in the classroom and in the subject
Participate in basic educational research
Critically reflect on their work to improve practice
Adapt to change
Values and ethics:
Care for and support the well-being of all learners
Maintain equity and fairness amongst learners and promote inclusive education
Continuously upgrade their own knowledge and skills
Interpersonal and intrapersonal abilities:
Collaborate and network with others including peers, head teachers, professional groups and parents
Communicate effectively with different audiences and using appropriate tools, including ICTs, and relevant forms of discourse
Lead and manage groups

Here the quest for quality teacher education in the institutions shapes and structures competences for teacher educators that extend to the mainstreaming EE/ESD. The starting point for the case of teacher professional development examined here was a review of the competences specified in the University of Zimbabwe (Table 18.1) in relation to the outcomes of the Rhodes University/SADC International Certificate Course in Environmental Education.

Working independently, but related to the broad competences in Table 18.1, the Rhodes University/SADC International Certificate Course in Environmental Education was developed to:

- Enhance the policy, institutional and contextual relevance of environmental education programmes and activities.
- Develop deeper understanding of unsustainability practices and associated environmental issues and risks and how to respond to them through educational mediating activities.
- Develop understanding of more sustainable alternatives and how to enhance them through educational mediating activities.
- Improve the use of educational methods and materials for mediation of better learning and more sustainable alternatives (SADC REEP, 2009).

The key characteristic of this course expressed in the outcomes that gave voice to the competences was the work-together/work-away tasks, a reflexive process of working in and out of practice where each participant would undertake a change project that focused on valued practice.

One of the authors of this chapter (Mukundu) attended the Rhodes University/ SADC International Certificate Course in EE and undertook a change project relating to the nutrition module of the biology component of the science education curriculum in the University of Zimbabwe, the context of her teacher education work. She sought to mainstream environment and sustainability education (EE/ESD) in the university curriculum by developing the capacity for mainstreaming these imperatives amongst her student teachers and into the schools and communities in which they were to work.

18.3 Mainstreaming ESE in the Faculty of Education

Even though this is a case of Mukundu's experiences in mainstreaming ESE, she did not engage in mainstreaming in isolation from the institutional practice and that of her colleagues. She realised that it would have been much easier for her to work in a community of practitioners with similar interests and in a receptive and supportive institution. At the time, some of her colleagues had also attended either the same course or similar courses that were offered through the SADC Regional Environmental Education Programme (SADC REEP). Her colleagues, thus, formed a loose community of practice which became a critical mass for mainstreaming environment and sustainability concerns and practices in the institution.

Through the SADC REEP network, the university also became a member of the UNEP Initiative, Mainstreaming Environment and Sustainability in African Universities (MESA), as well as a member of the Southern African Regional Universities Association (SARUA). SARUA, particularly, promotes mainstreaming of climate change and development education across its membership. SADC REEP also supported the MESA initiative of mainstreaming ESD by providing small funds specifically to support institutional capacity development activities. Capacity development was either through conducting change-oriented research activities or capacity-building workshops for the institutional staff.

Teacher educators including Mukundu, who had all received some training on mainstreaming ESE, thus teamed up and applied for funding which they used to facilitate a faculty-wide workshop on mainstreaming ESE. This expanded institutional capacity provided more supportive conditions for individuals such as Mukundu to innovate their change-oriented teacher education practice. Institutional capacity development was further enhanced by the participation of some educators in the International Training Programme (ITP) on Education for Sustainable Development. This training was provided by SADC REEP and Ramboll Natura, a Swedish NGO that was funded by the Swedish International Development Cooperation Agency (SIDA).

A further institutional capacity development opportunity emerged with the partnership between SADC REEP and the Swedish International Centre of Education for Sustainable Development (SWEDESD) that was extended to the Education for Strong Sustainability and Agency (ESSA) programme. With this expanded capacity, it was easier for the institution to take the lead in applying as a community of practice in ESE to be recognised by the United Nations University as a Regional Centre of Expertise (RCE) in ESD for Harare, the capital city of the country. What is evident across all of the developing processes briefly described here is that change in practice has been led by individuals and emerging groups who have developed the competences to work in communities of practice fostering change.

This section has detailed some of the contextual conditions influencing the capacity of teacher educators in this one institution. The following sections explore more deeply a specific capacity development course, reviewing how the course process and tasks came to influence practice.

18.4 The Rhodes University/SADC International Certificate Course in Environmental Education

Rhodes University and SADC REEP worked in partnership over 15 years to develop and conduct a 2-month work-together/work-away course for professionals in various education sectors including teacher education. This became known as the Rhodes University/SADC International Certificate Course in Environmental Education. It was a participatory course (no formal examinations) that sought to enable participants on the course to understand issues affecting society and the environment and to be able to deliberate and take action to promote more sustainable alternatives as well as solutions in their lived contexts by reflecting on what they could do in their own practice.

The course recognised that changed practices can emerge from deliberative interactions across two or more forms of habitus (Bourdieu, 1998), in this case the habitus of the workplace and that of the on-course contexts. The course had to link course work and contextual practices through a workplace-based change-oriented assignment, the change project. Course participants were expected to emerge from the course with enhanced capabilities, for their individual and social lives as well as in their professional practice. Capabilities in the course context were interpreted as the abilities that people have to make choices about what they value and how these could be developed through intersubjective reflexive deliberations in a course-supported process of learner-led change.

In order to facilitate such deliberations and actions, the course foregrounded reflexivity (Bourdieu, 1998). The process of supporting reflexivity is described at two levels. The first level is the course structure, while the second level is the detailed activities that participants contributed to and codeveloped during the course.

The Rhodes University/SADC International Certificate Course in EE had three developmental segments. The first segment was completed in the participant's workplace as a pre-course assignment. Participants, together with their seniors and colleagues, undertook an audit process to identify issues of concern for the course participant to develop a potential focus to be taken up in the course. The pre-course assignment was the first step in providing course participants with a tool to work collaboratively with institutional leadership and with colleagues in interrogating social-ecological and professional concerns towards opening an on-course conversation on how to respond to the concerns. While on the course, the emerging focus on a change project was to be developed as a response to be designed and framed in order to address particular concerns.

The next phase of the course was the on-course segment. Conceptual understanding of environment and sustainability issues was deepened through lectures, deliberation, practical activities and excursions. The on-course activities were covered in the following three units:

- Environmental issues and risks: This unit supported participants to reflect on issues identified in their contexts and to develop deeper knowledge of these issues and identify alternatives.
- Methods for mediating learning: This unit required reflexive engagement with methods of teaching and sharing information.
- Reflexive implementation of environmental education projects in communities of practice. This unit encouraged participants to think of working with others in their institution and in other institutions and to implement and evaluate their teaching practices, thereby promoting collaboration and wider forms of reflexivity at various levels.

The final part of the course involved implementation of their institutionally identified and self-defined 'change project' back in their institution. This was to be undertaken in collaboration with the institutional leadership and colleagues in the workplace, the same people who initially assisted in identifying the sustainability concerns and suggesting a possible change project. These colleagues came to constitute the workplace community of practice for each change project. What follows is an overview of the daily experiences course participants engaged in to continue to cultivate change through change-oriented learning.

18.5 Daily Programmatic Experiences

The daily on-course programme was divided into four sessions. During the morning session participants deliberated and reflected on the previous day's activities to review concepts and activities covered during the previous day that were relevant to them. This process provided an opportunity for participants to reflect on their understanding of the concepts and what that understanding meant for their intended change project.

The mid-morning session was intended to enhance conceptual understanding. This was commonly undertaken through exposition methods such as lectures from facilitators or by experts working on particular concepts such as climate change, policies and practice nexus and waste management. Videos illustrating sustainable practices were shown and discussed during this session. This work emphasised Vygotsky's view that language and dialogue are key to developing meaning of experience, a view that supports the need to enhance the language of environment and sustainability in the course through enrichment sessions that are content orientated and deliberative in nature (Hasan, 2004). The course set out to enhance the content knowledge of practitioners in order to give them the capacity to consider knowledge-informed individual and social action (Maxwell, 2012).

The last half of the second session each day was used for 'reading to learn'. Participants read through the core text and selected extra readings that related to their interests from the course file or literature searches. These readings were related to the concepts presented during the content enrichment session. They were intended to consolidate understanding of the concepts and also to enable participants to learn to read from more than one source in order to critically engage with the concept and be able to build a sound knowledge-informed argument to guide their expanding ESE work.

The third daily session was for developing and supporting change projects. In this session participants were scheduled, in turn, to present their change project to the whole group. It was mandatory for participants to make a PowerPoint presentation so that participants who were not familiar with the computer could develop these skills. In addition to sharing their change projects, presentation skills were critiqued. Peer assessment was encouraged as participants contributed in clarifying the change project. Using collegial critiques, the change project ideas evolved and became richer and more feasible as the course progressed.

During the last session each day, participants worked in regional knowledge exchange groups. In these groups, participants developed their professional skills in areas of their choice that were not necessarily covered in the course content but could be useful in their practice as educators implementing a change project in their institutional setting. The four elective groups were:

- Monitoring and evaluation of the course
- Documenting course activities through photographs and video recording
- Writing articles for publication in the Environmental Education Association of Southern Africa Bulletin
- The use of ICTs in course design

Some of these daily experiences had a profound influence on the practices of participants, particularly the variety of learner-centred teaching and learning methods used and the regional scope of the knowledge exchange amongst the groups.

Against this brief sketch of the developmental course progression, the following section of the chapter will present Mukundu's experiences as a course participant who implemented a change project.

18.6   Motivation to Participate in Professional Development in ESE

Upon joining the university in 2007, Mukundu worked with two colleagues on a SADC REEP-supported research project on Education for Sustainable Development that focused on sustainability issues around a school in a local rural settlement. As a teacher educator, her position in a university contributed to a concern for teaching, research and community service. One of her concerns became distinguishing what counted as powerful knowledge and as legitimate in a university course context where she had to generate tests and evaluate students. Through dealing with various knowledges, teacher educators are also in the position to establish knowledge relationships in society. With the intent of taking up these concerns, Mukundu enrolled in the Rhodes University/SADC International Certificate Course in EE.

In addition to engaging with the issue of quality education which her institution was striving for, the work with others on the course enhanced her capacity to effectively meet the performance expectations of her institution. She understood quality education to be evident in a form of education that made a positive impact on individuals and communities. It was not education for its own sake but a process that created experiences which enhanced individuals' lives by encouraging them to critique current conditions and explore better alternatives for the future. During the course she reported that an exploration of these questions had a positive influence on her practice more than others.

18.7   An Overview of Relevant On-Course Experiences

Mukundu valued the interaction in networked groups of different people from across the SADC subregion who worked on very similar environmental and sustainability challenges during the on-course sessions. As course participants, they interacted and shared different experiences so that she was able to relate issues experienced in other institutions to her own experiences in her institution. In addition, she identified activities that other institutions were working on that could also be conducted in her institution. She realised that learning new concepts in ESE helped to expand her social-ecological knowledge. This knowledge was necessary for developing different types of competences that she needed in her expanding ESE practice.

She found this constant reflexive engagement with her practice during the sessions very meaningful. Initially, she struggled to understand the notion of the change project as proposed in the pre-course assignment brief. She started off intending to embark on a reforesting project in the university grounds, a project that she and her colleagues had identified during the pre-course preparations. However, with more exposure during the on-course sessions, she realised that the project would be outside her existing teaching responsibilities. Learning support materials, on-course activities, tutor support and peer review enabled her to interrogate her practice and to craft a new change project based on her classroom practice. Her ideas shifted from reforestation to the inclusion of practical activities and excursions to enrich the teaching and learning of malnutrition within the module she taught on public health.

18.8   Implementation of the Change Project

On returning to her institution, she worked with her students to demonstrate growing the vegetables needed for a balanced diet. Here, she used innovative gardening practices she had experienced on the course. These were appropriate for her students who lived in urban areas like Harare, where homes do not have space for gardening and experience extreme shortages of water. Mukundu planned for her biology students to fill old tyres, old tins and old plastic bags with fertile soil and to plant different food crops. The containers were then to be placed in any available spaces within the college. The fact that the nutritional practices reach out to solve societal problems was closely associated with ESD learning processes which would strengthen educational quality and relevance. Learning biological concepts through practical activities would be meaningful to both the learners and the community she interacted with.

Due to challenges of limited space in her institution, Mukundu could not set up the innovative garden plots as she had planned. The challenge was thus taken into the living contexts of the students and out into the schools on teaching practice. Many expressed great enthusiasm and joy in being able to produce vegetables using innovative gardening with waste to provide fresh vegetables cheaply, widening choices of nutritive food through teaching about public health issues. As teachers, they could extend the talk about balanced diets by giving learners options on how to provide a simple solution through growing certain plants to prevent malnutrition. The learning process was practice based and resulted in the transformation of individuals and the way they lived. Even though the impact of such practice has not been evaluated, Mukundu believed she was contributing to quality education that was relevant to the biology students she taught through her small curriculum innovation project developed on the course.

18.9   Lessons Learned

Change of practice in teacher education is a complicated process and at times needs supportive teams for it to happen. Mukundu had to seek support from colleagues with similar capacity in order to raise the concern for mainstreaming environment and sustainability education at a faculty level and in her subject field. In the teaching of nutrition using the creative gardening techniques, she had taken students away from the usual concept-laden lectures that were the norm in their teaching. They had an opportunity to learn together in groups, to act on their ideas and to learn the core ideas in meaningful ways that related to the conditions of their daily lives and those of the students they were to teach. Student teachers thus collaborated to work with soil, manure and waste containers, materials that they would normally not use in a university classroom. The activities brought together a number of concepts ranging from conditions necessary for plant growth and development, graphing, the nature of soil, natural materials and nutrient cycles as well as types of crops that can be grown and how these can maintain good health in urban communities.

Student teachers engaged in much deliberation, negotiation and consensus building as they collaboratively engaged with the structures and discussed the various concepts, which they would not have done through a lecture. The practical gardening challenge was a key step in creating a platform for sociocultural learning where people learn through participating in groups (Edwards, 2014). Students were able to relate growing vegetables in their own homes and schools during the learning process as well as connect with the reality of their lived world and the possibilities thereof. In so doing, they were endowed with the capacity to pursue their personal and community goals around nutrition and poverty alleviation while interrogating their own practices and beliefs in science teaching and what the science they taught would mean to their own students (Adams, 2012). The change project approach engaged with in this way enhances the social relevance of science education and how it can contribute to sustainable development.

The interaction with various change projects during the on-course presentations provided fertile ground for the teacher educators to reorient teaching towards improving the teacher's role in the learning process. For example, the course's emphasis on active learning and participatory methods and the use of appropriate learning support materials were specific ways that enhanced Mukundu's understanding of her role as a teacher educator. In this way, learning processes through a change project approach managed to provide support for continuous academic teacher professional development. Follow-ups of students on teaching practice showed that some of them had established these innovative gardens. This was a sign that active learning processes were being cascaded to new learners while impacting on and addressing real situations.

It was useful for the participants to work with colleagues who attended the same course and those who attended other ESD courses such as the Swedish International Training Programme (ITP). Participating in the course created a new level of professional relationships with colleagues who had similar capacity and the heightened agency to enable the rest of the staff in a faculty to engage with ESD. This is the capacity that was used in the application for recognition with the United Nations University on the Regional Centre of Expertise (RCE) in ESD.

18.10 Conclusion

This account of the course and a change project is only one amongst many other very diverse initiatives characterising a course-supported mainstreaming of ESE in higher education through the curriculum, pedagogy and social innovation. The initial intention of mainstreaming activities using the change project was to change many teacher educators' mind-sets regarding how they recognise, describe and explain as well as teach issues relating to livelihood improvement. Change-oriented learning happened at various levels on the change project course. Firstly, the initial assignment facilitated Mukundu to think reflexively about her individual and institutional practice, together with colleagues. They seldom put heads together on such professional matters. She pointed out that transformation of her understanding that she used to develop the change project increased as she interacted with the course tools and processes. She implemented her curriculum-based change project in her practice but found collaborative and supportive conditions by institutional leadership and colleagues useful. Collaborative practice was further evident in her desire and ability to work in communities of practice. In addition to changing her teaching practice, the change project course-supported collaborative and democratic practice. Transformation of practice and of the individuals in a community of practice requires a receptive and supportive institution, but once the culture of practice gains momentum, it also transforms institutional practice and institutional relationships. The institution gets networked to other such institutions through supporting their change projects. Course-mediated change-oriented learning processes need to be structured to take into account how to facilitate reflexivity of the individual in an institutional setting and how that individual can be encouraged to interact with others to implement the valued social changes in and across communities of practice.

References

Adams, J. D. (2012). Community science: Capitalising on local ways of enacting science. In B. J. Fraser et al. (Eds.), Second international handbook of science education (24th ed.). London: Springer International Handbooks of Education.

Bourdieu, P. (1998). Practical reason: On the theory of action. Cambridge, UK: Polity Press.

Dembe'le', M., & Miaro-II, B.-R. (2013). Pedagogic renewal and teacher development in Sub-Saharan Africa. In B. Moon (Ed.), Teacher education and the challenge of development: A global analysis. London: Routledge.

Edwards, A. (2014). Designing tasks which engage learners with knowledge. In I. Thompson (Ed.), Task design, subject pedagogy and student engagement. London: Routledge.

Hasan, R. (2004). The world in words: Semiotic mediation, tenor and ideology. In G. Williams & A. Lukin (Eds.), The development of language: Functional perspectives on species and individuals (pp. 138–181). London: Continuum.

Kemmis, S., Wilkinson, J., Edward-Grove, C., Hardy, I., Grootenboer, P., & Bristol, L. (2014). Changing practices, changing education. Singapore, Singapore: Springer Science & Business Media.

Maxwell, J. A. (2012). A realist approach for qualitative research. London: Sage.

McGinnis, J. R., Hestness, E., Reidinger, K., Katz, P., Marbach-Ad, G., & Dai, A. (2012). Informal science education in formal teacher preparation. In B. J. Fraser et al. (Eds.), *Second international handbook of science education* (24th ed.). London: Springer International Handbooks of Education.

Southern African Development Community Regional Environmental Education programme (SADC REEP). (2009). *Course notes*. Howick, New Zealand: SADC REEP.

University of Zimbabwe. (n.d.). *Competences for lecturers*. Harare: University of Zimbabwe.

Chapter 19
Towards Professional Learning Communities: A Review

Sirkka Tshiningayamwe and Zintle Songqwaru

Professional learning communities (PLCs) have proven to be effective models of teachers' professional development. Therefore, for effective Education for Sustainable Development (ESD), PLCs should be considered in the planning of teacher professional development initiatives. This will enable teachers to address issues of quality teaching, learning and sustainable development issues. In the South African context, a 'professional learning community' is an emerging policy concept. This chapter thus provides a review of the emergence of teacher clusters in South Africa using a reflexive spiral model and praxis tasks. The chapter also draws from the Namibian and South African teacher cluster literature to comment on how teacher clusters could potentially be translated into PLCs to transform teachers' practices as they relate to ESD.

Increasingly, teacher collaborative models are used in helping teachers reshape their professional knowledge and change their classroom practices (Darling-Hammond, 1996; DuFour, 2004; Feger & Arruda, 2008; Jita & Ndlalane, 2009; Schmoker, 2006; Stoll, Bolam, McMahon, Wallace, & Thomas, 2006). In line with international trends, the *Integrated Strategic Planning Framework for Teacher Education and Development (ISPFTED) in South Africa (2011–2025)*, prepared by the Department of Higher Education and Training (DHET) and Department of Basic Education (DBE) (South Africa. Department of Higher Education and Training & Department of Basic Education (DHET & DBE), 2011) in South Africa, aims to promote teachers' continuing professional development through professional learning communities. This strategy is motivated by little impact that has resulted from various approaches and initiatives to teacher professional develop-

S. Tshiningayamwe (✉)
Environmental Learning Research Centre, Rhodes University, Grahamstown, South Africa
e-mail: sirkka.ts@gmail.com

Z. Songqwaru
Environmental Learning Research Centre, Rhodes University, Grahamstown, South Africa
e-mail: z.songqwaru@ru.ac.za

© Springer International Publishing Switzerland 2017
H. Lotz-Sisitka et al. (eds.), *Schooling for Sustainable Development in Africa*,
Schooling for Sustainable Development 8, DOI 10.1007/978-3-319-45989-9_19

ment over the years. It aims to ensure teachers' professional growth (ibid.; South Africa. Department of Basic Education (DBE) [DBE], 2015). De Clercq and Phiri (2013) observed that in South Africa, clusters are a version of PLCs. Many South African researchers have explored teacher clusters (Botha, 2012; De Clercq & Phiri; Jita & Ndlalane; Mokhele, 2013; Ndlalane, 2006; Ono & Ferreira, 2010; Steyn, 2013). These researchers observed that teacher clusters are meant to be structures for professional development but have not been used as intended (ibid.). These clusters appear to have been used as spaces for continuous assessment moderation (CASS) moderation only and not for professional development activities (Jita & Ndlalane). In the field of environmental education, teacher clusters in South Africa were established through the Learning for Sustainability project and the National Environmental Education Project for General Education and Training (NEEP-GET) (Du Toit & Sguazzin, 2000; Janse van Rensburg & Lotz-Sisitka, 2000). Teacher clusters were run using a spiral model (ibid.). The chapter thus provides a review of the emergence of teacher clusters in South Africa using a reflexive spiral model and praxis tasks. The chapter also draws from the Namibian and South African teacher cluster literature to comment on how teacher clusters could potentially be translated into PLCs to transform teachers' practices as they relate to ESD. This will be done by providing a literature review of the concept of PLCs and a review of teacher clusters in Namibia and South Africa. With an ESD focus, the chapter will provide a review of teacher professional development programmes in South Africa. It will conclude with an orientation of the concept of PLC in South Africa and explore a possible way forward.

19.1   Professional Learning Communities (PLCs)

In literature, PLCs are related to teachers' learning communities, teacher networks and critical friends' groups, but their nuanced differences and implications are not clear. The terms describe teachers engaging in similar activities and reflecting on their teaching practice. The focus of this chapter is the term 'PLC'. The concept has different shades of interpretation in different contexts, but there seems to be consensus among a number of authors that a conception of PLCs means 'a group of teachers sharing and critically interrogating their practice in an ongoing, reflective, collaborative, inclusive, learning-oriented, growth-promoting way to support innovation and knowledge sharing' (Stoll et al., 2006, p. 223). According to the ISPFTED policy, PLCs are defined as 'communities that provide the setting and necessary support for groups of classroom teachers, school managers and subject advisors to participate collectively in determining their own developmental trajectories, and to set up activities that will drive their development' (South Africa. Department of Higher Education and Training & Department of Basic Education (DHET & DBE), 2011, p. 14). These two definitions of PLCs highlight the role of teachers in PLCs. The definitions further highlight that PLCs require teachers to come together, to share and to reflect on their practices. Evident in the definition from the ISPFTED policy is that PLCs require teachers to be drivers of the PLC activities and that they should be capable to determine their own professional needs. Thus teachers need to be at the centre of the PLC activities (South Africa. Department of Basic Education, 2014). Thus, for PLCs to be effective models for teachers' practices as they relate to ESD, teachers need to be able to come together to interrogate, share and reflect on their ESD-related practices.

It is, however, worth noting that even though the concept of PLC is new, it is evident from the reviews of PLCs done by Stoll et al. (2006) that the concept of PLC has its roots in some older, well-known theories, such as Stenthouse's (1975) ideas of teacher as a researcher who plays an active role in curriculum development and Schon's (1983) notion of reflective practitioner. However, the concept only became popular in the 1990s, borrowing from 'learning organisation' theory in business as described by Senge (1990). The organisational structure of a PLC in education theory began to develop with the research conducted by Rosenholtz (1989). She found that teachers who were provided with opportunities for collegial collaboration learned from the experience and applied the knowledge gained in their classroom practices. PLCs are typically characterised by the following set of dimensions or attributes: shared beliefs, values and visions, reflective dialogue, inclusive membership, feedback to teacher, collaborative learning, shared practice, collective responsibility, mutual trust networks and partnerships (Darling-Hammond, 1996; DuFour, 2004; Feger & Arruda, 2008; Hord, 1997; Morrow, 2010; Stoll et al. 2006). These characteristics create a structure for continuing professional development through collaboration, continuous learning and meaningful learning. This may be through experience sharing, pedagogical analysis, observation, demonstration, feedback, experimentation, developing new methods and technical consultations from advisors (Borko, 2004). This implies that to improve teachers' ESD-related practices, there is a need to consider PLCs as models for professional development. These will create collaborative spaces that will allow teachers to continuously meet and share their experiences as they relate to ESD.

In the literature, PLCs are also associated with the theory of communities of practice. Wenger, McDermott and Snyder (2002) defined communities of practice as 'groups of people who share a concern, a set of problems, or a passion about a topic, and who deepen their knowledge and expertise in this area by interacting on an on-going basis' (p. 4). Thus, teachers who come together in the PLC have common interests, which in the case of ESD could be to improve their teaching practices related to the environment and sustainability content in the school curriculum. Teachers' PLCs are thus perceived as a framework for promoting and sustaining ongoing, effective professional growth (Morrow, 2010). PLCs inspire teachers to bring together their expertise, knowledge and enthusiasm in an effort to learn from one another (ibid.). The National Council of Teachers of English (2010) in the USA recognised PLCs as having the ability to bridge the gap between research and practice, to create spaces for addressing problems of practice and to foster transformative teaching. PLCs offer opportunities to tap teachers' tacit knowledge and make it public to be shared and critiqued (Schon, 1983). Strong PLCs are not only those in which new knowledge regarding content and pedagogy is acquired but also those in which existing assumptions about teaching and learning are challenged and critiqued

(Little, 2003). Thus continuing professional development initiatives should create spaces for teachers to critique and challenge each other's learning and teaching assumptions and practices.

Teachers should be responsible for their own continuing professional development, but support from different stakeholders is critical to enable them achieve their valued beings and doings in professional development initiatives (Burnette, 2002; Morrow, 2010). Huggins, Scheurich, and Morgan (2011) noted that 'without the inclusion of outside assistance in PLCs, collaboration simply cannot occur due to the lack of sufficient pedagogical and content knowledge within the community' (p. 85). This implies that in PLCs, teachers should be able to identify gaps in their content knowledge and other areas that hinder their successful implementation of ESD. Teachers should then be able to seek external support from different stakeholders, i.e. from subject advisors, higher institutions of education or non-governmental organisations, that will enable them to enhance their superficial implementation of PLCs. Lack of teachers' understandings of the interdependence and shared beliefs required for effective learning communities and difficulties in negotiating individual differences related to PLC activities can be challenges in PLCs (Westheimer, 1999).

In South Africa, like in Namibia, networks of teachers were termed clusters. Clusters are 'a group of teachers who meet together regularly with a facilitator or professional development mediator to work towards their own professional development within and away from the cluster meeting situation' (Du Toit & Sguazzin, 2000, p. 21). Schools use clusters as a tool to promote collaboration, reflection, sharing and continuous learning among teachers (Turkey, 2004) and to improve quality and relevance of the education in schools. These characteristics of teacher clusters are similar to some of the characteristics of PLCs given above. Like in Namibia, in South Africa, teacher clusters have largely been mandatory with a predetermined agenda from the Department of Basic Education (Dittmar, Mendelsohn, & Ward, 2002; Jita & Ndlalane, 2009). However, these clusters do not provide spaces for teachers to share and interrogate their practices; thus, even though teachers do meet regularly in clusters, they do not see them as spaces for professional development. However, research in the South African context does highlight the potential for transforming clusters into PLCs, making them spaces for professional development.

19.2 Teacher Clusters: The Namibian Perspective

The idea of clusters has been around in Namibia for some years, and several kinds of clusters have been formed at various times. The idea of teacher clusters was largely introduced by the Basic Education Project (BEP), but the growth of the system has mainly been driven by needs from within schools and regional education offices (Dittmar et al., 2002). The first clusters were introduced in 1996 in the Rundu education region. All the schools in the region were included in the cluster system with the aim of having a comprehensive framework to accommodate the needs of all the schools. Benefits which arose from the Rundu education region clusters led to the subsequent development of similar clusters in all other regions of the country (Dittmar et al. 2002). These clusters were supported by the Department of Basic Education district offices. Subject advisors in the regions were responsible for bringing together teachers teaching the same subjects. The establishment of clusters in Namibia was motivated by the need to improve the flow of learners from one school phase to another. It then gradually shifted in emphasis, more into a way of improving the management of schools, into improving communication between schools within clusters and between circuit offices and regional offices and into developing better supervision and training (Mendelsohn & Ward, 2001).

Clusters also aimed to improve teaching and learning practices by sharing resources, experience and expertise among staff (Dittmar et al., 2002). Clusters were also important in supporting teachers to improve their teaching methods and interpret the curriculum (ibid.). Many schools are far from other schools or circuit offices. Thus despite the intentions to improve levels of democratic participation and to decentralise the education system, too many decisions had to be taken at a head or regional office level, leading to delays, frustrations and low levels of ownership, accountability and responsibility. Schools needed more autonomy (Mendelsohn & Ward, 2001).

Some schools took advantage of the cluster system more than other schools. Schools that benefit most are the ones that are in greatest need of various kinds of support: for management purposes, teaching practices and examination requirements (Mendelsohn & Ward, 2002). Other schools, especially some of the large schools in urban areas, do not require such support to the same degree. They perceive the cluster system as more of an additional workload (Mendelsohn & Ward, 2001). Those teachers that benefit from clusters improve their classroom practice, have better ideas of what must be taught and benefit from ideas and materials that other teachers have developed (ibid.). The cluster system is also of great advantage to inspectors and managers. They benefit by having local management committees and cluster centre principals taking more responsibilities and decisions. Teachers support each other in subject groups and compile common schemes of work and examination papers. This improves standards and enables their learners to write common examinations (Dittmar et al., 2002). Teachers who were previously isolated in small schools working for years on their own with little professional contact with other teachers teaching the same grades and subjects became part of a larger support system (Mendelsohn & Ward 2001). Much of this isolation was broken by the formation of cluster-based subject groups; teachers created most of these groups themselves. The groups thus grew out of a need for more collegial support (Dittmar et al., 2002). Thus clusters have potential to meet the great demands for better quality education in Namibia (Dittmar et al. 2002). Among other benefits of clusters, the following have been noted by Dittmar et al. 2002:

- Principals and teachers learn from each other, sharing experiences and ideas, assisting each other with problems, consulting and co-operating at all levels.
- Teachers' morale and confidence are boosted, and their skills are developed as they work together to improve their teaching efforts within a supportive context.
- School visits from teachers or principals from within the cluster create a culture of sharing and mutual support.
- Advisory teachers can channel their inputs more effectively through cluster-based subject groups to reach all teachers within a given cluster.
- Cluster school members form a unified front to deal with issues, resulting in faster and more effective solutions that have greater ownership and local relevance.
- All participants develop greater competence as they learn to make decisions and take responsibility within their clusters.
- The cluster system provides a framework through which a more comprehensive and coordinated programme of training can be delivered efficiently at each cluster centre.

19.3   Teacher Clusters: The South African Perspective

In South Africa, there are official and unofficial teacher clusters. Official clusters are recognised by the Department of Education. Unofficial clusters are other kinds of clusters operating outside the official structures. They are 'alternative clusters' (Jita & Mokhele, 2012). The official clusters have good intentions to support teachers in their professional development but seem to limit teachers to continuous assessment moderation. Their agenda is often limited to curriculum implementation issues (ibid.). The official clusters try to give teachers the necessary physical, material and intellectual resources that the teachers would need in their clusters. Teachers do not have a choice of whether to attend the formal clusters or not. The formal clusters represent officialdom and, where possible, teachers deliberately stay away from the meetings (Jita & Mokhele, 2012).

Because of different needs, teachers form their own alternative clusters, of which departmental officials might not be aware (Jita & Mokhele, 2012). The formation of alternative clusters had for the most part been driven by the needs of the participating teachers and intermittent support, often from higher education institutions (HEIs) and/or NGOs (ibid.). Teachers recognised the alternative clusters to focus on the core issues of teacher development and classroom improvement (ibid.). It is this sharing and working collaboratively as a team to encourage teacher collegiality and learning that makes teachers participate in the alternative clusters.

Participation in the alternative clusters is voluntary and these clusters are self-directed. Teachers' participation in the alternative clusters is not mandatory, but teachers are happy to sacrifice their time to attend the meetings of their chosen alternative clusters even during weekends or school holidays. In the one example of the alternative clusters that Jita and Mokhele (2012) studied, they summarised the activities in that cluster as follows: The teachers would identify problem areas in their subjects and then arrange to meet in order to collaboratively plan and discuss ways of teaching the identified topics. The teachers would then bring in their learners on a Saturday for actual collaborative teaching of the identified topics. The arrangement enabled learners from the participating schools to attend the cluster lessons once every month. After the day's series of lessons, the teachers in the cluster would meet and reflect on each of the lessons. The collaborative teaching experiences and the collective reflection of the teachers supported and enriched each one of the participating teachers' classroom practices.

The development of alternative clusters in comparison to official clusters demonstrates teachers exercising their agency in terms of their participation (Jita & Mokhele, 2012). They seem to value clusters they have voluntarily joined and this is in line with what PLCs should be, where teachers collaboratively set out the agenda of what they want to get out of a PLC.

There has been substantive research on two projects that have contributed to the strengthening and establishment of clusters in South Africa, namely, the Learning for Sustainability project and the National Environmental Education Project for General Education and Training (NEEP-GET). The key focus of the projects was to strengthen teachers' environment and sustainability content knowledge. These projects and their professional development models will be discussed in the next section.

19.4 The Learning for Sustainability Project

The Learning and Sustainability pilot project was carried out in Gauteng and Mpumalanga between 1997 and 2000 (Janse van Rensburg & Lotz-Sisitka, 2000). The Learning for Sustainability project aimed at supporting the implementation of environmental learning in South Africa. The project drew heavily on the experience of the Namibian Life Science project which ran between 1991 and 2000. The project influenced the post-independence education system and has been one of the major projects that supported the cross-curricular integration of environmental education in the Namibian formal school system.

The Learning for Sustainability project followed a three-pillar approach of integrating environmental education in the curriculum: This consisted of teachers' development, curriculum development and materials development (Janse van Rensburg & Lotz-Sisitka, 2000). As part of the Learning for Sustainability project, a spiral model approach to teacher professional development was introduced (see Fig. 19.1). The spiral model was introduced out of concern that traditional professional development models (such as cascade model) used in South Africa were inadequate (Janse van Rensburg & Lotz-Sisitka, 2000). The spiral model promoted

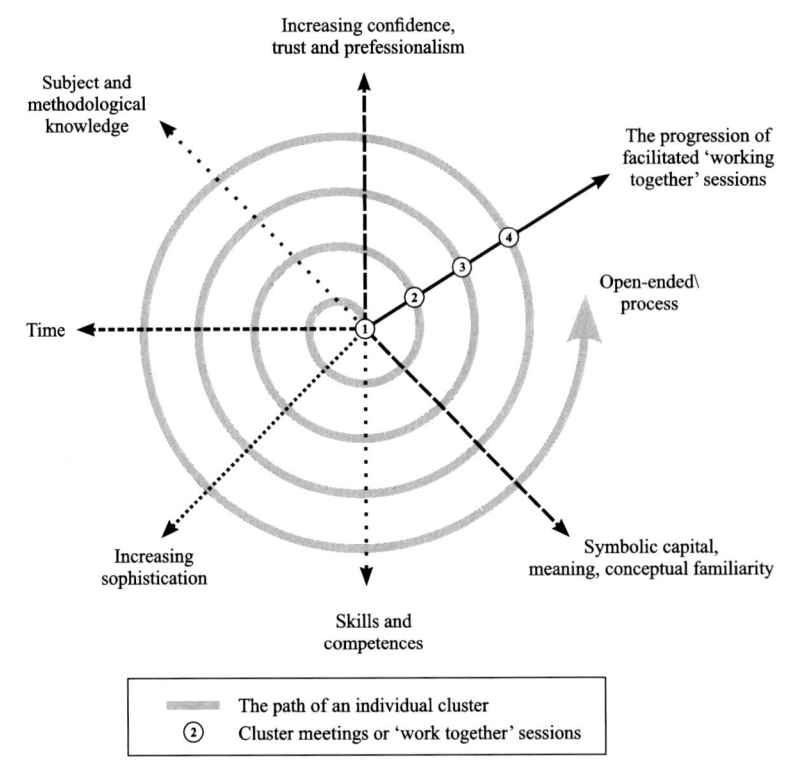

Fig. 19.1 The spiral model informed teacher professional development in the Learning for Sustainability pilot project (Janse van Rensburg & Lotz-Sisitka, 2000, p. 42)

increased sophistication of knowledge and practice over time that is achieved through a reflexive approach of working together (in the cluster) and then working away (in classrooms) and then reporting on practice back into ongoing cluster meetings.

The spiral model views professional development as 'a process that enables teachers to gain better understanding of their professional practice and reflect on it in light of policy over extended periods of time' (Squazzin & du Toit, 2000, p. 22). The spiral model enables teachers to contextualise the teaching of environmental content knowledge in the settings where they teach, as environmental issues may have different relevance in different contexts. Teachers also reflect on current practice and implement new ideas and then reflect on the implementation and get feedback from colleagues and facilitators. Doing this, they integrate theory and practice.

19.5 The NEEP-GET

In 2000, the National Environmental Education Project for General Education and Training (NEEP-GET) expanded upon the Learning for Sustainability project into nine provinces. The NEEP-GET was established by the Ministry of Education to support professional development of subject advisors and teachers (NEEP-GET, 2004). The NEEP-GET also aimed at strengthening environmental learning in the South African curriculum, within an outcome-based, learner-centred curriculum framework (ibid.). One of the key objectives of the project was to support professional development of educators in planning and implementing environmental learning programmes. The project worked with both curriculum support staff and teachers.

Professional development processes and programmes were structured around a spiral model within a cluster-based approach to professional development (Du Toit & Sguazzin, 2000). Groups of teachers (sometimes with curriculum staff supporters) met regularly to share ideas, perspectives and experiences related to curriculum development and professional development processes. The spiral model allowed for ongoing learning context through a series of 'work-together' and 'work-away' tasks, encouraging the development of educators as reflective practitioners. The project development process in the NEEP-GET was approached differently in the different provinces, shaped by various contextual factors, provincial organisations and management and geographic proximity of schools. A key feature of working within the process-based approach of the spiral model was the establishment and running of educator clusters. Below are some of the cluster models that were established in South Africa during the NEEP-GET, as summarised from NEEP-GET (2004).

- *Provincial cluster*: In some provinces, i.e. the Eastern Cape and the Gauteng province, both teachers and curriculum support staff clusters were established. Curriculum support staff clusters supported teachers and schools in district-based clusters. Teacher clusters in the provinces were coordinated and co-managed by the Department of Education district officials, HEIs and an NGO that were partners in the project. District cluster activities like approaches to lesson planning and portfolio development were informed by provincial clusters.
- *District-based clusters*. Curriculum support staff clusters were convened within the district structures of the province. These clusters established and worked with teacher clusters from schools in specific areas in the district. District-based clusters allowed educators to draw on locally available resources and services, e.g. local museums, nature reserves, local herbarium, waste management sites and environmental education centres, thus enabling teachers to focus on specific environmental issues within that specific part of the province.
- *School-based clusters*: The school-based clusters encouraged a whole-school development approach to environmental learning.

- *Learning area, grade or phase-specific cluster*: In some cases, learning area, phase or grade-specific clusters were established. These clusters allowed for the integration of environmental learning within specific learning area, phase or grade.
- *Lead teacher clusters*: In provinces where there was a shortage of curriculum support staff, lead teacher clusters were established. Lead teachers were expected to work with other teachers at their own school and other schools in close proximity, to support professional development and integration of environmental learning in the curriculum.

NEEP-GET (2004) observed that the cluster-based structures were seen to be valuable in professional development because of the following reasons:

- A focus of local context: Cluster groups established within a specific geographic district enabled participants to focus on prevalent local environmental issues in their own context.
- Creating learning spaces: For many educators, participating in ongoing cluster activities meant that they had the opportunity to interact with colleagues and share issues of mutual concern.
- Encouraged the development of relationships: Participating in cluster activities allowed educators to develop relationships as they engaged with processes of change in education. These relationships provided teachers with links to finding resources to support learning.

It was recommended by NEEP-GET (2004) that the setting up and sustaining of clusters should involve departmental officials and address the issue of accessing educators. Cluster activities should fit into teachers' busy schedules and timetables and should be linked to teachers' interests and priorities. Professional development programmes should be integrated into the in-service education and training (INSET) policies and programmes, so that participating educators are provided with an option of gaining credit for learning through these professional development programmes. Professional development programmes should also develop partnerships with the province. Jita and Mokhele (2012) observed that many of the alternative clusters sustained themselves through their existing networks with professional teacher organisations and the HEIs with which they were connected.

'NEEP supported the emergence of the Education for Sustainable Development Strategy in South Africa, but tended to concentrate more on piloting models for professional development of subject advisors at district level than on formal teacher education programmes' (Lotz-Sisitka, 2011, p. 33). Despite their many good practices and good policy frameworks, very little has been achieved in making a consistent and coherent impact in teacher education (ibid.). Fundisa for Change, a collaborative teacher professional development programme has been developed to enhance transformative environmental learning by establishing stronger institutionally located programmes within national policy and teacher development structures. The programme creates a platform for teachers to learn collaboratively and to form stronger works that creates opportunities for them to engage with each other outside of the training sessions. The programme enables teachers to share and reflect on their practice, and through this sharing and reflections, they will be able to identify areas where they need support from colleagues and any other experts.

19.6 Establishment of PLCs in South Africa

In South Africa, ISPFTED (2011–2025) was developed. The framework is viewed as part of an ongoing, dynamic planning process, which will continue to rely on the input of all teacher education and development stakeholders and through which the quality of teacher education and development will be improved over time (South Africa. Department of Higher Education and Training & Department of Basic Education (DHET & DBE), 2011). The primary outcome of the framework is the improvement of the quality of teacher education and development in order to improve the quality of teachers and teaching (ibid.).

The ISPFTED indicates that the establishment of PLCs will address the challenges teachers have experienced in accessing and receiving support, resources and continuing professional development opportunities close to where they live and work. PLCs will therefore be teacher support structures at provincial and district levels, thus enhancing teacher support at local level (South Africa. Department of Higher Education and Training & Department of Basic Education (DHET & DBE), 2011). The framework seeks to support continuous professional development of teachers so that they can adopt new orientations and approaches and to improve their subject content knowledge, pedagogic content knowledge, and practice and situational knowledge through a recognised accredited system of continuous development and through systems that support the establishment of PLCs. The key players in the establishment of PLCs are the provinces, districts, teacher organisations, subject-based professional teacher associations and, equally importantly, the teachers themselves (ibid.).

In order to successfully strengthen PLCs and establish new ones, the DHET and DBE will draw on available specialist knowledge of the specific focus areas, including expertise provided by NGOs and other specialist groups. The District Teacher Development Centres (DTDCs) and Provincial Teacher Development Institutes (PTDIs) will be established, with associated PLCs to provide ongoing professional development support for teachers (South Africa. Department of Higher Education and Training & Department of Basic Education (DHET & DBE), 2011). DTDCs will serve as local support sites for teachers to engage effectively with the course content, including independent, materials-based or online study, participation in formal or informal programmes and learning with colleagues and peers in PLCs. In provinces where DTDCs already exist, they will be strengthened to ensure that they meet the established norms and standards (ibid.).

The National Institute for Curriculum and Professional Development (NICPD) will support the work of PLCs by developing activities and materials that can help to stimulate their work. Teacher unions, as they play a key role in teacher professionalism and labour-related issues, have been tasked to promote teacher professionalism through advocating and supporting the establishment of PLCs and encouraging teachers to participate actively and meaningfully in these (South Africa. Department of Higher Education and Training & Department of Basic Education (DHET & DBE), 2011). PLCs will require external input but teachers should take control of their development. Diagnostic self-assessment tools will enable teachers to identify areas of weakness and use expertise within the PLC and appropriate CPD courses to address those. The framework has identified some activities that teachers could engage with in PLCs:

- Developing expertise in the analysis of learner results on evidence-based assessments such as Annual National Assessments (ANAs), among others, in order to determine teachers' own development trajectories
- Curriculum orientation activities, e.g. activities to develop understanding of, and the ability to use, the Curriculum and Assessment Policy Statements
- Learning how to interpret and use curriculum support materials such as the workbooks currently being developed and distributed to teachers and schools by the Department of Basic Education (DBE)
- Working together to learn from video records of practice and other learning materials (South Africa. Department of Higher Education and Training & Department of Basic Education (DHET & DBE), 2011, p.14)

Subject-based professional bodies like AMESA and programmes like Fundisa for Change that exist will be supported to enable the development and spread of PLCs. The framework proposes strengthening of existing PLCs and establishing of new ones. What is not clear is what exactly the framework refers to as a PLC. It is clear in the review that existing teachers' clusters in Namibia and South Africa share some PLC characteristics, but the nuances are not clear between clusters and PLCs. Can these clusters be PLCs? What would it take to make them PLCs? What is it about them that does not give them a 'PLC status'?

19.7 PLCs Going Forward

The South African Ministry of Basic Education held a national-wide colloquium on PLCs on 18–19 September 2014. Participants were from institutions of higher education, government departments, teacher unions, subject-based professional associations and CPD initiatives like Fundisa for Change who want to develop the functionings of the concept 'PLC'. Some of the key comments and questions from the colloquium were:

- More conceptual and practical work needs to be done before PLCs can be rolled out nationwide, and any guidelines on PLCs should *clearly describe the concept.*
- Do we only count PLCs that are truly teacher-led, data-driven and collaborative in nature? Or do we include all kinds of regular teaching meetings?
- Can PLCs be expected to become beacons of a collaborative, data-driven and self-driven view on professional development, if the education system in which they operate is prescriptive and bureaucratic?

- Should criteria such as teacher-driven, school-based, continuous and collaborative practice be key criteria without which a PLC is not a PLC anymore?
- Successful PLCs manages to instil a culture of '*us-ness*', teaching and learning become a collective responsibility for the school rather than an individual responsibility for the teacher.
- PLCs are inherently connected to the concept of *teacher agency*. How do we see the role of the teachers, as implementers of policy and curriculum or as education professionals? Is the current climate in South Africa conducive to teacher agency?
- Imposing PLCs on teachers and schools will result in command and control, compliance and resistance, nipping any potential of PLCs in the bud. Participation in PLCs should be voluntary, based on teachers' assessment of what constitutes valuable professional development for them.
- An important role for the Department of Basic Education lies in scaffolding and providing initial support to starting PLCs. The challenge lies in finding a good balance between prescription and support.
- The 80 h that are currently earmarked for professional development in teachers' schedules could be used for PLC activities.
- PLCs are primarily school-based structures to facilitate teaching and learning. Ideally, PLCs engage in a process of moving out of the local and back in to access and bring in external expertise in PLCs. (http://www.vvob.be/southafrica/)

All the key outputs have implications for initiatives like Fundisa for Change that want to strengthen the establishment of PLCs. They will have to determine how their activities address or respond to the comments and questions.

19.8 Fundisa for Change Programme and PLCs in South Africa

According to the framework, DHET and DBE will draw on available specialist knowledge of the specific focus areas to strengthen existing PLCs and to establish new ones. The expertise they will be drawing on will be from NGOs and other specialist groups. The Fundisa for Change initiative is a partnership programme of government, parastatals, NGOs in the environmental sector and HEIs committed to offering content-rich, pedagogically sound, continuous professional development courses for teachers (Lotz-Sisitka, 2011). The main objective of Fundisa for Change is to strengthen teachers' environmental and sustainability knowledge, including their pedagogical content knowledge and assessment practice. The Fundisa for Change programme is developing and piloting the concept of PLCs through research and by supporting ongoing interactions with teachers and ongoing professional development of teachers.

19.9   Conclusion

PLCs inspire educators to bring together their expertise, knowledge and enthusiasm in an effort to learn from one another. Programmes like Fundisa for Change need to explore how they can develop teachers' agency so teachers are able to drive their own professional development, can identify their needs and determine which activities will lead to their professional development. At the same time, these kinds of programmes should create spaces for teachers to participate collectively in their professional development. Currently Fundisa for Change does create a space for teachers to engage with each other and to share best practice. What still needs to be established is how the programme will strengthen existing PLCs and assist in establishing new ones.

References

Borko, H. (2004). Professional development and teacher learning: Mapping the terrain. *Education Researcher, 33*(8), 3–15.

Botha, E. M. (2012). Turning the tide: Professional Learning Communities (PLCs) to improve teaching practice and learning in South African public schools. *Africa Education Review, 9*(2), 395–411.

Burnette, B. (2002). How we formed our community: Lights and cameras are optional, but action is essential. *Journal of Staff Development, 23*(1), 51–54.

Darling-Hammond, L. (1996). The quiet revolution: Rethinking teacher development. *Educational Leadership, 53*(6), 4–10.

De Clercq, F., & Phiri, R. (2013). The challenges of school-based teacher development initiatives in South Africa and the potential of cluster teaching. *Perspectives in Education, 31*, 77–86.

Dittmar, F., Mendelsohn, J., Ward, V. (2002). *The school cluster system in Namibia: Framework for quality education.* Report for the Ministry of Basic Education, Sport and Culture and Basic Education Project of the Deutsche Gesellschaft fur Technische Zussamenarbeit: RAISON.

Du Toit, D., & Sguazzin, T. (2000). *A cluster approach to professional development support for teachers in South Africa: An illustrated proposal.* Johannesburg, South Africa: Learning for Sustainability Project.

DuFour, R. (2004). Schools as learning communities. *Educational Leadership, 61*(8), 6–11.

Feger, S., & Arruda, E. (2008). *Professional learning communities: Key themes from the literature.* Providence, RI: Education Alliance, Brown University.

Hord, S. M. (1997). *Professional learning communities: Communities of continuous inquiry and improvement.* Washington, DC: Office of Educational Research and Improvement.

Huggins, K. S., Scheurich, J. J., & Morgan, J. R. (2011). Professional learning communities as a leadership strategy to drive math success in an urban high school serving diverse, low-income students: A case study. *Journal of Education for Students Placed at Risk, 16*(2), 67–88.

Janse van Rensburg, E., & Lotz-Sisitka, H. (2000). *Learning and sustainability: An environmental education professional development case study informing education policy and practice.* Johannesburg, South Africa: Learning for Sustainability Project.

Jita, L. C., & Ndlalane, T. C. (2009). Teacher clusters in South Africa: Opportunities and constraints for teacher development and change. *Perspectives in Education, 27*(1), 58–68.

Jita, L. C., & Mokhele, M. L. (2012). Institutionalising teacher clusters in South Africa: Dilemmas and contradictions. *Perspectives in Education, 30*(2), 1–11.

Little, J. (2003). Inside teacher communities: Representations of classroom practice. *Teacher College Records, 105*(6), 913–945.

Lotz-Sisitka, H. (2011). National case study: Teacher professional development with an education for sustainable development focus in South Africa: Development of a network curriculum framework and resources for teacher education. *Southern African Journal of Environmental Education, 28*, 30–71.

Mendelsohn, J., & Ward, V. (2001). *A review of clusters of Schools in Namibia.* Report for the Ministry of Basic Education, Sport and Culture and Basic Education Project of the Deutsche Gesellschaft fur Technische Zussamenarbeit: RAISON.

Mokhele, M. (2013). Empowering teachers: An alternative model for professional development in South Africa. *Journal of Social Sciences, 34*(1), 73–81.

Morrow, J.E. (2010). *Teachers 'perceptions of professional learning communities as opportunities for promoting professional growth.* Unpublished doctoral (Education) study, Graduate School Appalachian State University.

National Council of Teachers of English. (2010). *Teacher learning communities.* A policy research brief. James R. Squire Office of Policy Research.

Ndlalane, T. C. (2006). *Teacher clusters or network as opportunities for learning about science content and pedagogical content knowledge.* Unpublished Doctoral Thesis. South Africa: University of Pretoria.

NEEP-GET. (2004). *Cluster-based approaches to professional development.* Howick, South Africa: National Environmental Education Project for General Education and Training/ Share-Net.

Ono, Y., & Ferreira, J. (2010). A case study of continuing teacher professional development through lesson study in South Africa. *South African Journal of Education, 30*, 59–74.

Rosenholtz, S. J. (1989). *Teachers' workplace. The social organization of schools.* New York: Longman.

Schmoker, M. (2006). *Results NOW: How we can achieve unprecedented improvements in teaching and learning?* Alexandria, VA: ASCD.

Schon, D. A. (1983). *The reflective practitioner.* New York: Basic Books.

Senge, P. M. (1990). *The fifth discipline: The art and practice of the learning organization.* London: Century Business.

South Africa. Department of Basic Education. (2014). Report on the colloquium on professional learning communities-held at the Department of Basic Education, Pretoria. Draft. 18–19 Sept 2014.

South Africa. Department of Basic Education (DBE). (2015). *Professional learning communities- a guideline for South African schools.* Pretoria, South Africa: Department of Basic Education.

South Africa. Department of Higher Education and Training & Department of Basic Education (DHET & DBE). (2011). *Integrated strategic planning framework for teacher education and development in South Africa.* Pretoria, South Africa: DHET/DBE. 2011–2025.

Squazzin, T., & du Toit, D. (2000). *The spiral model: New options for supporting the professional development of implementers of outcomes-based education.* Johannesburg, South Africa: Learning for Sustainability Project.

Stenthouse, L. (1975). *An introduction to curriculum research and development.* London: Heinemann Educational Books.

Steyn, G. M. (2013). Building professional learning communities to enhance continuing professional development in South African schools. *Anthropologist, 15*(3), 277–289.

Stoll, L., Bolam, R., McMahon, A., Wallace, M., & Thomas, S. (2006). Professional learning communities: A review of the literature. *Journal of Educational Change, 7*, 221–258.

Turkey, I. (2004). An online social constructivist tool: A secondary school experience in the developing world. *Turkish online Journal of Distance Education, 9*, 3–14.

Wenger, E., McDermott, R., & Snyder, W. M. (2002). *Cultivating communities of practice: A guide to maintaining knowledge.* Boston: Harvard Business School Press.

Westheimer, J. (1999). Communities and consequences: An inquiry into ideology and practice in teachers' professional work. *Education Administration Quarterly, 35*(1), 71–105.

Chapter 20
Development of 'Fundisana Online': An ESD E-Learning Programme for Teacher Education

Shepherd Urenje, Wolfgang Brunner, and Andrew Petersen

This chapter shares insights into the development and use of Fundisana Online, an ESD e-learning teacher education programme, in teacher professional development for mainstreaming education for sustainable development. We describe how SADC REEP and SWEDESD are working in partnership with teacher education institutions across the SADC region to develop the regional ESD course, Fundisana Online. We describe the collaborative process in building the Fundisana course elements. These elements include key competences for sustainable development, relevant content within and beyond subject borders, effective pedagogic methods to support the learning process and the use of a blended learning approach to integrate pedagogy and technology.

Between 2012 and 2014, the Southern African Development Community (SADC) Regional Environmental Education Programme (REEP) and the Swedish International Centre of Education for Sustainable Development (SWEDESD) collaborated in mainstreaming education for sustainable development (ESD) in teacher education. SADC REEP and SWEDESD are the lead organisations coordinating the Education for Strong Sustainability and Agency (ESSA) Programme, a multi-year collaborative partnership between 42 universities and teacher education institutions within the SADC region (SWEDESD, 2014) (Fig. 20.1).

S. Urenje (✉) • W. Brunner • A. Petersen
Swedish International Centre of Education for Sustainable Development (SWEDESD),
Uppsala, Sweden
e-mail: shepherd.urenje@swedesd.uu.se; andrew.petersen@uct.ac.za

© Springer International Publishing Switzerland 2017
H. Lotz-Sisitka et al. (eds.), *Schooling for Sustainable Development in Africa*,
Schooling for Sustainable Development 8, DOI 10.1007/978-3-319-45989-9_20

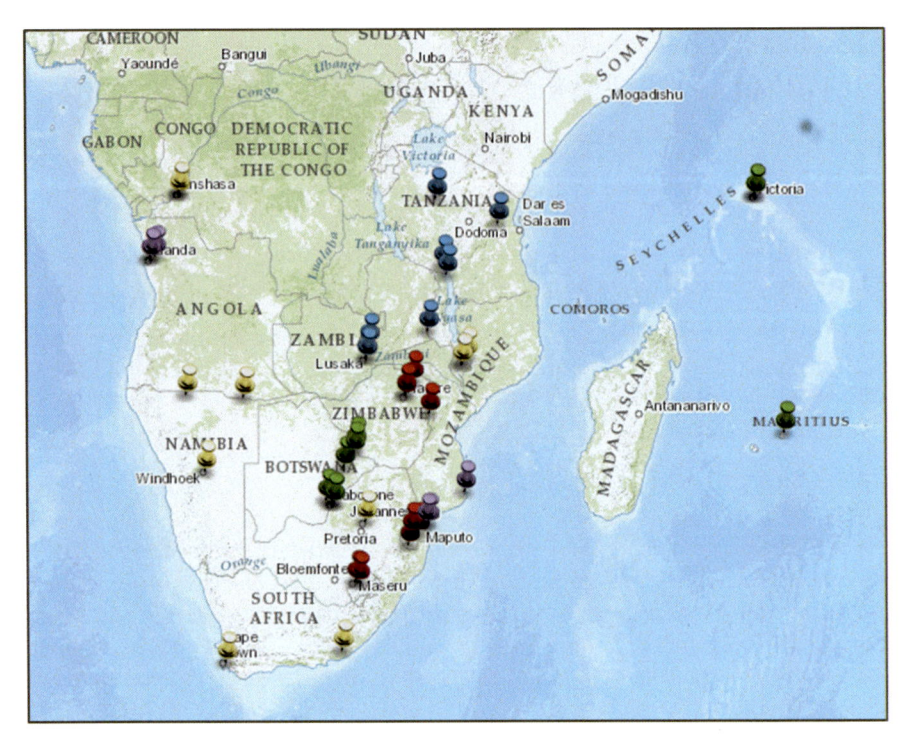

Fig. 20.1 Universities and teacher education institutions engaged in the ESSA programme

20.1 The Aim of the ESSA Programme

The aim of the ESSA programme is to allow teacher educators and their institutions to introduce innovative methods and relevant content related to education for sustainable development in their syllabuses and working practices. In order to create desired change processes, the programme has:

- Developed and tried out teaching materials and classroom examples
- Developed and tried out strategies for dialogue on institutional levels with deans and rectors
- Developed and tried out workshop programmes for teacher educators
- Engaged 42 universities and teacher education institutions, together with 83 teacher educators in change projects on implementing ESD at their institutions

A critical analysis of the 42 change projects initiated in the partnership was conducted by the University of Botswana (Ketlhoilwe, Silo, & Boikhutso, 2014). The evaluation revealed that 60 % of the institutions were grappling with integrating environment and sustainability into their curriculum. Further discussions between SWEDESD and SADC REEP arrived at a decision to offer an online ESD course, Fundisana Online, to support current and future partner institutions in the desired integration process. These outcomes, together with the fundamental environment and sustainability demands in the SADC region, have been used as the backdrop for the development of Fundisana Online. In a number of Southern African languages, the term Fundisana means 'learning from and with each other'.

The Fundisana Online course therefore emerged as a natural continuation and expansion of this partnership and is meant to become a regional course for mainstreaming ESD in teacher education. A team of practising teacher educators from Belvedere Technical Teachers' College, Zimbabwe; Mauritius Institute of Education; University of Botswana; University of Cape Town, South Africa; and Uppsala University, Sweden, are leading the course development. The course builds upon the rich and fertile ground of EE/ESD that already exists in the SADC region. The existing regional networks and competences, together with the outcomes of the ESSA programme, are used in the build-up of the Fundisana Online course. Box 20.1 shows the ESSA programme specifications that have guided development of the Fundisana Online course.

Box 20.1: ESSA Programme Specifications for ESD Course
The course should:

- Bridge the gap between teacher education and classroom practice.
- Present an integrated and holistic view on ESD that could attract participants regardless of subject background.
- Present an integrated view on the mutually dependent dimensions of ESD representing relevant content, effective methods and desired competences.
- Speed up and facilitate the mainstreaming of ESD in teacher education at SADC universities and teacher education institutions.
- Respond to demands from both the SADC protocol on education and from teacher educators in the region.
- Respond to the international demands expressed in the UNESCO world conference on the environment (Rio + 20) declarations and the upcoming Global Action Programme on ESD.

In the next sections, we explain how the course was developed in order to satisfy these specifications and guidelines. We start by explaining the cyclic process of developing classroom examples and follow this with identification of the essential dimensions of ESD to be incorporated in the teaching and learning.

20.1.1 The ESSA Cyclic Process Development of Classroom Examples

When developing the classroom examples, we used exactly the same methods that teacher educators expect to see their students using in the classroom. This was not an easy task as it required us to demonstrate how to combine the theory of education and content. We used a cyclic interactive approach where each cycle led to a deepening and widening of the course units and also clarified how they are connected and dependent on each other. This holistic and systemic approach has facilitated our work and guided the strategies we used. During this process it also becomes obvious that in the developing content and methods that strengthen ESD competences, we needed to include substantial changes at different policy levels. This led to the consultation with the leaders in teacher education and the running of five workshops with teacher educators from 42 institutions in the 14 SADC countries. The process started with a collaborative discourse which attempted to answer the following questions:

- What are the new demands that science and society put upon learners, teachers and teacher educators?
- What type of lessons would we like to see in the classroom as a response to these new demands?
- How do these new demands from science and society and the new contents, method and approaches for the classroom influence the way we educate teachers?

We have come back to these three fundamental questions at each stage of the course development process with different actors in the education system. Figure 20.2 (adapted from Coghlan & Brannick, 2001) is a summary of the events that have taken place between 2012 and 2014 to develop and fine-tune cutting-edge examples for integrating ESD in teacher education.

In cycle 1 we participated in a dialogue with partner institutions. The first stage was the development of classroom examples utilising the experience of teacher educators from Swedish and SADC teacher education institutions which resulted in the 2012 publication, *The parts and the whole: A holistic approach to environment and sustainability education*. This publication is being used by participating institutions as a starting point in discussing the integration of ESD into teacher education, and it was used as the main document for discussions in cycle 2.

Cycle 2 was the dialogue we had with the deans of faculties and colleges of education. In 2013 and 2014, SADC REEP and SWEDESD coordinated two workshops for leaders in teacher education from participating universities and colleges of education. These meetings mandated SWEDESD and SADC REEP to coordinate practical efforts of integrating ESD in their institutions. Two teacher educators from each institution were seconded as project managers for mainstreaming ESD in their institutions.

Cycle 3 involved dialogue with teacher educators. Five cluster workshops were conducted in 2013 and 2014 with teacher education institutions from all 14 SADC countries as the first step in institutionalising the ESD into teacher education.

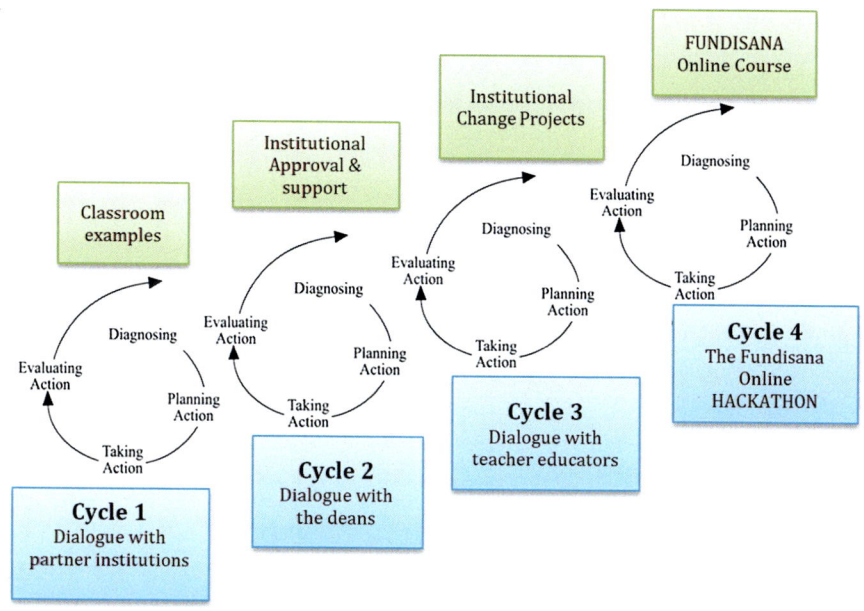

Fig. 20.2 Development cycle (Adapted from Coghlan and Brannick, 2001)

Following the workshops, the participants initiated change projects for testing and implementing ESD approaches into their institutions.

Fundisana Online in cycle 4 is a result of the analysis of the most pressing needs being addressed by the change projects. The online course will create possibilities for a faster and more effective way of mainstreaming ESD in teacher education with the potential to reach teacher educators in the entire SADC region.

20.1.2 Essential Dimensions of Teaching ESD

In this combined effort of building the units, elements and approaches in the online course, we have aimed to find a condensed way of describing the relevant skills essential for teaching and learning in a rapidly changing world and the implications these processes should have in teacher education. During this process we have continuously turned back to the interconnected and mutually dependent dimensions of ESD. These included representing relevant content, clarifying effective methods and working with desired competences to frame the process as illustrated in Fig. 20.3.

Each cycle in this iterative process has led to a deepening and widening of the perspectives, and it has also clarified how they are connected and dependent on each other. This holistic and systemic approach has facilitated our work and guided the strategies we used. During this process it has also become obvious that in the development of content and methods to strengthen EE/ESD competences, we also have

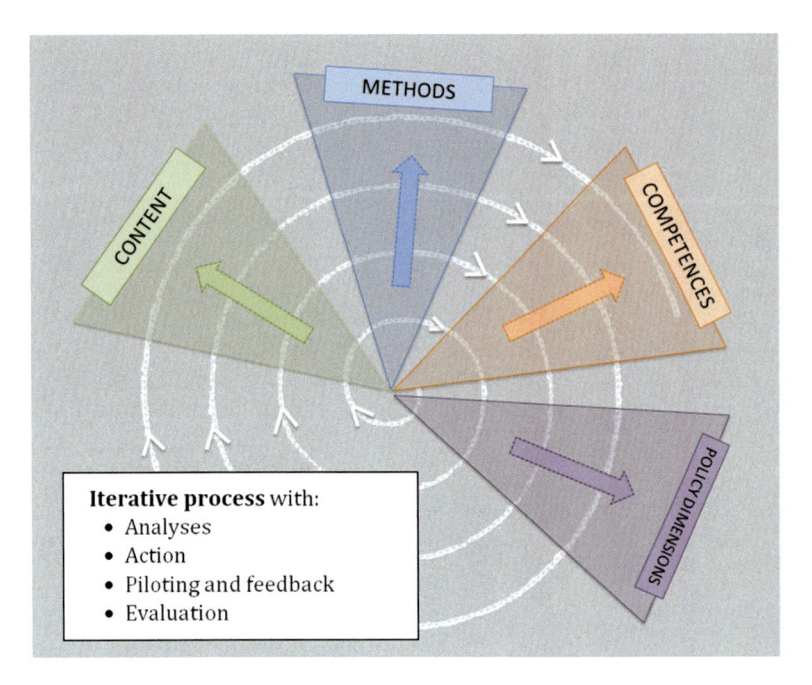

Fig. 20.3 Essential elements for ESD

to include substantial changes at different policy levels. In our effort to empower the teacher with the 'relevant skills' essential for teaching and learning in a world changing so rapidly, we continuously made use of the interconnected and mutually dependent dimensions of ESD representing desired competences, relevant content and the most effective methods for teaching and learning for sustainability. We have called this combination the ESD Navigation Tool (see Fig. 20.6).

20.1.3   Essential Competences

The sustainability challenges in Southern Africa informed the content and methods of teaching and learning that are essential in developing the specific competences needed for sustainable living. The debate on the type of key competencies to be considered as essential for graduating students has been ongoing for a number of years. The concept of competencies is seen as an essential landmark for orienting teaching and learning for sustainable development. In our quest to identify the key competencies essential to support teacher education, we have relied on the work of Wiek, Withycombe and Redman (2011) who synthesised five essential competencies shown in Box 20.2.

> **Box 20.2: Key Competences in Sustainability**
> **(Adapted after Wiek et al. 2011)**
>
> - **Systems thinking competence**
> - The ability to see, understand and relate the different parts in a system –
>   and how these parts together connect issues to come up with a whole
>   picture
> - **Anticipatory competence**
> - The ability to critically analyse and evaluate current situations in view of
>   predicting and envisioning future scenarios and their possible outcomes
> - **Normative competence**
> - The ability to collectively demonstrate an understanding of values and
>   principles in view of negotiating and integrating these in your vision and
>   practice of sustainability
> - **Strategic competence**
> - The ability to collectively design and implement interventions and to
>   enable and manage change processes towards sustainability issues
> - **Interpersonal competence**
> - The ability to create an environment that enables people to learn from and
>   with each other. The ability to motivate, enable and facilitate collaborative
>   and participatory learning processes regarding sustainability issues
>
> **Action competence is embedded in each of the above competences.**

We have used these competences as a guideline for describing desired competences and outcomes for the teacher education system which is being catered for by Fundisana Online. In the design of the online course and in the presentation of classroom examples and themes, we used this framework of competences and when necessary worked with participants to 'translate' these into concepts, tasks and desired outcomes adapted to the expected users.

20.1.4 Relevant Content

The content of the online course has emerged from our collaborative effort to analyse necessary skills and competences for the transition towards a sustainable society. We have tried to find a condensed way of describing what 'relevant skills for a rapidly changing world' could embrace and translated that into a number of key concepts that together could form an understanding of sustainability (Fig. 20.4).

These concepts create a deeper awareness of being connected to our life-supporting ecosystems and demonstrate necessary elements in building a fair, just and equal society. Many of them also hold 'holistic dimensions' in the sense that they amalgamate perspectives from different subject areas or bring together natural and social sciences. The crucial task is to arrange these holistic concepts into

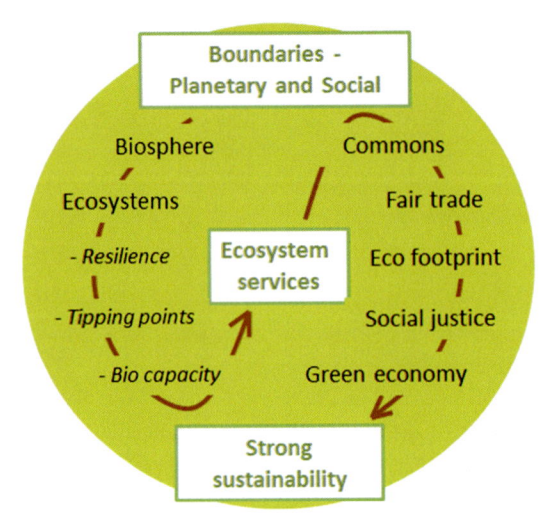

Fig. 20.4 Relevant content

themes/units in such a way that they hold relevant sustainability dimensions and at the same time can be taught with classroom examples that have their origin within traditional subjects.

20.1.5   Effective Methods

Although we acknowledge that there is a place for didactic instruction, we highlight the need for agency. In the Fundisana Online course, we will use methods with the intention to facilitate a pedagogical shift in teaching and learning from *transmissive* and *authoritative* approaches towards *participatory* and *transformative* approaches. To facilitate this transformation, we have tried to find examples and approaches that encourage:

- Reflexive thought and action (which encourages metacognition) (Flavell, 1979)
- Collaborative learning (which emphasises socio-constructivist learning) (Vygotsky, 1978)
- Making connections
- Learning as inquiry
- Enabling empowerment, transformation and emancipation (which emphasises action competence) (Jensen & Schnack, 1997)

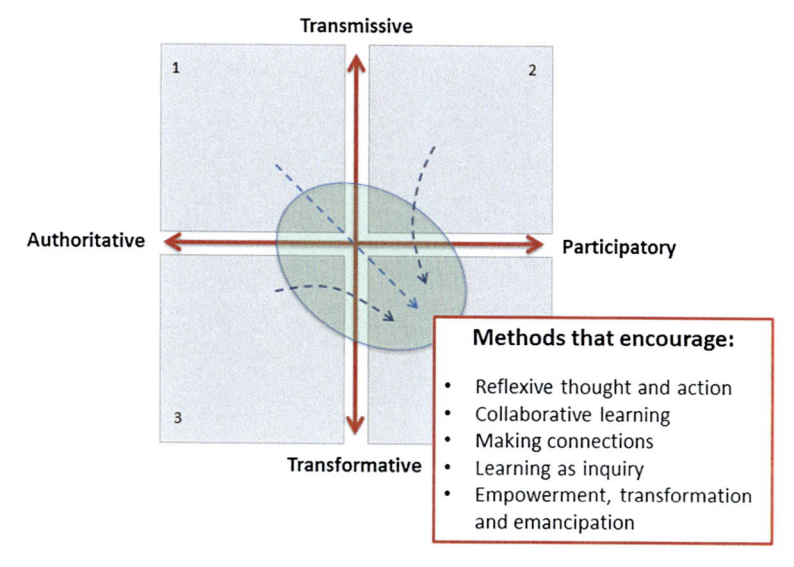

Fig. 20.5 Effective methods

By combining certain content with selected methods, we aimed to develop the specific competences that are necessary for environmental and sustainability education. The course also carefully blended theory of learning and the practice of teaching and learning in such a way that teacher educators could use the same methods in their own teaching as they wished to see teachers using in their classrooms. These are the same methods that we believe have the potential to steer teaching and learning (through teacher education) from a dominance of transmissive and authoritarian perspectives to more transformative and participatory approaches, as shown in Fig. 20.5.

20.1.6 The Fundisana ESD Navigation Tool

The interconnected and mutually dependent dimensions of ESD, i.e. essential competencies, relevant content and effective methods, have been built into the Fundisana Online course navigation tool as shown in Fig. 20.6.

The Fundisana ESD Navigation Tool is the result of combining the three fundamentals described above:

- The outline mapping of essential **competences** required for contributing to the development of a sustainable society sustainability
- The locally and more widely relevant **content** required to respond to environmental degradation, social instability and lack of equity
- The most effective pedagogy or **methods** that have the potential to steer teaching and learning from being transmissive and authoritarian to transformative and participatory

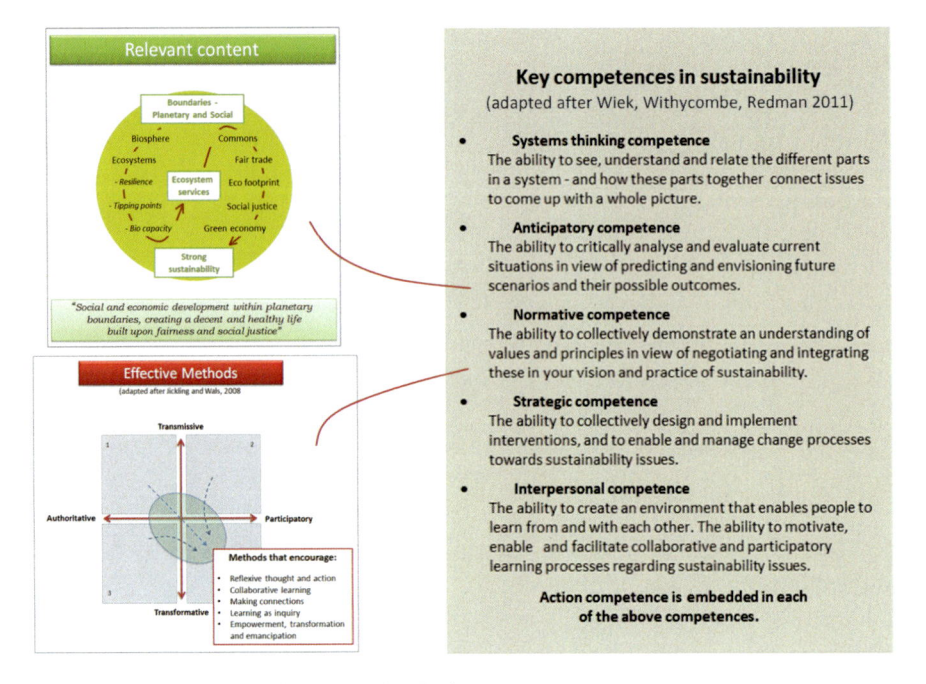

Fig. 20.6 The Fundisana ESD Navigation Tool

We have worked with this navigator in the development of the course units and classroom examples and to inspire the course participants to refer to it in their comments and reflections.

20.1.7   Course Structure and Course Units

We have also used the ESD navigator to blend examples and course units so that they lead to the desired competences. The course has six units shown in Fig. 20.7.

- **Systems thinking:** simplified models to support a basic understanding of systems and how systems thinking competences can be applied in the classroom.
- **Social sustainability and resource flow:** demonstrates how to engage learners in discussions on sustainability and to problematise the use of limited resources.
- **Risk and uncertainty:** highlights some local and global challenges that have emerged due to unsustainable behaviour of different kinds.
- **Eco-footprints and commons:** uses the concepts of *eco-footprints* and *commons* to discuss how to share the resources in a fare and sustainable way.
- **Global citizenship and agency:** discusses ethics and values that deal with sustainability issues and how to teach to create agency among the learners.

- **Self-reflection and application:** participants use the Fundisana ESD navigator to reflect on their learning processes and come up with well-structured plans on change projects they would like to launch at their institutions.

The basic purpose of the course is to support and to strengthen the teacher educators and their students' competence to integrate ESD in their teaching. The long-term aim is thus to ensure that, when student teachers leave the college or university, they will have a basic competence to combine and transform their subject knowledge and their knowledge of ESD into their teaching and learning. The course culminates in the participants conducting change projects in their institutions.

Figure 20.7 shows the change project as the final unit of the Fundisana Online course. The student teacher is given the opportunity to apply the knowledge and skills acquired in the preceding units. The change project approach is based on the assumption that participating institutions are committed to implement ESD in their syllabuses and working practices and that they are willing to invest their own resources to achieve these changes. The participants' change projects will address their own teaching on ESD or other relevant ESD challenges within their institution.

In the design for Fundisana Online, we were aware that university and college structures do not change easily and that institutional procedure sometimes is a hindrance. The participants will have to adapt to the local conditions and be realistic in their change projects. A stepwise introduction of ESD could be one way of doing it:

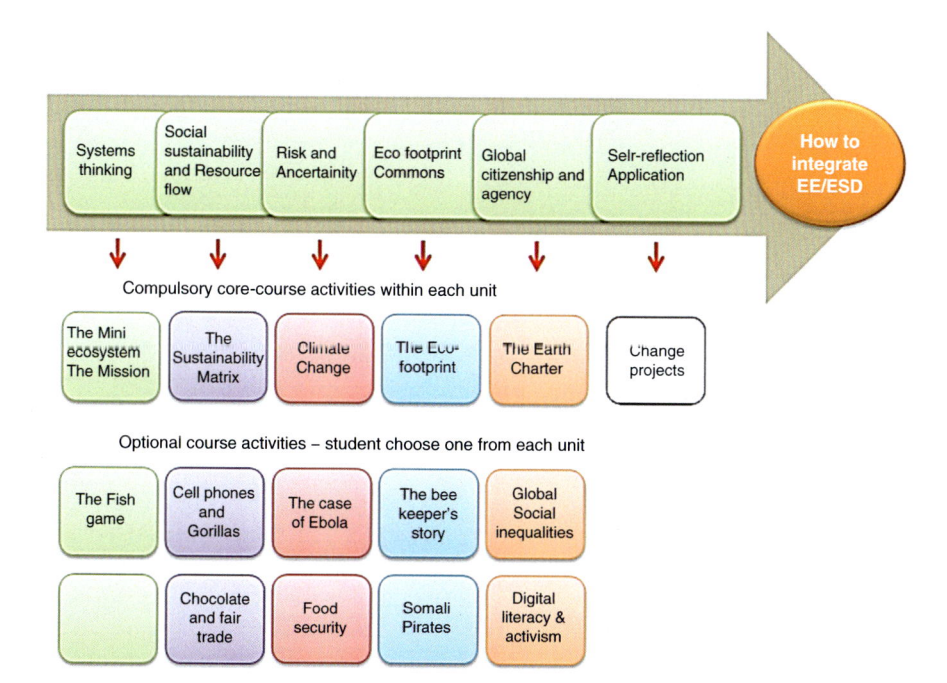

Fig. 20.7 The six course units

- A first step could be just to change *single lessons* in an existing lesson sequence.
- A second step could be to bring in *new ESD elements* in an existing course.
- A third step could be in collaboration with partners to come up with a *syllabus reformation* where the ESD perspectives are integrated.

The change project could also be carried out by producing *a new lesson example* that later on could be integrated directly into the course or become part of the 'bank of good examples' attached to the course. In that way the participant will become part of a strong 'community of ESD practice' (Wenger, 1998) that continuously develops within the course. The Fundisana ESD navigator can be used as a guide in this process.

20.2   Discussion

Fundisana Online is designed to provide a blended learning experience. It combines (blends) traditional classroom methods at the institution with an e-learning platform, blended learning and a hybrid teaching and learning methodology which represents a much greater change in basic technique than simply adding computers to classrooms. Fundisana Online represents a fundamental change in the way students and teacher educators approach teaching and learning experiences. While still being in a 'brick-and-mortar' lecture room, face-to-face classroom methods are combined with computer-mediated activities to transform and improve the quality and relevance of the learning process. The teacher educator provides face-to-face support on a flexible basis through activities such as small-group instruction, group projects and individual tutoring. E-learning creates flexibility for students, who can have access to substantial components of their courses anywhere, anytime and often in any format. Fundisana Online is designed to use less classroom time since students have access to their course materials, lectures and readings online.

The University of Cape Town (UCT) has used the Fundisana Online model to integrate ESD into mainstream education in an elective course for pre-service teachers participating in the Postgraduate Certificate in Education (PGCE) course. The aim of the course was to strengthen the students' capacity for integration of ESE and climate change issues into the curriculum, and it included carefully scaffolded e-learning progression. The second example, also at UCT, was environmental learning for in-service teachers. In this case the Fundisana Online approach was modelled in two instances, namely, the science, technology and society module and energy and change. Massive open online courses (MOOCs) are situated within an inward-outward continuum according to Czerniewicz, Deacon, Smail and Walji (2014). We have explored both options. The integration of the Fundisana Online course into our existing courses is an example of an inward-facing course, whereas the Fundisana Online course currently being developed for in-service teachers is an example of an outward-facing MOOC. Both these interventions can be regarded as 'distributed flip courses' which include blended learning (Stanford, 2014, cited in

Czerniewicz et al., 2014). Overall, Fundisana Online can be considered to be a meta-MOOC with the inside courses nested as small private online courses (SPOCs) with a common MOOC – this can be viewed as a Russian doll-nested model (Czerniewicz et al., 2014). We have the potential to offer formal online credit-bearing courses (e.g. our science, technology, society and the environment as part of CPTD) and non-formal credit-bearing courses which use the crowd-based blended learning model and can be considered to be an emerging MOOC (Czerniewicz et al., 2014).

The use of blended learning environments is still developing; hence, more research is needed to gain a comprehensive understanding of perceptions and problems that teacher educators and student teachers face when integrating pedagogy and content knowledge into the Fundisana Online learning environment, the strategies they employ to address these problems and how they use the blended learning tools to overcome these challenges (Caravias, 2014). In general, scepticism of digital and online learning is widespread because many technology-enabled educational practices are still largely untested, and their educational utility and value remain in question (Hidden curriculum, 2014). Critics of blended learning also question whether the practice can provide students with enough personal attention, guidance and assistance from teachers. In cases where students may not be self-directed, self-disciplined or organised enough to learn effectively without regular supervision, the Fundisana Online model of blended learning can be ineffective.

Considering that many teacher education institutions in SADC countries are either not or are badly connected digitally, institutions that need it most may find Fundisana Online inaccessible. In this case an offline option is preferred with online activities for specific sharing and discussion tasks. Critics also question the extent to which the mentors for the student teachers have received or will receive adequate training in how to instruct their students effectively in a blended learning context given that the teacher educators are not 'digital natives' (Kennedy, Judd, Churchward, Gray, & Krause, 2008). Teacher educators will be required to use new technologies and possibly more sophisticated instructional practices.

20.3 Conclusion

This chapter has given insights into the development of the Fundisana Online course to support teachers to mainstream ESD effectively. The Fundisana Online blended learning approach offers a unique opportunity to integrate pedagogy and technology fully. While e-pedagogy may not be separated from any other pedagogy, research and evaluation literature suggests that new modes of teaching and learning are emerging through the use of online networks, access to remote experts and, more recently, mobile technologies (Learning and Development Learning Development Centre, 2014). The Fundisana Online course is based on the view that teacher education supported and facilitated by ICT has the potential to effect and sustain effective teaching and learning approaches that will:

1. Enable student teachers to make *connections* by making it possible for them to enter and explore new learning environments, overcoming barriers of distance and time. Students will develop capabilities to make connections across learning areas as well as to home practices and the wider world.
2. Facilitate *shared learning* by enabling students to join and/or create communities of learners that extend well beyond their institutions (national) and countries (international).
3. Assist in the creation of *supportive learning environments* by offering resources that take account of individual, cultural or developmental differences.
4. Enhance *opportunities to learn* by offering students virtual experiences and tools that *save them time, allowing them to take their learning further.*
5. Take students through active learning processes where they build knowledge through the inquiry-based manipulation of digital artefacts such as online tasks, simulations, games or microworlds.
6. Enable interactive learning where students build knowledge through inquiry-based collaborative interaction with students from other institutions at home and abroad. The course will give teacher educators an opportunity to become co-learners as well as facilitators.
7. Encourage critical thinking and reflexive thought processes by giving feedback to each other. When students complete a task, they will be required to reflect and comment on at least one other person's work before the system allows them to proceed to the next task.

By opting for the blended learning approach, Fundisana Online will use ICT to supplement traditional ways of teaching and in addition open up new and different ways of learning. The Fundisana Online technology will prepare students and teacher educators for the future, thereby increasing the opportunities to interact, to learn about their work and to foster collaboration. The technology will also help build links with other teacher education communities and enable input from practising teachers.

References

Caravias, V. (2014). *Teachers' conceptions and approaches to blended learning: A literature review*. Victoria, Australia: Swinburne University of Technology.

Coghlan, D., & Brannick, T. (2001). *Doing action research in your own organisation*. London: Sage.

Czerniewicz, L., Deacon, J., Smail, A., & Walji, S. (2014). Developing world MOOCs: A curriculum view of the MOOC landscape. *Journal of Global Literacies, Technologies, and Emerging Pedagogies, 2*(3), 122–139. Retrieved from http://joglep.com/index.php/archives/voulme-2-issue-3/.

Flavell, J. H. (1979). Metacognition and cognitive monitoring. A new area of cognitive-development inquiry. *American Psychologist, 34*(10), 906–911. doi:10.1037/0003-066X.34.10.906.

Hidden curriculum. (2014). In S. Abbott (Ed.), *The glossary of education reform*. Retrieved from http://edglossary.org/hidden-curriculum

Jensen, B. B., & Schnack, K. (1997). The action competence approach in environmental education. *Environmental Education Research, 3*(2), 163–178.

Kennedy, G. E., Judd, T. S., Churchward, A., Gray, K., & Krause, K. (2008). First year students' experiences with technology: Are they really digital natives? *Australian Journal of Educational Technology, 24*(1), 108–122.

Ketlhoilwe, M. J., Silo, N., & Boikhutso, K. (2014). ESSA change projects evaluation report; Teacher education institutions in the SADC Region. Visby, Sweden: Evaluation Report, Swedish International Centre of Education of Sustainable Development (SWEDESD).

Learning Development Centre. (2014). *The changing face of e-pedagogy.* Coventry: University of Warwick. Retrieved from https://www2.warwick.ac.uk/services/ldc/resource/interactions/celi/chap7/article1/dempster.pdf.

SWEDESD. (2014). The EESA programme. In L. S. Vygotsky (Ed.), *Mind and society.* Cambridge, UK: Cambridge University Press. Retrieved from www.swedesd.se.

Vygotsky, L. S. (1978). Mind and society. Cambridge: Cambridge University Press.

Wenger, E. (1998). *Communities of practice: Learning, meaning and identity.* New York: Cambridge University Press.

Wiek, A., Withycombe, L., & Redman, C. L. (2011). *Integrated Research System for Sustainability Science*, United Nations University and Springer. Retrieved July 16, 2014, from http://www.e3washington.org/upload/profile/resources/file-268.pdf.

Printed in the United States
By Bookmasters